THE POETRY OF VICTORIAN SCIENTISTS

A surprising number of Victorian scientists wrote poetry. Many came to science as children through such games and toys as the spinning-top, soap-bubbles, and mathematical puzzles, and this playfulness carried through to both their professional work and writing of lyrical and satirical verse. This is the first study of an oddly neglected body of work that offers a unique record of the nature and cultures of Victorian science. Such figures as the physicist James Clerk Maxwell toy with ideas of nonsense, as through their poetry they strive to delineate the boundaries of the new professional science and discover the nature of scientific creativity. Also considering Edward Lear, Daniel Brown finds the Victorian renaissances in research science and nonsense literature to be curiously interrelated. Whereas science and literature studies have focused mostly upon canonical literary figures, this original and important book conversely explores the uses to which literature was put by eminent Victorian scientists.

DANIEL BROWN is Professor of English at the University of Southampton.

CAMBRIDGE STUDIES IN NINETEENTH-CENTURY
LITERATURE AND CULTURE

General editor
Gillian Beer, *University of Cambridge*

Editorial board
Isobel Armstrong, *Birkbeck, University of London*
Kate Flint, *Rutgers University*
Catherine Gallagher, *University of California, Berkeley*
D. A. Miller, *University of California, Berkeley*
J. Hillis Miller, *University of California, Irvine*
Daniel Pick, *Birkbeck, University of London*
Mary Poovey, *New York University*
Sally Shuttleworth, *University of Oxford*
Herbert Tucker, *University of Virginia*

Nineteenth-century British literature and culture have been rich fields for interdisciplinary studies. Since the turn of the twentieth century, scholars and critics have tracked the intersections and tensions between Victorian literature and the visual arts, politics, social organisation, economic life, technical innovations, scientific thought – in short, culture in its broadest sense. In recent years, theoretical challenges and historiographical shifts have unsettled the assumptions of previous scholarly synthesis and called into question the terms of older debates. Whereas the tendency in much past literary critical interpretation was to use the metaphor of culture as 'background', feminist, Foucauldian, and other analyses have employed more dynamic models that raise questions of power and of circulation. Such developments have reanimated the field. This series aims to accommodate and promote the most interesting work being undertaken on the frontiers of the field of nineteenth-century literary studies: work which intersects fruitfully with other fields of study such as history, or literary theory, or the history of science. Comparative as well as interdisciplinary approaches are welcomed.

A complete list of titles published will be found at the end of the book.

THE POETRY OF VICTORIAN SCIENTISTS

Style, Science and Nonsense

DANIEL BROWN

CAMBRIDGE
UNIVERSITY PRESS

CAMBRIDGE UNIVERSITY PRESS
Cambridge, New York, Melbourne, Madrid, Cape Town,
Singapore, São Paulo, Delhi, Mexico City

Cambridge University Press
The Edinburgh Building, Cambridge CB2 8RU, UK

Published in the United States of America by Cambridge University Press, New York

www.cambridge.org
Information on this title: www.cambridge.org/9781107023376

© Daniel Brown 2013

This publication is in copyright. Subject to statutory exception
and to the provisions of relevant collective licensing agreements,
no reproduction of any part may take place without the written
permission of Cambridge University Press.

First published 2013

Printed and bound in the United Kingdom by the MPG Books Group

A catalogue record for this publication is available from the British Library

Library of Congress Cataloguing in Publication data
Brown, Daniel, 1961– author.
The poetry of Victorian scientists : style, science and nonsense / Daniel Brown.
pages   cm. – (Cambridge studies in nineteenth-century literature and culture ; 83)
ISBN 978-1-107-02337-6 (hardback)
1. English poetry – 19th century – History and criticism.   2. Scientists' writings.   3. Literature and science – Great Britain – History – 19th century.   I. Title.
PR595.S33B76   2012
821'.80936–dc23
2012020239

ISBN 978-1-107-02337-6 Hardback

Cambridge University Press has no responsibility for the persistence or
accuracy of URLs for external or third-party internet websites referred to
in this publication, and does not guarantee that any content on such
websites is, or will remain, accurate or appropriate.

*Mihály Juditnak, kedves nagynénémnek,
és öcsémnek, Brown Raymondnak, a valódi tudósnak.*

Contents

List of illustrations	*page* x
Acknowledgements	xi

1	Professionals and amateurs, work and play: William Rowan Hamilton, Edward Lear and James Clerk Maxwell	1
2	Edinburgh natural philosophy and Cambridge mathematics	39
3	Knowing more than you think: James Clerk Maxwell on puns, analogies and dreams	59
4	Red Lions: Edward Forbes and James Clerk Maxwell	89
5	Popular science lectures: *'A Tyndallic Ode'*	110
6	John Tyndall and 'the Scientific Use of the Imagination'	142
7	*'Molecular Evolution'*: Maxwell, Tyndall and Lucretius	164
8	James Joseph Sylvester: the romance of space	183
9	James Joseph Sylvester: the calculus of forms	207
10	Science on Parnassus	234

Notes	263
Works cited	287
Index	302

Illustrations

1. A page from Edward Lear's sketchbook, Rose-ringed Parrakeet and visitors to the Zoological Gardens, *c.* 1830. MS Typ 55.9 (60), Houghton Library, Harvard University. *page* 21
2. Insignia of the Metropolitan Red Lion Association, early 1850s. Imperial College, London. 90
3. Frederick Pollock's 1865 Valentine for Tyndall, 'The Ice Flower'. Royal Institution of Great Britain, London. 125
4. A page from P. G. Tait's Scrapbook, 'Pigmies are pigmies still – though perched on Alps'. James Clerk Maxwell Foundation, Edinburgh. 146
5. James Clerk Maxwell, Stereograms: Lines of Curvature of Ellipsoid, Concyclic Spherical Ellipse. Add.7655.v.i.11 1–2, Cambridge University Library. Reproduced by kind permission of the Syndics of Cambridge University Library. 256
6. *Venus Physica*, engraving by Gilliam van der Gouwen of a print by Jan Goere, from Lucretius, *De Rerum Natura* (London: Tonson, 1712). Reproduced by kind permission of the Syndics of Cambridge University Library. 259

Acknowledgements

I would like to thank my parents, Aniko and Kevin Brown, for all their help and patience, and also my brother Raymond. Gillian Beer kindly read an early draft of some chapters on Maxwell, and emerged to observe that they needed 'aerating'. I would accordingly like to dedicate all the bubbles in the book to her. I am grateful to Russell Hays for his attentive reading of a late draft, to my colleague Nicolas Damjanovich, who offered an astute commentary on a version of Chapter 3, and to Aedan Pope and Mark Ioppolo, for checking some passages on mathematics from Chapters 8 and 9. I would like to thank my editors Linda Bree, Maartje Scheltens, Josephine Lane, Jodie Hodgson and Anna Zaranko at Cambridge University Press, along with the Press's readers of the typescript, Alice Jenkins and Peter Pesic, for their useful and improving suggestions. While the study has tended to keep to itself over the years, its writing has been facilitated by the kindness, and various stimulating and amusing milieux, furnished by Hilary Fraser and Nick Burton, John and Gillian Beer, James Garner and Charlotte Squire, Russell Hays, Neale and Mary Hays, Rudolf Balázs, Chancellor Von' Burnickis, Anna Novotny, Áronson Gáborné, Frank Callanan, Patrick Healy, Gail Jones, Patience Hook, Burt and Barbara Main, Gillian Jenkins, and my 2011 'Modernism' classes.

This project benefitted from an Australia Research Council 'Discovery' Grant. Citations from the James Joseph Sylvester papers are used by permission of the Master and Fellows of St John's College, Cambridge. I am also very grateful to the following institutions for allowing me access to their manuscript collections and kind permission to reproduce material: Cambridge University Library; the Bodleian Library, Oxford; the James Clerk Maxwell Foundation, Edinburgh; Imperial College, London; and the Royal Institution of Great Britain, London.

An earlier version of parts of Chapter 7 was published in *Minds, Bodies, Machines, 1770–1930*, edited by Deirdre Coleman and Hilary Fraser (2011).

CHAPTER 1

Professionals and amateurs, work and play: William Rowan Hamilton, Edward Lear and James Clerk Maxwell

Having met William Wordsworth for the first time during a visit to the Lake District in September 1827, the 22-year-old Irish mathematician and astronomer William Rowan Hamilton wrote to his sister Eliza of the 'burning thoughts and words' that drew the two together as fast friends.[1] Hamilton became Wordsworth's 'poetic disciple',[2] sending and addressing many of his poems to him, while the eminent romantic poet, who was then fifty-seven, compared his new friend to Samuel Taylor Coleridge, describing them as 'the two most wonderful men, taking all their endowments together, that he had ever met'.[3] Three months before their meeting, while still an undergraduate, Hamilton had been made Andrews Professor of Astronomy at Trinity College Dublin, a position that included with it the title of Astronomer Royal for Ireland and the directorship of the Dunsink Observatory, where he would live. Although he kept this position for the rest of his life, his most important contributions to science would be not in astronomy but in mathematics, principally in the field of quaternions that he largely invented.

Looking back to his youth, Hamilton wrote to his friend and fellow astronomer, mathematician and poet John Herschel in 1847 that 'it would really seem to have been at one time a toss-up, whether I should turn out a rhymer or an analyst [i.e., a mathematician working with continental calculus]'. Wordsworth encouraged his interests in science. The main theme of Wordsworth's sustained campaign of criticising Hamilton's verse is summed up by the charge he makes in his first letter on the subject, that 'the workmanship (what else could be expected from so young a writer?) is not what it ought to be'. Wordsworth pushed Hamilton to make a choice between the two vocations, although not in favour of poetry:

You send me showers of verses, which I receive with much pleasure, as do we all; yet have we fears that this employment may seduce you from the path of Science which you seem destined to tread with so much honour to yourself and profit to others. Again and again I must repeat, that the composition of verse is infinitely more of an art than men are prepared to believe, and absolute success in it depends upon innumerable *minutiæ*, which it grieves me you should stoop to acquire a knowledge of.[4]

In such letters to Hamilton, Wordsworth consistently stresses the need for what is effectively a professional commitment to the practice of poetry, a dedication he assumes that science also requires.

Wordsworth came to Ireland as Hamilton's guest in August 1829, staying with him and his sisters at the Observatory. Eliza kept a journal of the visit, which documents a particularly intense discussion that her brother had with Wordsworth about poetry and science. It appears to have been prompted by Wordsworth's recitation of a provocative passage from Book IV of *The Excursion* (1814) as part of his argument that current positivist, utilitarian and materialist applications of science, as Eliza records him saying, 'waged war with and wished to extinguish Imagination in the mind of man'.[5] Science is characterised in the poem as a slavishly positivist brute, which, the poet allows, could nonetheless be admitted to the realm of poetry and the imagination, but only in a chastened form and on a strictly temporary, probationary and subordinate basis:

> Science then
> Shall be a precious visitant; and then,
> And only then, be worthy of her name:
> For then her heart shall kindle; her dull eye,
> Dull and inanimate, no more shall hang
> Chained to its object in brute slavery;
> But taught with patient interest to watch
> The processes of things, and serve the cause
> Of order and distinctness, not for this
> Shall it forget that its most noble use,
> Its most illustrious province, must be found
> In furnishing clear guidance, a support
> Not treacherous, to the mind's *excursive* power.[6]

Science is addressed as a Caliban that can be freed from its enslavement to materialism, tutored and refined to join for a time Ariel and Prospero's realm. Invited to lend its strength to the imagination, it must first overcome a vicious native propensity to undermine such poetic vision. Soon after Wordsworth's return to England, in October 1829, Hamilton formally renounced his poetic ambitions. Indeed, risking the charge of performative contradiction, he reported his decision to the muse directly and courteously in his poem 'To Poetry'.[7] He continued to write poetry privately, often as an adjunct to his science, for the rest of his life.

Hamilton's friendship with Wordsworth, and the parity he gave to his practices of poetry and science in the 1820s, are telling of, and indeed emblematic for, an historical moment of transition in British culture, in which poetry, beginning its gradual decline in power and prestige with the waning of romanticism, meets with ascendant science. His choice of science over poetry came on the eve of a decade in which British science strove forthrightly to become professional. This occurred primarily through the efforts of the British Association for the Advancement of Science (BAAS). Established at a meeting organised by the Scots natural philosopher Sir David Brewster at York in September 1831, the Association was the outcome of a public debate about the decline of science in Britain. Led initially by Herschel and the Scots geologist John Playfair in the late 1820s, then Brewster and the mathematician Charles Babbage, much of the discussion focused upon the inadequacies of the Royal Society. These had been brought to the fore during the heated 1830 campaign for the Society's president, principally through Babbage's provocative *Reflections on the Decline of Science in England, and on Some of Its Causes*, but also by the physician Augustus Granville's *Science Without a Head: or, the Royal Society Dissected*. Drawing upon their experience as Fellows, since 1816 and 1817 respectively, Babbage and Granville criticised the Society as a bastion of aristocratic class privilege, ill suited to recognising scientific merit and facilitating scientific research.[8]

Anxious to distinguish themselves from the old order represented by the Royal Society, the delegates to the early meetings of the BAAS accordingly felt, as the Cambridge polymath William Whewell recalls in 1834, 'the want of any name by which we can designate the students of the knowledge of the material world collectively'. Early efforts to clarify the amorphous principle of science that prevailed at the time made appeals to a broad range of interested parties, including Coleridge, who attended the 1833 meeting at Cambridge. Whewell records his contribution to the discussion: '*Philosophers* was felt to be too wide and too lofty a term, and was very properly forbidden them by Mr. Coleridge, both in his capacity of philologer and metaphysician.' Coleridge was, as Trevor H. Levere observes, keen to distinguish philosophers from 'those who merely studied the material world', a generic identity that is caricatured by such further suggestions from the meeting 'for these *naturæ curiosi*' as '*nature-poker*, or *nature-peeper*'. Whewell's account of the 1833 meeting credits Coleridge's observation with prompting 'some ingenious gentleman' (actually Whewell himself) to momentously distinguish the new professional with the coinage '*scientist*'.[9] While this term would not receive general acceptance until the early twentieth century, the need that was felt for it during the early 1830s marks a decisive shift in the conception of professional science.

It was largely through the efforts of the BAAS that, as Jack Morrell and Arnold Thackrey observe, 'the word "science" took on new and narrower meanings in the 1830s and 1840s', so that it applied to a specialist, purportedly superior, form of understanding, which for the Victorians 'came to be seen as the intellectual progenitor of technology, the guarantor of God's order and rule, the proper way of gaining knowledge, and the key to national prosperity and international harmony'.[10] Radical reforms to the Cambridge Mathematical Tripos course during this time also yielded Britain's first professional science qualification, which, as the following chapters document, furnished the foundation for a new research culture, indeed a renaissance, in British mathematics and physics during the mid-Victorian period.

The last chapter of Babbage's *Reflections on the Decline of Science* is entitled 'Suggestions for the Advancement of Science in England', while the book reprints as an appendix a report on the 1828 meeting of the *Gesellschaft Deutscher Naturforscher und Ärzte*, the body that the *naturphilosophe* Lorenz Oken had begun in 1822. Prompted by Babbage's book and modelling itself on the German organisation, the BAAS responded to the growth and increasing diversity of the sciences by arranging itself in discrete sections, 'each with its own president and committee, and indicated by letters', as the biologist T. H. Huxley reports from the 1851 Ipswich meeting: 'For instance, Section A is for Mathematics and Physics; Section B for Chemistry, etc; my own section, that of Natural History, was D.'[11] The alphabetical arrangement was deliberate and hierarchical. Much as the incipient discipline of English literature was at the time rallying around the figure of Shakespeare, the British Association traced its pedigree to Isaac Newton and made his disciplines of mathematics and physics the paradigm for its science, twinning and honouring them as Section A. This was largely the work of Whewell, who coined the word '*Physicist*' in 1840, for the scientist who investigates 'the ideas of force, matter, and the properties of matter'.[12] Physics was taken as the type for the new professional science, so that as an 1839 report in the periodical *British Critic* observes of the BAAS, 'Physical science, the science, in other words, of matter and material things, now arrogates in effect the name "science" to itself.'[13] At the original York meeting, Whewell suggested that an eminent representative of each area of science should report on 'the present state of the science',[14] which in subsequent meetings became the sectional president's address. Each meeting of this 'parliament of the scientific world',[15] as Whewell described it in 1833, would appoint an overall president, who would present an address on the current state and progress of British science.

Reviewing the BAAS in his address as Secretary to the 1834 meeting at Edinburgh, James David Forbes, the newly elected professor of natural philosophy at Edinburgh University, describes 'the perfectly *unique* character of this Association, and the high aims to which its efforts are directed'. He notes that no other British scientific body, including the Royal Society, had made such an 'attempt to guide in any specific direction the investigations of their members, or to form any school of science for the initiation of fresh inquirers'. Unlike the continental models that inspired it, the BAAS not only met annually to offer research papers, but resolved that it 'should continue to operate during the intervals of these public assemblies, and should aspire to give an impulse to every part of the scientific system, to mature scientific enterprise, and to direct the labours requisite for discovery'. Various committees were established by the Association to promote the welfare of science in Britain and by its sections to commission particular programmes of research, which would be funded by members' dues and report back to the meetings. Even at this early date, its work in professionalising and regularising scientific practice included devising 'instructions for conducting uniform systems of observation', and disseminating them 'in the New World', while Forbes also notes a grant to establish a committee on standard measurements, 'the Numerical Constants of Nature and Art', to be led by Babbage. Furthermore, 'the Association has fulfilled its pledge of stimulating Government to the aid of science', with £500 having been advanced by the Chancellery for 'Greenwich Observations'.[16]

The argument that informed Brewster's push for a new scientific association made common cause with literature and the arts, observing, for instance, that 'the sciences and the arts of England are in a wretched state of depression' and noting 'the indirect persecution of our scientific and literary men by their exclusion from all the honours of the State'.[17] Furthermore, many provincial scientific societies at the time treated literature as integral to moral and natural philosophy. Despite such arguments and precedents,[18] the BAAS chose not to include literature amongst the sections, a decision that effectively institutionalises the separation of poetry and science at the inception of British professional science. It mirrors the opposition between the two that Wordsworth also makes, which similarly became institutional with the growth of his reputation amongst the Victorians as the greatest and definitive romantic poet, and the consequent consolidation of his thought within a British Romantic ideology that shaped the emergent discipline of English literature, originally through F. D. Maurice's teaching of the new subject at King's College London from 1840 to 1853, in working men's institutes, in schools and from the 1890s at Oxford and other universities.[19]

The adoption by the rising discipline of English literature of a Wordsworthian and Coleridgean romantic ideology was also facilitated by the BAAS's blanket rejection of metaphysics, despite Hamilton's strong advocacy of it, as 'merely ideal'.[20] The new professional science repudiated romantic science and its harmonious, indeed integral relations with metaphysics and poetry. In the early 1850s the young physicist James Clerk Maxwell, having completed a degree at Edinburgh University that had strong elements of both experimental science and metaphysics, was frustrated to find that his subsequent Mathematical Tripos studies at Cambridge offered little scope for philosophical inquiry. While at Cambridge, he accordingly turned to writing poetry as the private medium in which he further developed the epistemological grounds for his scientific practice and, through a critique of the Tripos system, his intellectual ethics.

Victorian and twentieth-century academic philosophy, other than British Idealism, has, like science, demonstrated little interest in Romantic metaphysics, which accordingly found refuge in English literature, where it has continued the Wordsworthian battle against positivist, utilitarian and materialist science. Consistent with this Romantic ideology, studies in Victorian science and literature have until quite recently dwelt exclusively upon the responses of canonical literary figures to geology and evolutionary biology rather than mathematics and physics, while conversely poetry and literary criticism by Victorian scientists have been largely ignored.[21]

SCIENTISTS AND ARTISTS

The neologism *'scientist'* was formed at the 1833 meeting of the BAAS not only out of respect for Coleridge's proprietorial use of the term 'philosopher', but, as Whewell notes, 'by analogy with *artist*'. He elaborates this formal relation in 1840, writing in the preface to his *Philosophy of the Inductive Sciences* that 'as an Artist is a Musician, Painter, or Poet, a Scientist is a Mathematician, Physicist, or Naturalist'.[22] Indeed Whewell also wrote poetry and so may, as both his analogy and balanced phrasing suggest, have found similarities and correspondences to exist between the activities of the artist and those of the scientist, much as his friend Hamilton did. While the formal analogy by which Whewell arrives at his coinage bequeaths a nominal trace of the artist to the new Victorian 'scientist', the relation between the two becomes substantive in the work of some of the age's greatest scientific practitioners. Hamilton asserts definitively the kinship he sees to exist between science and poetry in 'To Poetry' as, referring nostalgically to an original prelapsarian unity of poetry and science, the 'joint abode' of the 'Spirit of Beauty' and her 'sister Truth', he advises the former that 'my life be now / Bound to thy sister Truth by solemn vow'.[23]

Hamilton took the opportunity in his 1831 Introductory Lecture on Astronomy at Trinity College to discourse upon 'the latent imagination that is involved in the processes of Science'. He begins by reciting and recommending to his students the passage from *The Excursion* that Wordsworth had read to him during his stay at the Observatory. While the BAAS would effectively make Newton the patron saint of their new British science, Hamilton describes him in his lecture as an artist, whose powers of imagination 'caused many ideal worlds to pass before him', an image that may have informed the lines that Wordsworth added to his description of the great natural philosopher in the 1850 version of *The Prelude*, as 'a mind for ever / Voyaging through strange seas of Thought, alone'.[24] Hamilton provides a more personal and radical introduction to astronomy, and the relations he sees science to have to poetry, in his early poem 'Ode to the Moon under Total Eclipse'. Written in 1823 a few weeks before his eighteenth birthday, the 'Ode' remained a favourite with Hamilton throughout his life. His friend and first biographer Robert Perceval Graves records that 'among his papers are several copies of it in his own handwriting, showing that even to the last year of his life he attached a special value to it'.[25] The poem offers a perspective on science that Hamilton evidently felt able to affirm throughout his distinguished career.

Dated 'July 23', the day of the eclipse it records, the 'Ode' is prefaced with a text that provides the empirical datum for all that follows: '*The moon under Total Eclipse is not invisible, but of a dark red colour.*' The poem observes that this raw phenomenon has itself been eclipsed by various hypotheses:

> O queen of yon ethereal plain,
>   With slow majestic step advancing,
> 'Mid thine attendant starry train,
>   The subject waves beneath thee dancing,
> As Dian moves through Delian Shades,
> Above her circling Oread maids:
> Why had that crimson red
> Thy lovely brow o'erspread?

Oh! wherefore that portentous gloom,
Meet for the tenants of the tomb?

Say is it but a passing cloud
   Far in some higher sphere,
Which thus around thee winds its shroud,
   While all the heaven is clear:
When all the stars are brightly burning,
Each in his wonted orbit turning?

Or wizard from his murky cell,
   Who bows thee to his power,
By magic word and muttered spell
   In this, Night's witching hour?

Or is it, as the sages say
   Versed in celestial lore,
Our Earth athwart Light's pathless way,
   Which bars it from thy shore:
Whose shadowy cone, with noiseless pace,
Through the infinity of space,
Hath darkly crossed thine orb on high,
And dimmed it to our wondering eye?

On thee the Nations gaze,
With looks of wild amaze,
   And anxious ask what means the sign:
What dread disaster nigh
Is boded by thine eye
   Lowering with aspect thus malign?

Surveying his subject, Hamilton acknowledges that historically most of its sources have been literary, and that his scientific discipline of astronomy has its roots in mythology and poetry. The opening stanza draws upon allegories by Edmund Spenser, William Shakespeare, Walter Raleigh and their contemporaries, which figure Queen Elizabeth I as the moon goddess Astraea, to dramatise the eclipse as a Renaissance masque. While the regal moon's power over its 'subject waves' makes a scientific observation about the tides, this case of action-at-a-distance also licenses the succeeding hypothesis, in which the moon is in turn subjected to eclipse, as 'By magic word and muttered spell' a wizard 'bows thee to his power'. With its invocation of courtly politics and fate, a 'portentous gloom', the Elizabethan allegory also introduces classical hermeneutic concerns with the eclipse as an omen, as 'Nations' ask 'what means the sign[?]', figuring it as an evil 'eye' looking down upon them: 'Earthquake or famine, sword or fire, / Is menaced by that look of ire.' The scientific account of the lunar eclipse, which explains that the Earth's 'shadowy cone / ... / Hath darkly crossed thine orb on high', is placed unobtrusively amongst the other hypotheses. Its proponents are described as 'sages', an archaic designation that suggests an affinity with the 'wizard' of the earlier stanza. 'Versed in celestial lore', their science is cast as simply another tradition, another mode of explanation, which, like lyric poetry, tells of a transcendent sphere. Its scientist sages are like bards, being 'versed' in a lore that is presumably transmitted orally. The poem sees phenomena to signify richly and reverberantly, and correspondingly to accommodate and require a range of hermeneutics. 'I unite', Hamilton writes to Eliza in a letter on an earlier eclipse, 'in some degree the poet with the astronomer.'[26]

Each of the explanations that Hamilton's 'Ode' proffers *correlates* with its phenomenon, but only one can be said in fact to *cause* it, and so be distinguished as a scientific principle. It is a measure of the transcendental epistemic authority that scientific explanations have since acquired that merely to juxtapose them with 'poetic' glosses, as Hamilton's verses do, is enough to make the latter appear nonsensical. However, while such discriminations are easily, and often patronisingly, made in hindsight, they are much more problematic in practice. For Victorian scientists in particular, for whom many new disciplines, and indeed new conceptions of science itself, were being formed, while such established disciplines as mathematics and physics were increasingly pursuing speculative fields of inquiry, the difference between science and nonsense was not always clear-cut. Indeed the Victorian period marks a renaissance not only in research science but in nonsense literature. That the great inventors of Victorian nonsense Edward Lear and Lewis Carroll also practised science, the former as a natural history illustrator, the latter as a mathematician, suggests that the relation between the two fields is not merely contingent. Jean-Jacques Lecercle argues in his *Philosophy of Nonsense* that the genre borrows the themes of 'exploration and taxonomy' from natural history, and, while he focuses upon Carroll, acknowledges that the discipline 'is a direct source for [Lear's] nonsense'.[27] The following sections of the chapter examine the legacy for Lear's nonsense verses and drawings of the innovative natural history work he did during the transformative decade of the 1830s.

NATURAL HISTORY AND HISTORIC NONSENSE

While writing flawed poetry may, as Wordsworth suggests to Hamilton, be an occupational hazard for scientist-poets, ostensibly 'bad' verse becomes professional, indeed canonical, with Lear's limericks. The liminal nature of both Lear's scientific work as a natural history illustrator and his literary work as a writer of nonsense contrast neatly with Hamilton's classical scientific credentials in mathematics and astronomy and the high art inspirations and aspirations of his poetry, his 'fondness for classical and for elegant literature'.[28]

Lear's brief but spectacular career as a natural history lithographer began in 1830, the year he turned eighteen and started working on his *Illustrations of the Family of Psittacidae, or Parrots*, two folios of which were published later that year, with the monograph itself appearing, albeit incomplete, in 1832. The first book of imperial folio lithographed birds published in Britain, it immediately gained Lear international recognition as an ornithological draughtsman. The genre of the grand natural history plate pioneered in the *Psittacidae* monograph is best known through the examples published by John Gould, who took from Lear the folio format and use of lithography, and indeed employed him during the 1830s. Gould included plates by Lear, often without acknowledgement, in *A Century of Birds Hitherto Unfigured from the Himalaya Mountains* (1830–2), *A Monograph of the Ramphastidae, or Family of Toucans* (1834), *The Birds of Europe* (1832–7), *The Birds of Australia* (1837), *Icones Avium* (1837) and *A Monograph of the Trogonidae, or Family of Trogons* (1858–75). Due to his deteriorating eyesight, however, Lear had to leave such work in 1837, turning instead to the complementary hypermetropic exercise of landscape painting, and to the nonsense writings and drawings for which he is best known.

One of the admirers of Lear's *Psittacidae* monograph, Edward Stanley, the President of the London Zoological Society and (from 1834) the thirteenth Earl of Derby, employed Lear from 1832 until 1837 to make an illustrated catalogue of the mammals and birds kept in his private zoo at Knowsley Hall, near Liverpool. Lear lived with Stanley and his family for much of this period, during which time he made paintings of about a hundred of the creatures and, to amuse the children in the family, began writing and drawing his nonsense works, including the 128 limericks that survive in the Knowsley album. Seventeen of his paintings of Stanley's animals were published in *Gleanings from the Menagerie and Aviary at Knowsley Hall* in 1846, the same year that he gathered many of his limericks from this productive period in his first *Book of Nonsense*, which he dedicates to his patron's 'great-grandchildren, grand-nephews and grand-nieces'. Their parallel histories suggest that his natural history and nonsense were for Lear compatible, even comparable, activities.

While Lear's flourishing as a natural history illustrator dates from the beginnings of Victorian professional science, modern nonsense, which he largely defined, developed concurrently as its ostensible opposite, an apparently unmeaning other to what became, in subsequent decades, the paradigmatic epistemic regime. Nonsense appears to offer a childish respite from the seriousness of science and the adult world it shaped, of intellectual, technological and economic progress encumbered with the various existential challenges that flowed from such ideas as geological time, Darwinian evolution and entropy. Despite their innovative nature, Lear's lithographs became outmoded scientifically as the practice of natural history they participate in was transformed, largely by Huxley and the other members of Section D of the BAAS, into the modern life sciences. Combining careful scientific observation with artistic refinement and expressivism, Lear's hand-coloured lithographic plates, 'a form of visual natural theology',[29] found their subscribers and other patrons amongst the wealthy, often aristocratic, Anglican scientific amateurs and connoisseurs that the emerging professional science of the BAAS was defining itself against.

Akin to the natural theological conceit of the Book of Nature, the natural history lithographs published by Lear and Gould imply a conception of nature as Creation, in which art accordingly coincides with science. Much as Whewell, Herschel and their peers saw their scientific writings as replicating parts of the original Book of Nature and drawing out its truths, the natural history plate similarly presupposes that man, made in the Creator's image, can grasp and pay tribute to the Creation by effectively re-creating it in his own artefacts. As attentive as they are to the empirical details of nature, such plates subordinate them by degrees to ideal principles of species type and aesthetics. Furthermore, while natural history illustrations have long been regarded as the province of art historians and have only recently been studied also for their scientific meanings and currency,[30] the original credentials of Lear's plates have been further diminished by being read through the more celebrated nonsense works, as expressions of personal quirkiness rather than scientific record.

Lear's *Psittacidae* was the first illustrated monograph devoted to a single family of birds. It was also distinguished by its practice of drawing its subjects not only from the preserved remains of specimens, in this case those held in the Royal Zoological Society Museum, but also directly from living examples, particular birds in the Regent's Park Zoological Gardens, which had opened in 1828, two years after the death of its original proponent, the colonial governor Sir Stamford Raffles. Lear would work in the birdhouse, making sketches of the parrots and taking notes about their colours and careful measurements of their wingspans, beaks and the length of their legs.[31]

As a means of recording and identifying the physical characteristics of species, the natural history illustration has obvious advantages over the written descriptions that usually accompanied them. Lear was renowned for the precision and detail of his work, crucial scientific values that led him to unwittingly recognise a new species of Macau. The accuracy with which he reproduced the specimen that was at the time presented at Regent's Park Zoo as the Hyacinthine Maccaw allowed Charles Lucien Bonaparte, the ornithologist nephew of Napoleon, to recognise it in 1856 as a new species, which he accordingly named Lear's Macau (*Anodorhynchus leari*). Another he named Lear's Cockatoo (*Lapochroa leari*).[32] The scientific achievement of Lear's work on parrots in carefully delineating species was also recognised by the Linnaean Society. Nominated the day after the first two folios of the *Psittacidae* were published in 1830, he was made an Associate of the Society two years later, when he was twenty. Lear clearly states his scientific concern with carefully discriminating and documenting the species of the *Psittacidae* in a letter to Sir William Jardine in 1834, where, explaining that financial losses from his project meant that he could not complete it, he urges the publisher to produce a further book on parrots: 'I am very anxious to see this publication – and particularly, Sir William, I hope you will clear up several demi-groupes which are at present very obscure – P[sittacidae]. viridis – the Genus Brotogeris – & the white winged Parakeet – Murinius & others are among those I allude to.'[33]

Lear informs Jardine that not only will there be no further plates for his *Psittacidae*, but 'neither will there be – (from me) any letterpress'.[34] The form of the natural history text he recognises here coincides with that of his nonsense alphabets, where the entry for each letter consists of a picture of an object, in most cases an animal, followed by a concise description of it. Akin to his alphabets, the nonsense botanies usually consist of a definitive picture and a mock-Linnaean binomial title, such as 'Bottlephorkia Spoonifolia' and 'Manypeeplia Upsidedownia'. They hark back to the early speculative days of taxonomy, when Linnaeus's sexual criteria for classifying plants by their flowers was elaborated in verse by Erasmus Darwin as *The Loves of Plants* (1789). Here, a possible model for Lear's 'Cockatooca Superba',[35] the lily *Gloriosa Superba*, co-exists with the botanic unicorn, an 'intensely yellow' woolly, four-stemmed 'Fern' with a 'flexile neck' reputed to come from China, the *Wunderkammer* classic *Polypodium Barometz*: 'Eyes with mute tenderness her distant dam, / Or seems to bleat, a *Vegetable Lamb*.'[36] Indeed, going back further, to the origins of botanical science in distinguishing the edible from the inedible, the 1870 'Nonsense Botany' is presented by Lear, in a prefacing 'Extract from the *Nonsense Gazette*, for

August, 1870', as one of the scientific 'communications from our valued and learned contributor, Professor Bosh, whose labours in the fields of Culinary and Botanical science, are so well known to all the world'.[37] The historical origins of botany in concerns with the edible are acknowledged here, rather as in his ode on the eclipse Hamilton recollects the beginnings of astronomy in myth and portents. While the plants of the nonsense botanies are created *ex nihilo* with the invention of their name, from which the drawings and commentaries follow, many of the alphabets propose conversely that letters originate with and embody their objects.

Lear's nonsense alphabets assert a mimetic hypothesis that proposes that each letter 'was once' the object that is now designated by a word beginning with it: 'R was a rattlesnake.'[38] As if parodying the plethora of mid-Victorian theories about the origin of language, this immanentist principle suggests an iconic or hieroglyphic theory, or perhaps, taking a cue from Friedrich Max Müller, a mythological one, in which, like Daphne's transformation into the tree, animals and allied objects are petrified as particular letters. It also gestures toward the onomatopoeic theory of the origin of language that Max Müller made light of as the 'Bow-wow' theory a decade earlier in his *Lectures on the Science of Language*.[39] Both the percussive sound of the rattlesnake's tail and the creature's venomous prerogative are distilled in the trilled sound of the letter it becomes: 'r! / Rattlesnake bite!'[40]

The owl is the best known of the animals from the earlier natural history plates that Lear identifies with particular letters in his alphabets: 'O was once a little owl.' Above this text is a simple drawing of the bird, one of many in Lear's nonsense works that correspond to his earlier illustrations of owls belonging to the genus *Bubo* for Gould's *Birds of Europe*. The letter 'O' is rendered typographically in Lear's text as an oval, which describes in outline the shape of both the 'little owl' and its eyes, the most identifiable physical attribute of the animal, as they are represented in his accompanying sketch. As if to encourage this comparison, a capital and a lower case example of the letter hover abstractly above and below the drawing. The owl's counterpart in another nonsense alphabet, 'O was an oyster', similarly lends its form to the letter, as the concluding equation of the two makes clear: 'o! / Open mouth'd Oyster!' As well as acknowledging the oval shape of the creature, this attribution also suggests the round, open-mouthed, vowel sound that extends and gives substance to the empty form described by the letter, an effect that is more pronounced in the case of the owl. Below Lear's sketch of the owl is a catalogue of the creature's attributes, which is rendered in a terse verse equivalent of the note-form favoured by many natural history texts: 'Prowly, / Howly, / Owly / Browny fowly.'[41] The full soft vowel sound that puffs up the word 'owl', rather as Lear does the delicate chest feathers of the Eagle Owl (*Bubo Maximus*) in his lithograph of the bird, is reiterated and foregrounded with the enunciation of each defining attribute of the creature, an oddly synaesthetic resonance that gives further credence to his nonsense mimetic theory of the origin of the alphabet.

Like the nonsense botanies and alphabets, Lear's limericks each consist of an illustration with a description below it. Indeed, by giving the distinctive physical characteristics, behaviours and location of its creatures in a pithy and ostensibly factual verbal account and a descriptive drawing, the limerick could be mistaken for a burlesque version of the natural history text. Their subjects are usually identified with a crude binomial, first by age and gender as a generic human type, and then by the place they inhabit, a peculiar habitat parallel to those of animals:

> There was an Old Man of Dundee,
> Who frequented the top of a tree;
> When disturbed by the crows, he abruptly arose,
> And exclaimed, 'I'll return to Dundee.'[42]

Their anecdotal nature gives Lear's limericks another similarity to contemporary natural history texts, where in the absence of more extensive empirical evidence such brief narratives often featured in descriptions of species. This was especially the case during the great age of imperial and taxonomical expansion in the latter half of the eighteenth century and the first half of the nineteenth, when the locations and behaviour of many creatures were known only through a few individuals and impressions. The Comte de Buffon's forty-four volume *Histoire Naturelle* (1749–88), for example, includes copious anecdotes, as do works by Georges Cuvier, and E. T. Bennett's *Gardens and Menagerie of the Zoological Society Delineated* (1830–1), where Lear's first illustrations appeared, when he was nineteen. Curious stories also prevail in popular Victorian accounts of natural history in, for example, the *Penny Magazine*, Samuel G. Goodrich's *Peter Parley's Tales of Animals* (5th edn, 1835) and Francis Buckland's series *Curiosities of Natural History* (1857–72).

Although the limerick went unnamed until the close of the nineteenth century, it has a long history in oral traditions of folk and children's verse, becoming fixed as a published form only a few years before Lear began to renew it. Reputedly the first book of limericks, *The History of Sixteen Wonderful Old Women*, appeared anonymously in 1820, while Lear apparently first encountered the genre in Richard Scrafton Sharpes's *Anecdotes* and *Adventures of Fifteen Gentlemen* (1821). The limerick can be usefully contrasted with the similarly matched pictures and verses of children's chapbooks. Insofar as both forms were aimed at a juvenile readership, the chapbooks draw instructive examples from the readers' peer group, while the limericks allow no such moral identification with their subjects, as they focus upon the idiosyncrasies of peculiar adults. Usually viewed as too old and eccentric to learn from their behaviour, or to be blamed for it, such characters present objects not of moral censure and instruction, but of curiosity and mirth.

Lear evidently preferred the limerick's tolerance, indeed its celebration, of eccentric individuality and senseless behaviour over the punitive insistence in chap-books and other current children's literature upon discipline and conformity, an attitude that he allocates to the unspecified societal 'they' who in many of his verses persecute his anomalous protagonists. He is notorious as something of a social outsider, whose personal idiosyncrasies critics and biographers often compare with the subjects of his nonsense work. Indeed the author encourages this identification in 'How pleasant to know Mr Lear'[43] and many of his other verses and his drawings, which attribute to their subjects his love of animals and such physical features as long noses, round bodies and spindly legs, elements that are staples for his later caricatures of himself. It is not difficult to see how such a character would sympathise with the incompletely socialised nature of children, for whom social mores are apt to appear as novel and puzzling dictates, and that he would come to reproduce and indulge such perceptions by recording in his limericks the absurd and arbitrary behaviour of various adult characters, an affinity that the ostensibly child-like crudeness of the accompanying drawings would seem to confirm. The limerick allowed Lear to present individual and social phenomena not morally but with a pseudo-scientific objectivity, so that his body of work in this genre achieves something akin to a mock social science.

HOLOTYPES

Lear's natural history prints participate in an enlightenment science of reasonable literalness. Providing positivist descriptions that specify taxonomical classifications, their rationalist dedication to clarity and order marks their continuity with his nonsense work. As Lecercle observes, 'Nonsense is a meaning-preserving activity: its implicit goal is to save meaning by maintaining the correspondence between signifier and signified.' Lear's nonsense botanies and alphabets, albeit each working from opposite directions, were earlier seen to achieve this goal; the botanies literally extrapolating the signified from the signifier, while the alphabets conversely derive each letter from a concrete object. Like enlightenment science, nonsense is predicated upon an antipathy to the semantic instability of metaphor, which disrupts such straightforward relations. They are united in a common front against this fundamental principle of poetry. Lecercle writes that nonsense combats the metaphoric indeterminacy of language with its 'matter-of-factness',[44] the quality that gives Lear's limericks a pseudo-scientific authority.

Lear's natural history illustrations demonstrate several ways in which the defining characteristics of a species can be brought to the fore. So, for example, in the illustration of the Australian marsupial *Macropus Parryi* that he drew in 1834 for an article by E. T. Bennett in *The Transactions of the Zoological Society*, a magnified side view of its jawbones is embedded in the grass on which the animal stands, primarily to show its distinctive third incisor.[45] In one of the woodcuts he contributed to Bell's *History of British Quadrupeds*, the Greater Horse-shoe Bat obligingly presents itself in a direct frontal view, so that both the face and body can be seen clearly, with one wing extended and the other folded.[46] Such contrived ways of making prominent certain identifying features tend toward caricature, an emphasis through exaggeration that marks an affinity with Lear's early limericks and the drawings that accompanied them. Apt to present a creature's defining attributes as odd and arbitrary, the nonsense alphabets demonstrate this link between Lear's natural history work and his anthropology in the limericks:

> O, was an Owl
> Who objected to light
> When he made a great noise
> Every hour in the night
> O!
> Noisy old owl![47]

The taxonomical propensity for caricature was endemic to British natural history at the time Lear began working on his *Psittacidae*. It is demonstrated clearly and influentially by the illustrations that Frederick Nodder and later his son Ralph made for George Shaw's *Naturalist's Miscellany* (1790–1813). These hand-coloured copperplate engravings catalogue the defining characteristics of each creature directly and overtly, often to rather whimsical effect, as for instance in the lumbering toad-like frog, *Rana Australiaca* (Heleioporus australiacus) with its ungainly gait, bulbous yellow-spotted black body, and leering eye and mouth (so that, as Shaw comments, 'Its rarity must therefore apologize for its deformity').[48] Such distortions may be smoothed out artistically and appear more natural in the grander more finished forms of the large natural history folios, much as do the extended limbs and other expressive anatomical deformities in some mannerist paintings. Nevertheless, elements of caricature can also be discerned in such lithographs, not least in the pioneering examples produced by Lear, which take as their subjects some of the more curious, ostensibly comic or awkward, types of animals, such as parrots, flamingos, toucans, owls, kangaroos, monkeys and tortoises, some of which lend themselves to nonsense treatments in his verses, pictures and alphabets.

The representation of each species by an individual animal was, at the time Lear was making his illustrations, entrenched in British natural history practice through the name-bearing specimen or holotype. This is usually a preserved example from which a species is first described, so that it accordingly provides the definitive reference for all subsequent identifications. Shaw can take much of the credit for establishing this practice in Britain, as he based the descriptions of species in the *Naturalist's Miscellany* upon the Natural History collections of the British Museum, with which he worked as the zoologist. The name-bearing reference for a species may consist of syntypes, multiple specimens that belong to its original description, or include the secondary form of the allotype, a specimen of the opposite sex from the holotype, as is often required in describing bird species. As noted earlier, Lear used specimens from the Zoological Society Museum for his *Psittacidae* folios, and would also have been provided with such preserved remains in his work for Gould. The taxidermist for the Zoological Society from 1828 and its superintendent of Ornithology from 1833, Gould also procured further specimens from around the world for his lithographs, leaving his work with the Society in 1838 for an eighteen-month collecting expedition to Australia. Lear's illustration of the *Macropus Parryi* was, as the inclusion of its jaw bones indicates, drawn from its holotype, 'a single specimen' that had recently been presented to the Society by Sir Edward W. Parry.[49] Name-bearers for species could also be furnished by drawings, such as those in the British Museum's Natural History collection by Sydney Parkinson, who accompanied Sir Joseph Banks on James Cook's *Endeavour* expedition to the Pacific in 1771, only to die on the voyage. The type of the frog *Rana Australiaca* is similarly based on Nodder's 1795 illustration for the *Naturalist's Miscellany*.[50] It is a tribute to Lear's scientific prowess that Bonaparte made the artist's lithograph of the Hyacinthine Maccaw the holotype for the species he discriminates as *Anodorhynchus leari*.

By being tethered to a holotype each species is effectively described as an individual writ large, a conception that allows it to be represented not so much as a caricature but anthropomorphically as a character. The artist's reconstructive work from the remnants of the name-bearer, which were often poorly preserved and usually consisted of bones, principally skulls, and pelts, provided ample opportunity for such construal. Lear's *Macropus* is accordingly given a sweet rather comical facial expression that contrasts dramatically with the bare bones of the animal's jaws alongside it, from which this expression was presumably extrapolated. While Lear's practice of drawing and painting from living animals lends itself to the sympathies and embellishments of portraiture, his most celebrated bird illustration, the magnificent lithograph of the Eagle Owl,[51] is nonetheless telling for being drawn from a taxidermed specimen.

The Eagle Owl plate invites a reading of it as an allegory of the relation Lear sees animals to have to their human percipients. Many of his birds look out to the viewer; the toucans curiously and drolly, the flamingos somewhat aloofly, the Long-billed Parakeet with a sagacious and philosophical expression, while the cockatoos can be affable or coyly flirtatious. Ann Colley observes that the Eagle Owl's eyes 'tenaciously hold the viewer's and reduce the importance of the onlooker', and she argues further that in all of Lear's ornithological illustrations 'there is an exchange which protects the bird's integrity and diminishes the observer's power'.[52] Lear's Owl gazes austerely into our world, like an avian Olympia. It challenges the viewer's conventional prerogative to assimilate its being into our categories and desires, rather as Edouard Manet's provocative painting *Olympia* (1863) does by giving its eponymous female nude a bold outward gaze of her own, which challenges the convention in painting, prevalent since the Renaissance, of such subjects offering themselves demurely to the viewer's gaze. Ostensibly fixed upon some prey it sees, the Owl's poised gaze at once asserts its independent being whilst also suggesting the possibility of some relation to the viewer.

The Owl's gaze can be compared to the extravagant and playful poses of some of Lear's toucans, or his Great Auk, depicted paddling vigorously whilst eating a fish. Each such expressive and characteristic gesture shows that in composing his lithographs Lear was not simply working from life or its preserved remains, but painting portraits. Such distinctive behaviours are treated here not as mechanistic biological principles but rather, in the manner of portrait painting, as discretionary gestures that constitute and disclose individual character. His animals are endowed with an autonomous subjectivity, a quality that appears anthropomorphic only through the assumption, entrenched in modern philosophy since Descartes, that human beings have the monopoly of such being, that it distinguishes us from the rest of nature.

A sub-group of the limericks asserts an ontological parity, indeed the coalescence, of man and animal, with the two being depicted in the accompanying drawings parallel to one another, often facing each other, the fundamental situation of Lear's early natural history work. His drawings for such limericks do not present the physical features of their animals anthropomorphically, in the manner of such contemporaries as J. J. Grandville and John Tenniel. Rather it is their humans who conversely assume a resemblance to the creatures, with the two often adopting the same physical attitudes and stances. Examples of human protagonists represented in this way include 'an Old Man with an owl' who sits on a fence and drinks bitter ale with his fellow creature, 'an old person of Crowle, / Who lived in the nest of an owl', 'an old man of El Hums', who 'with the other birds round' picks crumbs 'off the ground', and 'That affable person of Nice' 'Whose associates were usually geese'. In further examples of his limericks, such as 'That instructive old man in a Marsh' who 'sang songs to a frog', and 'an old person of Skye, / Who waltz'd with a Bluebottle fly', animal and human face each other as mirror images. In the former instance the old man and the frog lean toward one another in the same shape and stance, with elongated hand and webbed fingers almost touching, while in the second the two hold together hand and insect foot, with the man wearing large round glasses, tails over a horizontally striped upper garment and black trousers and shoes. Their direct gaze, and that also between 'an Old Man who said "Hush!"' and an oversized bird in a bush in another such drawing,[53] recall Lear's earlier equivocal ornithological lithographs, in which, consistent with his preference for drawing from live models, the singular characters of the animals engage with their viewers through a range of direct gazes, in subtle defiance of the taxonomy that is imposed upon them. Such drawings of Lear's human characters with their animal friends offer emblems for Gilles Deleuze and Felix Guattari's idea of 'becoming animal',[54] while their interactions illustrate this idea directly, for each is a shared activity of becoming. These limericks are also true to Lear's early intense experience working on the *Psittacidae* monograph; 'for the last 12 months', he writes in October 1831, 'I have so moved – thought – looked at, – & existed among Parrots – that should any transmigration take place at my decease I am sure my soul would be very uncomfortable in anything but one of the Psittacidae'.[55] Lear's benign mirroring images of humans meeting and merging with their animal others also resonate with early and mid-Victorian preoccupations with biological developmentalism and the gathering debate over what Huxley describes as 'Man's Place in Nature', the relation of human beings to the other organisms. Indeed they offer suggestive and perhaps comforting counterparts to caricatures of Man either confronting or resembling a (usually simian) animal, which with the advent of Darwinism became staple tropes in *Punch* and other popular magazines.[56]

'MAN'S PLACE IN NATURE'

The sketchbooks that Lear used to make his preliminary studies of the birds at Regent's Park for the *Psittacidae* monograph include caricatures of visitors to the zoo, who evidently stared not only at the usual inhabitants of the parrot house but also at the artist enclosed with them making his drawings. The best known of these sketches began as a study of the Rose-ringed Parakeet, which is drawn perched on a branch, with its head lowered, so that its gaze is directed downward (Figure 1). Lear has added a round-bodied man below the bird, who looks upwards to meet its gaze. This pencil drawing depicts the basic confrontation of animal and man that Lear experienced most intensely and thoughtfully as an ornithological draughtsman working from live subjects, and also witnessed in the zoo visitors' attitudes to the birds. The drawing indicates Lear's identification with the Parrakeet rather than the visitor, who represents a type that stares at both of them alike. His professional eye is incorporated, with the bird's curiosity, in the sharp retaliatory gaze that the parrot directs at the gawping visitor.

The parrot's gaze illustrates an observation that the poet Gerard Manley Hopkins makes in his notes on nature some thirty-five years later: 'What you look hard at seems to look hard at you.'[57] The tension between man and bird dramatised in Lear's drawing not only poses the fundamental question addressed by his professional work, of what we can know of such creatures, but also suggests a rationale for the nonsense works. Like the Eagle Owl of Lear's lithograph, the Parrakeet asserts its autonomy forthrightly with its gaze. Indeed it prevails here, simplifying – effectively interpellating – the zoo visitor as a caricature, a suggestive prototype for the unwieldy round figures that become a staple of the limericks. There are smaller sketches of minor figures scattered around the drawing of the parrot and his viewer, including another rotund man and a further type developed in the limericks, an exaggeratedly elongated individual, who walks with a stick, as does also a stooped old man who appears in front of him. The picture indicates that the conception and iconography of the human figures in the limericks was encouraged by his natural history work, as, identifying with the birds in his illustrations, Lear scrutinises their human visitors.

Figure 1 A page from Edward Lear's sketchbook, Rose-ringed Parrakeet and visitors to the Zoological Gardens, c. 1830. MS Typ 55.9 (60), Houghton Library, Harvard University.

While the animals of Lear's early natural history illustrations are represented respectfully and sympathetically as each having an autonomous being and a dignified bearing, the nonsense works usually figure human beings conversely, not only as caricatures but in the way that both Christian and modern philosophical thought habitually distinguish animals, as creatures that, far from being free-willed agents, are subject to the fixed laws of nature, both their own and that of their physical environment. The Nonsense Botany's 'Manypeeplia Upsidedownia', in which a series of people hang by their feet like snowdrops from the leaning stem of a plant, is an emblem, indeed almost a manifesto, of this inversion, a systematic reversal of our conventional ideas about ourselves in relation to other organisms. Another such specimen, 'Phattfacia Stupenda', in which a large round, quite mad-looking, human face, akin to that of the Yonghy-Bonghy-Bo,[58] springs from a sparsely leaved stem, similarly illustrates this conception of human nature as fixed and subject, whilst also recalling the obese characters that exemplify this condition in the limericks.

The drawings for Lear's early limericks tend to depict their characters' subjection to fixed laws emblematically as a slapstick battle with gravity, the fundamental force of external nature. Indeed most represent some instance of instability, and several a prelude to a fall. Many protagonists are endowed with large bellies and heads, so that some, such as the elderly habitués of Calcutta, Cheadle and Bromley, and a Young Lady of Turkey, are confined to the ground.[59] Others, including an Old Man of Nepaul and another with a beard, are depicted falling off horses or into the water, perched uncertainly on chairs or in trees, or balancing on clogs, at a casement, or on one foot whilst reading Homer, before jumping off a cliff.[60] An Old Person of Spain sits in a pose of perpetual falling, with his feet and his hands in the air, while others remain resolutely horizontal.[61] Many have their arms stretched out alarmingly straight behind their heads, often to counterbalance such features as a large nose, a prominent pin-like chin, or a beard with birds living in it, while an Old Person of Tring is bent forward by the weight of a very large nose-ring.[62]

At their most optimistic and joyous, Lear's characters voluntarily engage with the force of gravity, happily heightening the conflict with it by exercising their own muscular forces in countervailing movements and by balancing. This takes the form of dancing or simply walking gracefully, both of which are usually seen in the limericks to follow from an affinity with birds, the natural masters of defying gravity. They are a variant of the category of characters noted earlier who enjoy symmetrical and coincident relations with animals. The 'Old Man, on whose nose, / Most birds of the air could repose' is pictured as having a very long nose upon which a series of birds are happily perched as he walks along, smiling, thin, upright and perfectly balanced with his arms outstretched, in stark contrast to the ungainly gait of his peers in other limericks. The 'Young Lady whose bonnet, / Came untied when the birds sate upon it' is similarly depicted happily carrying several birds on her large bonnet whilst balancing balletically on the toe of one shoe, with the other leg and arms outstretched, echoing the poses of the birds that fly around her, and like them prevailing over gravity. She has the audacious and insouciant poise of Lear's Toco Toucan (*Ramphastos Toco*), a large bird pictured toying mischievously with gravity, as it leans far forward on its tree branch with its tail feathers extended, its wings slightly akimbo and its head and large yellow beak raised.[63] In a similarly elegant pose to the Young Lady with the bonnet, albeit one that is reminiscent of fencing rather than dancing, the 'Young Lady of Hull, / Who was chased by a virulent Bull' is pictured countering this particular force of nature with a smile, one foot forward, one arm in the air and another pointing a spade at the animal.[64]

Lear's drawings for *A Book of Nonsense*, cruder and starker than those for the later limericks, present many of their protagonists as unwieldy lumpy creatures that sit precariously, lurch forward, or else share their author's 'disposition to tumble here & there',[65] like helpless atoms. His apparently satirical reductionism resonates with Victorian doctrines of biological evolution, which applied deterministic models of animal nature and more radical forms of materialism to describe human beings. Indeed in the 1830s, Lear's work was brought together tangentially with that of Charles Darwin, through his unacknowledged contribution of lithographic plates to the volume on birds (1838–41) for the *Zoology of the Voyage of the H.M.S. Beagle*, which Darwin had enlisted Gould to produce. As Jonathan Smith recounts, Gould inadvertently precipitated the young scientist's theory of island speciation by identifying various Galapagos finches and mockingbirds as distinct species and correlating them with their physical isolation on various islands. By recording these distinctions and correspondences, Gould's volume on birds marks a crucial point at which Darwinian biology would lead the life sciences away from the natural historical and natural theological cosmos that his folios rendered so grandly and entrepreneurially, to place the discipline on its modern professional footing. While Lear makes his own obscured contributions to this volume, it is, as was suggested earlier, rather in his nonsensical departures from natural history that his work offers affinities with the new biology. The simplified pictures used for the limericks suggest a pastiche of the utilitarian illustrations favoured by the emerging life sciences, which, encouraged alike by the palaeontologist Richard Owen's comparative anatomy and Darwin's mechanism of natural selection, became over the course of the century increasingly minimal and monochromatic, functionalist and diagrammatic. In the period stretching from Robert Chambers's *Vestiges of the Natural History of Creation* (1844) through to Darwin's *Descent of Man* (1871) and its aftermath, the odd parallel universe of Lear's verses and drawings offered a surreal articulation of popular Victorian preoccupations with 'Man's Place in Nature'.

As well as making his characters ostentatiously subject to (or otherwise engaged with) the physical law of gravity, they are each also defined scientifically by an idiosyncratic causal principle. In contrast to the various moral entities that populate novels of the time, the protagonists of Lear's limericks are usually defined by peculiar tragi-comic chains of causality, various involuntary compulsions, impulses, customs, habits, accidents and arbitrary perceptions and reflexes that determine their interactions with the phenomenal world and their fate:

> There was an old Person whose habits,
> Induced him to feed upon Rabbits;
> When he'd eaten eighteen, he turned perfectly green,
> Upon which he relinquished those habits.[66]

The drawing for this limerick is, as Lisa Ede observes, ambiguous.[67] It depicts a stolid big-bellied man in a state of some alarm, apparently in the act of relinquishing his habits, as a cheeky-looking rabbit appears to be escaping from his mouth, and an earlier escapee to be hurtling through the air above a group of its peers, who all appear happy and knowing.

While such human creatures have difficulties with various forces of gravity, habit and compulsion, they are most subject to the arbitrary form of the limerick, the ultimate laws by which they come into being and act. The peculiar rhyme scheme, which in the first two lines of the example cited above foredooms the old Person to his rabbit-habit, is dominated by the overarching law that (in most cases) makes the first and last lines end with the same word or phrase, a formal gesture of enclosure and futility that precludes the possibility of real change and growth. The old Person who in the final line of his limerick 'relinquished those habits' that were introduced at the close of the first, does so through the intervening bout of Pavlovian conditioning, he is not the reflective hero of a *bildungsroman*.

KNOWLEDGE FROM NONSENSE

The representation of a species as an idiosyncratic individual, which proceeds from the synecdochal principle of the holotype and its proxy and occasional instance, the natural history plate, is turned on its head in Lear's limericks and their drawings, which reciprocally cast the eccentric individual as a species unto her- or himself, apparently denuded of moral autonomy and agency. Each such protagonist is fixed by a generic nature and snugly contained within its own world:

> There was an Old Man of Peru,
> Who never knew what he should do;
> So he tore off his hair, and behaved like a bear,
> That intrinsic Old Man of Peru.[68]

The limerick concludes preposterously that the indecisiveness and consequent random behaviour of the Old Man is authentic and noumenal, that it constitutes the thing-in-itself. The summary description of his nature as 'intrinsic' highlights the distinctive and closed quality of the respective ontologies that define the human creatures of Lear's limericks. As was suggested earlier, this hermeticism is marked formally by the enclosing rhyme that the first line makes with the last, which usually takes the form of a simple repetition of the name of the place, or less often a physical or behavioural peculiarity, that defines the person. Flanked by these terms, each of which rhymes identically with the second line, the limerick faces inwards, to its intrinsic world, at the heart of which is the internal rhyme of the third line, a formal innovation that comes of compressing the third and fourth lines of the five-line 1820s limerick, which has an AABBA rhyme scheme. It is a form that, like the descriptions that accompany natural history illustration, is dedicated to fixing the distinctive identity of its object.

By placing repetition instead of rhyme at the end of its final line, Lear's limericks enact the prosodic equivalent of tautology in logic. This is another way in which nonsense sides with science against poetry. 'Tautology is', Lecercle observes, 'the best form of anti-metaphor: simple repetition prevents shift of meaning, and Nonsense shows a strong relish for analytic truths.'[69] The closing line of the 'Old Man of Peru', which adds to the generic title of its protagonist a distinctive adjective, is typical of Lear's limericks, and demonstrates a further way in which they function hermetically, and indeed pseudo-scientifically, namely as nonsense deductions, mock-syllogisms. This is a function, a formal recognition, of the scientific principle of causality that the limerick subscribes to so rigidly and absurdly, as the distinctive nature of the verse's object, specified by its habitat and behavioural trait, is made to yield its consequence.

Lear's limericks each consist of three categorical propositions. The first rhyming couplet, which occupies the opening lines, and the second, contained by the third line, form a pair of premises that yield ostensibly new knowledge through the new adverb or adjective that in the conclusion often conditions the formulaic name of its singular species:

> There was an Old Person of Hurst,
> Who drank when he was not athirst;
> When they said, 'You'll grow fatter,' he answered, 'What matter?'
> That globular Person of Hurst.

The implicit middle term in this limerick is the idea of drinking when not thirsty (which the second premise declares leads to obesity), the minor term is the 'Person of Hurst' and the major term the attribute of obesity, of becoming 'globular'. As this example demonstrates, the rigorous causality of the limericks (e.g., excessive consumption of liquids results in obesity) lends itself to such sequential quasi-syllogistic reasoning. The repetition of the minor term in the final line is varied by the application of the major term, the conditional adjective or adverb, which is usually more than the simple reiteration of the major premise, being rather an inference from it that yields either new knowledge or a nonsensical pretence to it.

The major terms of Lear's mock syllogisms are often not only concise but strikingly arbitrary renditions of their major premises, as occurs with the Old Man of Peru, whose distinctive behaviour, of tearing out his hair and behaving like a bear, is distilled to the attribute 'intrinsic', rather than, for example, dismissed as inexplicably irrational, simply mad. Apparently feted to yield new knowledge, the adverb or adjective introduced in the concluding line often appears to be a momentous addition to the generic title of the limerick's subject:

> There was a Young Person of Crete,
> Whose toilette was far from complete;
> She dressed in a sack, spickle-speckled with black,
> That ombliferous person of Crete.[70]

The Young Person of Crete is contained both by her distinctive sack and the corresponding coinage 'ombliferous'. For Lecercle coinages are the complement to matter-of-factness in the 'double strategy' that nonsense uses to counteract metaphor, for they are 'metaphors gone berserk', a way of destroying metaphor by 'push[ing] its undecidableness to the limit'.[71] This consideration does not, however, mean that such neologisms are mere gibberish. Positioned as an inference drawn from the third line, the predicate 'ombliferous' suggests a correlate for the Young Person's condition of being held within her sack, as does also the neologism's suggestively mimetic sound. In its first syllable, the open sound of the initial vowel, the 'O' that is in its shape also open like the mouth of a sack, is deepened and gently closed by the consonants that follow it, much as in the accompanying drawing the sack is carefully tied around the Young Person's neck with string in a bow. The sack is the *differentio specifica*, the specific difference that distinguishes the Young Person of Crete from other organisms, while the Latinate suffix that completes the neologism renders this quality generic. The coinage suggests a nonsense taxonomy, a conceit that Lear develops more completely in his Nonsense Botany, with such pseudo-Latinate, mock-Linnaean binomials as the 'Phattfacia Stupenda' encountered earlier, and 'Tickia Orologica'.[72] Such coinages give a recklessly *a priori* description of genus and species, preposterously allowing language to dictate reality, to bring distinctive organisms into being by fiat, through a nonsense *logos*. The *OED* observes that the English suffix '-ferous' is used especially in natural history, and gives such instances as carboniferous, cocciferous and fossiliferous. Its Latin origin, *fer-us*, meaning wild or savage, a sense familiar from such English words as feral and ferocious, extends hyperbolically the initial coy reference to the dress protocols of this Young Person, 'Whose toilette was far from complete'. It applies to the sack she wears, which is distinguished in the manner of natural history specimens by its markings, being pied like a bird's egg, 'spickle-speckled with black'.

Like the nonsense binomials of the botanies, the neologism 'ombliferous' marks the 'Young Person of Crete' as both specific, indeed singular, as it applies uniquely to her, and also generic, for its Latinate adjectival formulation indicates a class of beings. Similarly, in the coinage that distinguishes 'The Scroobious Snake' in one of the nonsense rhymes, the use of the Latin suffix advises that the creature is characterised by, or full of, the quality designated by the first syllable, which, similar to that of 'ombliferous', enacts through the longer but more contained sound of its doubled vowel a version of enclosure, the curtailment of his serpentine reach, for he 'always wore a Hat on his Head, for / fear he should bite anybody'.[73] His scrupulous scroobious nature distinguishes him from the rattlesnake of the nonsense alphabet. Such neologisms highlight Lear's treatment of his subjects as reifications of language and the sole specimens of a class, singular inhabitants of the preposterous taxonomies that compose the huge counterfactual science of his nonsense works.

COUNTERFACTUALS

The contemporary philosopher David Lewis has drawn attention to the relations that exist between our conceptions of causation and counterfactual thought, imaginary representations of possible situations that do not actually exist in our world. He begins his 1973 essay 'Causation' by observing that in defining a cause the eighteenth-century Scots philosopher David

Hume inadvertently does so 'twice over', first by specifying a regular succession of events, *'an object followed by another, and where all the objects, similar to the first, are followed by objects similar to the second'*, but then, in ostensibly restating his definition, he makes cause dependent upon counterfactuals: 'Or, in other words, *where, if the first object had not been, the second never had existed.*'[74] 'When you entertain a counterfactual,' Seahwa Kim and Cei Maslen note, 'you imagine or make-believe that the antecedent is true, and consider what else has to be (or is) also true.' The antecedent can be regarded as a hypothesis,[75] a *'what-if'*, a function that Maxwell clarifies with his principle of the 'thought experiment'.

Introduced in his 1871 book *Theory of Heat*, the 'thought experiment' has since become a staple of physics and philosophy and is used widely in other disciplines. For its original example Maxwell builds upon the work he did from the late 1850s on the diffusion of gases, in which he theorises the physical properties of gas 'on dynamic principles', by analogy with the mechanics of elastic solid particles, as rapidly moving and constantly colliding. A hypothetical agent, which his fellow physicist William Thomson later dubbed 'Maxwell's Demon', is tasked with mediating the flow of gas through an aperture between two containers of gas to direct warmer (and hence more fast moving) molecules into one side and cooler (more slowly moving) molecules to the other, in apparent contravention of the second law of thermodynamics, which asserts that all energy in the form of heat moves to distribute itself evenly throughout space:

> we have seen that the molecules in a vessel full of air at uniform temperature are moving with velocities by no means uniform, though the mean velocity of any great number of them, arbitrarily selected, is almost exactly uniform. Now let us suppose that such a vessel is divided into two portions, A and B, by a division in which there is a small hole, and that a being, who can see the individual molecules, opens and closes this hole, so as to allow only the swifter molecules to pass from A to B, and only the slower ones to pass from B to A. He will thus, without expenditure of work, raise the temperature of B and lower that of A, in contradiction to the second law of thermodynamics.[76]

This experiment demonstrates that the second law is not the strict invariant law that its title suggests, but a statistical principle.

Maxwell's thought experiment brings into being an imaginary sub-microscopic world in order to trace its consequences. As well as describing such scientific counterfactuals, the term 'make-believe', which Kim and Maslen use for the way that we attribute truth to imagined antecedents, evokes ideas of children's play. This association has come to form another branch of contemporary thinking about counterfactuals, principally through the work of the developmental psychologist Alison Gopnik. Indeed she and her colleagues develop Lewis's work on counterfactuals and causal thinking to furnish an evolutionary rationale for human childhood, an unproductive period of dependency that is inordinately long in our species compared to that of others. Gopnik argues that rather than relying exclusively upon a set of instincts, native propensities for certain behaviours that enable other animals to function as adults relatively quickly, our extended period of childhood allows us to develop supple imaginative ways of learning that yield causal theories about how the physical, psychological and social worlds work. Such theories, which are seen to provide the grounds by which individuals can both respond to change and impose it upon the object world, are developed by children through their imaginative play, as they explore the consequences of various counterfactual representations they create, such as an imaginary friend, a soup-eating dragon, a world where everyone walks backwards, or teddy bears taking tea.

The causal sequences of Lear's limericks correspond suggestively to this account of play, and may explain children's recognition and enjoyment of them, and indeed the ticklish or uncanny resonances with obscure childhood experiences they can have for adults. As was noted earlier, the illustrations and opening lines to Lear's limericks typically define a character with a few distinctive attributes, as a hypothesis, the consequences of which the remaining lines explore and disclose. As inconsequential as these consequences often are, they nonetheless represent a conceptual advance, an achievement that is often registered in the adjective or adverb added to the protagonist's title at the close of the limerick, as was noted earlier of 'That globular Person of Hurst' and 'That ombliferous person of Crete'. By citing actual place names and allocating idiosyncratic individuals to them, the limericks enumerate 'possible worlds', Lewis's idea of causally isolated universes, in which particular counterfactuals are made actual.[77]

The structural affinity between the hypothetical speculations of scientists and the imaginary play of children, the shared protocols of 'make-believe' alluded to earlier, is noted and its implications developed by Gopnik and Andrew N. Meltzoff in their essay 'The Scientist as Child'. The respective modes of counterfactual play by which children and scientists develop theories of the world find common ground in the genre of the limerick that Lear originally devised for the children at Knowsley Hall, while he was there as resident natural history painter. Like Maxwell's thought experiment, the limerick presents a distillation of the scientific hypothesis. Walter Benjamin describes Edgar Allan Poe's story 'The Man of the Crowd' as an 'X-ray picture of a detective story', as losing 'the drapery represented by crime' it pares the genre down to 'the pursuer, the crowd, and an unknown man'.[78] Lear's limerick can be appreciated similarly as an X-ray of the scientific hypothesis, as shorn of any pretence of purpose or sense it presents the raw speculative audacity of imaginative play in a simple causal sequence.

As the means of considering and anticipating possibilities, and therefore working out how to act in new situations, counterfactual thinking is especially requisite in times of great change. It is accordingly unsurprising to find that it features prominently during the period of modernity. Theoretical concerns with counterfactuals occur during the eighteenth and nineteenth centuries in new philosophical interests in causation, the invention of the concept of ideology, and philosophical and literary preoccupations with the imagination. Not only does research science develop rapidly at this time, but also, parallel to it, the new social thought experiment of the novel, which by positing a particular set of characters and circumstances generates lines of narrative consequences.

Having read Brontë's *Villette* in 1853, the year of its publication, Maxwell wrote to his friend and biographer Lewis Campbell with an account of the novel that describes not the plot but its premises, an outline of the principal characters that furnish the defining conditions for its counterfactual experiment:

It is autobiographic in form. The *ego* is a personage of great self-knowledge and self-restraint, strength of principle and courage when roused, otherwise preferring the station of an onlooker.

Then there is an excellent prying, upright, Jesuitical, and successful French school directress; a fiery, finical, physiognomic professor, priestridden, but taking his own way in benevolence as in other things, etc. etc.[79]

An engine for new knowledge, *Villette* progressively discloses and clarifies the consequences of such complex and distinctive qualities for individual psychology and social relations.

Hamilton regards science and poetry as forms of counterfactual thinking that create theories of the world. Furthermore, like Lear, he suggests that they can be traced to childhood play. His early autobiographical poem 'The Enthusiast', which dates from January 1826, attributes its heroic protagonist with both 'the bright gift of Poetry' and that of Science, which it traces to the reveries of childhood:

> He had been gifted, too, with sterner powers.
> Even while a child he laid his daring hand
> On Science' golden key; and ere the tastes
> Or sports of boyhood yet had passed away

> Oft would he hold communion with the mind
> Of Newton.

Hamilton received a demanding and idiosyncratic education, being taught from infancy classical, oriental and modern European languages by his uncle James, who raised him and ran the diocesan school at Trim. Precociously intelligent and engaged by mathematics at an early age, Hamilton's 'boyhood' play is indeed likely to have extended to imaginative 'communion with the mind / Of Newton'. As both scientist and poet the Enthusiast is defined by a capacity for bold counterfactual thought: 'He would create new worlds.'[80] Hamilton's communion with Newton evidently continued into adulthood, for, as was noted earlier, in the 1831 Introductory Lecture on Astronomy he presents the great natural philosopher as akin to the artist in his proliferent ability to create such worlds:

Then Newton came; he felt that power not less than beauty was an object of intellect, that the unity of law, as well as that of form, could make the Infinite, One; he framed therefore a universe of energies; or rather, as the mind of an artist calls up many forms, he meditated on many laws and caused many ideal worlds to pass before him: and when he chose the law that bears his name, he seems to have been half determined by its mathematical simplicity, and consequent intellectual beauty, and only half by its agreement with the phenomena already observed. While, therefore, I do not pretend that the Newtonian philosophy is likely to make men better painters, or sculptors, or poets, than if it never had been invented, I yet consider the structure of that philosophy as bearing much analogy to the productions of painting, sculpture, and poetry, and as being not less than they an intellectual and imaginative creation, having properly only an ideal truth, though charming partly by resemblance. The world which Newton constructed was like the outward world; but had it not been so, he might still have chosen to contemplate it.

Here, as in his 'Ode to the Moon under Total Eclipse', Hamilton defends the value of counterfactual play for its own sake, of arriving at theories of the world imaginatively, against the requirement that such conceptual structures prove veridical: 'For imagined possibility affects us otherwise than believed reality: the interest of the *has been*, the *is*, and the *will be*, differs from that of the *may* and the *might*; and both these interests are combined in physical science in its perfection.'[81]

'THE PHILOSOPHY OF TOYS'

Hamilton describes science as a distracting infatuation in a letter he wrote to his cousin Arthur in May 1823, while preparing to take the entrance examinations for Trinity College Dublin: 'The time I have given to Science has been very small indeed; for I fear becoming again infatuated with it, and prefer giving my leisure even to less valuable reading, if it can be connected in any way with Classical literature.'[82] The classical curricula that dictated the schooling of well-born boys for most of the nineteenth century contained little science, other than a few Books of Euclid's *Elements of Geometry*. Scientific pursuits were effectively relegated to leisure hours and the domestic realms of play and amusement, to be taught by parents and siblings through didactic dialogues and novels. Examples of the former include Jeremiah Joyce's much reprinted *Scientific Dialogues* (1800–3) and Jane Marcet's immensely popular *Conversations on Chemistry* (1806) and *Conversations on Natural Philosophy* (1819). A pair of pioneering novels on science written for children by Maria Edgeworth, a friend of both Hamilton and Marcet, and her father Richard, about the siblings Harry and Lucy (1801; 1813), amply demonstrate the domestic nature of the genre and its concerns. Maria observes in the introduction to the third volume, which she wrote in 1825, after the death of her father, that these books were 'the very first attempt to give any correct elementary knowledge or taste for science in a narrative suited to the comprehension of children, and calculated to amuse and interest, as well as instruct'.[83] The physician John Ayrton Paris dedicates the early editions of a children's novel he wrote to Edgeworth, whilst choosing not to credit his own authorship, a deficit he compensates for by giving the book a long informative title: *Philosophy in Sport made Science in Earnest; Being an Attempt to Illustrate the First Principles of Natural Philosophy by the Aid of Popular Toys and Sports* (1827).

Addressing 'parents and preceptors' in his preface to *Philosophy in Sport*, the anonymous author mischievously declares a swords-into-ploughshares approach to teaching, implicitly contrasting his pedagogy with disciplinarian models of formal education: 'Imagine not, however, that I shall recommend the dismissal of the cane, or the whip ... with the one I shall construct the bow of the kite, with the other I shall spin the top.' In the following extract from the first chapter of the novel, Mr Seymour, the father of the children Louisa and Tom who are to be inducted into 'the philosophy of toys' over their summer holidays from school, explains his idea to his old friend, the 'Rev. Peter Twaddleton, Master of Arts ... Fellow of the Society of Antiquaries' and one-time Fellow of Jesus College, Cambridge:

'I have long thought,' said Mr Seymour, 'that all the first principles of natural philosophy might be easily taught, and beautifully illustrated, by the common toys which have been invented for the amusement of youth.'

'A fig for your philosophy!' was the unceremonious and chilling reply of the vicar. 'What have boys,' continued he, 'to do with philosophy? Let them learn their grammar, scan their hexameters, and construe Virgil; it is time enough to inflict upon them the torments of science after their names have been entered on the University boards.'

By the close of the chapter Twaddleton has been brought around to his friend's idea, and is retained as an approving observer for the sunny season of experiments with toys that the novel recounts. Twaddleton gives to the venture the imprimatur of the cleric and the Cambridge Fellow, albeit one who had 'only succeeded in obtaining the "*wooden spoon*", an honour which devolves upon the last of the "*junior optimes*'", that is, the third or lowest class of graduates.[84] Not only the vicar, but all the adult characters provide commentaries on the experiments and children's play, which they draw principally from classical sources, but also canonical British literature and history, to demonstrate to the reader the intrinsic compatibility of science with both the classics and national culture.

Maxwell knew Paris's book. In a letter dated 8 July 1853, he reports to the Reverend C. B. Tayler, the uncle of his friend 'Freshman Tayler' at Cambridge, that the publisher 'Macmillan was talking to me to-day about elementary books of natural science, and he had found the deficiency, but had a good report of "Philosophy in Sport made Science in Earnest", which I spoke of with you.'[85] A seventh edition of the book, 'Revised and Considerably Enlarged', and dedicated to Paris's friend Michael Faraday, the Fullerian Professor of Chemistry at the Royal Institution, was published in 1853. A draft essay Maxwell wrote in 1877, 'On the Dimensions of Physical Quantities', suggests that the book remained with him in subsequent decades. The essay includes a passage on viscosity that in its collocation of poetic and scientific terms harks back to the discussion of soap bubbles in *Philosophy in Sport*, drawing in particular upon a set of four lines that Paris quotes from Samuel Rogers's 1793 poem 'On a Tear': 'Thus the superficial tension of liquids is the force which as Rogers tells us "moulds a tear".'[86] The young Maxwell's interest in particular scientific toys, which is documented by his old friend Campbell, indicates that he also knew this popular book from childhood.

Mainly remembered for inventing the thaumatrope in 1824, Paris introduced this toy three years later in the first edition of his book, and indeed marketed it very profitably. The device consists of a simple disc with pictures on either side (for example, a parrot on one side and an empty cage on the other), which when spun around its diameter, appear to coincide in a single picture (a parrot in a cage). Campbell remembers the juvenile Maxwell 'twirling his magic discs'.[87] Subsequent editions of Paris's book also describe the phenakistoscope, invented by the Belgian physicist Joseph Plateau in 1832, which similarly depends upon persistence of vision to furnish the illusion of a moving image, 'an improvement on the thaumatrope' that, Campbell writes, Maxwell had 'in full operation before 1839', his eighth year. Consisting of a large disc that spins about its transverse axis, it has a series of evenly placed slits near its outer edge, through which the viewer sees reflected from a facing mirror a succession of gradually varied images placed regularly on the other side of the disc, thereby giving the illusion of movement. Campbell catalogues a variety of such discs that the young Maxwell drew himself, many of them exploring scientific themes: 'the tadpole that wriggles from the egg and changes gradually into a swimming frog; the cog-wheels moved by the pendulum, and acting with the precision of clockwork . . . intricate coloured patterns, of which the hues shift and open and close as in a kaleidoscope'.[88]

Play is for Maxwell literal and abiding, its continuities with his childhood palpable, as he retained a delight throughout his life in such toys as the zoetrope, the 'dynamical top'[89] and his favourite, the devil-on-two-sticks, a wooden reel that is thrown and spun in the air using a string drawn between two hand-held sticks. 'Games or make-believe are', Kim and Maslen observe, 'one species of imaginative activity: specifically, they are exercises of the imagination involving props.'[90] Maxwell provides for such play in the plans he makes in 1850, when he was nineteen, to explore natural philosophy: 'Common Optics at length; and for experimental philosophy, twisting and bending certain glass and metal rods, making jellies, unannealed glass, and crystals, and dissecting eyes – and playing Devils.' This parity between scientific experiment and play is similarly demonstrated in 1856, when he commissioned a scientific instrument maker to construct for him both a stereoscope and a new 'devil'.[91] Throughout his life, Maxwell played devils; a direct, bodily and immediate way of engaging with fundamental phenomena of physics. This Parisian principle provides an important premise for the new Cambridge Mathematical Tripos course Maxwell shaped from 1871 as the first Cavendish Professor of Experimental Physics, for, as he declares in his inaugural lecture, 'we may find illustrations of the highest doctrines of science in games and gymnastics, in travelling by land and by water, in storms of the air and of the sea, and wherever there is matter in motion'.[92]

J. J. Thomson, one of his successors as Cavendish Professor, makes some connections between Maxwell's toys and his scientific discoveries in the essay he wrote for the 1931 volume of essays commemorating the centenary of his subject's birth. Speaking of the 'devil', he writes that:

in fact it is a home-made gyroscope with all the paradoxical properties of that instrument. He attained great skill with it, and no doubt it led him to the construction of his dynamical top, by which he demonstrated in a striking way the properties of bodies in rotation. Another toy which attracted him in boyhood and to which later on he also gave a scientific application was the zoetrope or wheel of life. Long afterwards he used it to represent the way two circular vortex rings play at leap-frog with each other. This is I think the first application of the principle of the cinematograph to scientific purposes.[93]

Maxwell begins his 1857 paper 'On a Dynamical Top' by observing that 'The mathematicians of the last age, searching through nature for problems worthy of their analysis, found in this toy of their youth, ample occupation for their highest mathematical powers.' Following their example, he proceeds to deduce *some suggestions as to the Earth's motion*[94] from the behaviour of the improved spinning-top he invented.

Another of his toys that Thomson notes, the zoetrope, invented in 1834 by William Horner, dispenses with the phenakistoscope's need of a mirror by transferring the picture sequence to the inside of a revolving cylinder that is similarly viewed through slits placed evenly near its upper edge. Maxwell improved the illusion by placing convex lenses in place of the slits. He also made the toy the subject of a problem for the 1869 Tripos examination.[95] 'Among the many optical contrivances designed by Professor Maxwell' that William Garnett alludes to in the biography he wrote with Campbell, only one other is cited, 'the real-image stereoscope', which he again improved with the use of lenses.[96] The physicist Horace Lamb, who was one of Maxwell's students at Cambridge in the early 1870s and an occasional visitor to his home, recalls that 'He had two toys which he would sometimes bring out to entertain fresh visitors. One was the "dynamical top" . . . The other toy was a form of ophthalmoscope which he had independently invented, probably in connection with his researches in Dioptrics. He was wont to demonstrate the use of this on himself and his friends, including his dog.'[97]

His friend from childhood, Campbell was uniquely well placed to appreciate the continuity between Maxwell's early play and his later science:

The Galloway boy was in many ways the father of the Cambridge man; and even the 'ploys' of his childhood contained a germ of his life-work. Indeed, it may be said that with him, despite the popular adage, 'Work when you work,' etc., play was always passing into work and work into play. In twirling his magic discs, his mind was already busied about the cause of optical phenomena. He plied the devil-on-two-sticks with the same eager industry, and with the same simple enjoyment, with which he afterwards spun his dynamical top. And amidst his profoundest investigations, whether about the Rings of Saturn or the Lines of Force, or the molecular structure of material things, the playful spirit of his boyhood was ever ready to break forth.[98]

As Paris has his protagonist observe, '*Play* and *work* – *amusement* and *instruction* – *toys* and *tasks* – are invariably but most unjustifiably employed as words of contrast and opposition.' He also has Seymour cite the following dictum from the beginnings of British science: '"To enter into the kingdom of knowledge," said Lord Bacon, "we must put on the spirit of little children."'[99]

In 1873, Maxwell wrote a review for *Nature* of scientific essays by Plateau, the inventor of the phenakistoscope, on another children's game: soap-bubbles. Maxwell begins by observing that 'On an Etruscan vase in the Louvre figures of children are seen blowing bubbles', and then identifies himself and his readers with them:

Our admiration of the beautiful and delicate forms, growing and developing themselves, the feeling that it is *our* breath which is turning dirty soap-suds into spheres of splendour, the fear lest by an irreverent touch we may cause the gorgeous vision to vanish with a sputter of soapy water in our eyes, our wistful gaze as we watch the perfected bubble when it sails away from the pipe's mouth to join, somewhere in the sky, all the other beautiful things that have vanished before it, assure us that, whatever our nominal age may be we are the same family as those Etruscan children.[100]

The woodcut by George Cruikshank of two children blowing bubbles that heads Paris's chapter on soap-bubbles and liquids (and is reproduced on the dust-jacket of the present book) may, like the passage on Rogers's poem, have stayed with Maxwell, readying him for the identification he makes here of scientists with such children.[101] Indeed, the woodcut offers a more concrete source for Maxwell than the image from the Etruscan urn he cites, which is of doubtful authenticity and drawn from a description that Plateau quotes in his book.[102]

Scientists are characterised in Maxwell's review essay by Nietzsche's criterion of maturity, as having reacquired 'the seriousness we had towards play when we were children'.[103] They are dedicated to the fragile and elusive poetry of phenomenal nature:

Here . . . we have a book, in two volumes, octavo, written by a distinguished man of science, and occupied for the most part with the theory and practice of bubble-blowing. Can the poetry of bubbles survive this? Will not the lovely visions which have floated before the eyes of untold generations collapse at the rude touch of Science, and 'yield their place to cold material laws'? No, we need go no further than this book and its author to learn that the beauty and mystery of natural phenomena may make such an impression on a fresh and open mind that no physical obstacle can ever check the course of thought and study which it has once called forth.

Plateau is presented as a poetic seer, a characterisation that is enhanced, as it is for Homer and Milton, by his blindness, the 'physical obstacle' that resulted from a series of experiments in which, as Maxwell explains, 'he exposed his eyes to an excess of light'.[104]

Poetry is one of Maxwell's great ludic forms. Science converges with children's games, while their songs are recognised as rudimentary poetry, in '(Cats) Cradle Song, By a Babe in Knots', a late poem that Maxwell wrote for Peter Guthrie Tait, his closest colleague and friend. Beginning with the lines, 'Peter the Repeater, / Platted round a platter', the poem describes Tait's pioneering research work into knots as child's play.[105] According to his friend W. N. Lawson, Maxwell was 'constantly producing' his 'quaint verses' while they were studying together at Cambridge, 'and bringing [them] round to his friends, with a sly chuckle at the humour, which, though his own, no one enjoyed more than himself'.[106] Oddly neglected and even dismissed (a recent study by Jason Rudy asserts that they 'are neither entirely serious nor of the sort that rewards extended reflection'[107]), the poems that Maxwell wrote about science offer a unique record of the creative play and critical perspectives of the greatest physicist of his age, whose achievements are commonly ranked with those of Newton and Einstein.

As the most momentous of its scientists, and the most accomplished of the poets among them, Maxwell is central to the following study of poetry by Victorian scientists. The focus upon research, which was inspired by the German model of science, combined with the public interest in its phenomena and findings, encouraged by such popular lecturers as Michael Faraday and John Tyndall at the Royal Institution and J. H. Pepper at the Royal Polytechnic Institution, saw mid-Victorian physics shaped by imperatives of discovery and novelty. For Maxwell this meant that his discipline, more so than the rest of science, was hemmed in by nonsense. As the liminal form of science, nonsense draws out new and audacious speculations and discoveries that affront established ideas, only to be subsequently accepted into its mainstream, while as its liminal case, its unmeaning other, nonsense demarcates the boundaries of scientific knowledge. Maxwell accordingly uses nonsense to mock and exclude those ideas and practices that he judges to have only the pretence of being scientific, such as spiritualism, 'Scientific Materialism', and popular science. He enjoys the nihilistic high-stakes play of nonsense, the sport of entertaining and establishing the counter-intuitive hypotheses that since his time physics has become notorious for, in the knowledge that, as history demonstrates, their credibility is likely to be revoked by future discoveries, that they too will come to be recognised as literal nonsense. Writing in 1783, the poet William Cowper observes that 'The *vortices* of Descartes gave way to the gravitation of Newton, and this again is threatened by the electrical fluid of a modern. One generation blows bubbles, and the next breaks them.'[108]

CHAPTER 2

Edinburgh natural philosophy and Cambridge mathematics

Maxwell often signed his letters to Tait with the formula '$\frac{dp}{dt}$', which, as Tait's colleague and biographer C. G. Knott explains, '$=JCM$', an 'expression for the Second Law of Thermodynamics' that was devised by William Thomson.[1] With Maxwell's adoption of it, the formula becomes a pun on his initials and his mortality, the second law of thermodynamics being unique amongst the laws of physics in asserting time's arrow, as over time energy for work becomes degraded and dissipated as unusable heat. The working scientist will become a cold body. His scientific achievements will be superseded by new discoveries and theories. In his 1959 Rede Lecture on 'The Two Cultures and the Scientific Revolution', C. P. Snow argues that the second law of thermodynamics has a cultural importance on a par with '*a work of Shakespeare's*'.[2] By making his signature, the seal and symbol of his selfhood, at once a scientific principle and his *memento mori*, Maxwell does indeed draw from the second law a poise and poignancy, a fusion of punning play and metaphysical gravity, that offers to bridge the parallel that Snow makes of such science with Shakespeare. It implies an urgency to Maxwell's scientific work, which at the time of his premature death in 1879, at the age of forty-eight, had long been prodigious and profound.

Consistent with his delight in the 'devil' and other toys, Maxwell's pun demonstrates his (signature) lightness and playfulness. As an undergraduate at Cambridge, studying for the Mathematical Tripos from 1850 to 1854, Maxwell was invited to join the 'Apostles', a select group of up to twelve students who met around the reading of a member's essay, written on a topic of his choosing. Maxwell's essays for the Apostles accordingly offer a valuable record of the, mainly philosophical, questions that most exercised him as an undergraduate. They are, Campbell writes, 'real indications of the writer's speculative tendencies, and are most characteristic of the activity and fullness of his mind, of his ironical humour, and of his provoking discursiveness and indirectness of expression'.³ Maxwell begins his 1856 essay for the Apostles 'Are there Real Analogies in Nature?' by observing a reciprocal relation between the pun, which is rarely taken seriously as a form of knowledge, and the more generally credited epistemological principle of the analogy:

> In the ancient and religious foundation of Peterhouse there is observed this rule, that whoso makes a pun shall be counted the author of it, but that whoso pretends to find it out shall be counted the publisher of it, and that both shall be fined. Now, as in a pun two truths lie hid under one expression, so in an analogy one truth is discovered under two expressions. Every question concerning analogies is therefore the reciprocal of a question concerning puns, and the solutions can be transposed by reciprocation. But since we are still in doubt as to the legitimacy of reasoning by analogy, and as reasoning even by paradox has been pronounced less heinous than reasoning by puns, we must adopt the direct method with respect to analogy, and then, if necessary, deduce by reciprocation the theory of puns.⁴

More than an entertaining gambit to engage his audience, these jocular opening remarks define Maxwell's poised orientation to the problem of knowledge, which he recognises here as both metaphysical and cultural.

For Lecercle, as Chapter 1 observed, nonsense opposes the normative workings of metaphor, which it identifies with semantic instability. Nonsense flanks metaphor on one side with forms of literalism, such as tautology and 'matter-of-factness', and on the other with coinages that precipitate its 'undecidableness'. While the preceding chapter argued that such liminal phenomena are encouraged by Lear's practice of natural history illustration and exemplified in his nonsense work, they do not exempt the median instance of the metaphor, nor the more speculative sciences that employ it, from the charge of nonsense. Modern physics, a milieu of field theory, ethers, energy and mathematical models, lends itself to the use of analogies, with their attendant risks of semantic duplicity, conflation, and indeed febrile imaginings. Maxwell delights in the semantic equivocation of analogy, which he, like Lewis Carroll, sees to have its reciprocal and more overt case in the pun.

While there can be no quarrel with the trivial axioms of tautology, new knowledge arises less certainly and securely through the comparison of different terms, a problem that Maxwell poses drolly with the radical instances of the pun and the analogy. Both teeter on the brink of nonsense, as in each case the relation between their respective twin terms, understood literally, is unmeaning. The statement from Shakespeare's *Romeo and Juliet* that 'Juliet is the sun'⁵ is literally false, for the disparate terms equated here have in themselves little in common beyond the simple predicates of existence and extension. The punning quibble on the word 'son' in the opening lines of his *Richard III* similarly casts the king as the 'sun': 'Now is the winter of our discontent / Made glorious summer by this son of York.'⁶ To read 'son' and 'sun' literally and discretely makes minimal or no sense, the meaning of the lines occurs only in the punning convergence of these terms, which is facilitated and enhanced by the knowledge that York's heraldic emblem is the Sun in Splendour. In such instances analogies and puns snatch sense from the literalist jaws of nonsense.

An avid reader of Carroll's Alice books, Maxwell wrote the following note, in mirror-writing, to remind Tait to send him a copy of *Through the Looking-Glass*: 'Why have *you* forgotten to send Alice. We remain in Wonderland till she appears.'⁷ As if courteously returning this reflective homage, Carroll's book demonstrates Maxwell's dictum about the reciprocity of analogies and puns through its revival of 'dead' metaphors. So, for instance, in 'The Garden of Live Flowers', the figure of the flower-bed, the metaphoricity of which has been lost sight of as it passed into common usage, furnishes a pun that explains why it is that Alice had not heard flowers talk before; '"In most gardens," the Tiger-lily said, "they make the beds too soft – so that the flowers are always asleep."' Similarly, at dinner with Alice and the White Queen later in the book, the Red Queen, having formally introduced Alice to the prospective courses, a leg of mutton and then a plum pudding, has them removed from the table as 'it isn't etiquette to cut anyone you've been introduced to'.⁸ The pun becomes the conduit through which the literal meaning that originally furnished a now forgotten metaphor is retrieved and applied directly to the new meaning. Instances of sense that were once twinned in a metaphor are subsequently reunited in a pun that yields a teasing example of nonsense. Maxwell uses hermeneutic tropes of hiding and discovery to describe the pun as 'two truths ... hid under one expression' and the converse case of the analogy, in which 'one truth is discovered under two expressions'. Truth is extracted, actualised, from analogies and puns by the mind's intrinsic powers of wit and play, so that hermeneutic success is like getting a joke, the happy resolution of a puzzle.

PUNS AGAINST PEDANTRY

The institutional hostility to the pun that Maxwell dramatises in his 'Analogy' essay is represented in Paris's *Philosophy in Sport* by Twaddleton, whose 'antipathy to an English pun was so extravagant as to be ridiculous'. Twaddleton's disgust at punning is traced to his experience at Cambridge, and in particular to 'his frequent intercourse with the Johnians, a race of students who have, from time immemorial, been identified with the most profligate class of punsters'.⁹ Maxwell pointedly identifies Peterhouse, where he spent his first term at Cambridge, with a Twaddletonian prohibition on punning. Garnett writes in his 1879 obituary of his colleague and friend that 'Clerk Maxwell's first term in Cambridge does not seem to have been a very happy one ... Finding himself comparatively without friends at the end of the term, he ... migrated to Trinity on December 14th, 1850' where he 'found congenial spirits'.[10]

The opening sentence of the 'Analogies' essay brings to its most poised ironic expression an opposition that preoccupied Maxwell at Cambridge, between inflexible rules and free play: 'In the ancient and religious foundation of Peterhouse there is observed this rule, that whoso makes a pun shall be counted the author of it, but that whoso pretends to find it out shall be counted the publisher of it, and that both shall be fined.' This solemn statute holds within it the risk of its own infringement, as it declares that not only the perpetrator but the perceiver of the pun will be punished. Indeed to observe the second part of their rule by recognising a pun, the prerequisite for identifying its 'author', would make the college authorities its 'publisher', and so equally liable to punishment. However, the college's venerable rule-bound ethos means that it would not be amenable to this implication after all, as its pompous literalism precludes it from recognising any such instance of word-play. Freely announcing and indeed universalising its authors' and enforcers' complete lack of imagination and humour, the statute discounts claims to having discovered a pun as mere pretence, for it stipulates that only 'whoso pretends to find it out shall be counted the publisher of it'.

Maxwell makes fun of college rules and their policing in his 1854 verse 'Valedictory Address to the D[ea]n'. A parody of Robert Burns's poem 'John Anderson', the poem is written in dubious honour of the Senior Dean, John Alexander Frere, who was leaving the university to become Rector of Shillington. Frere was notorious for sending finicky 'note-lets' to students for the slightest neglect of college protocols, which Maxwell parodies here:

> Why he was out of College,
>   Till two o'clock or near,
> The Senior Dean requests to know,
>   Yours truly, J. A. Frere.[11]

Maxwell came to Cambridge in October 1850 after studying natural philosophy for three years at the University of Edinburgh. His impatience with the English college system and its rules can be explained by his earlier experience at the Scots university, where, as one of his contemporaries records, undergraduates were treated 'less as students than as young men ... with no prospect of chapel in the morning, and with no fear of being shut out at night'.[12] This difference is understood by Maxwell as one of not only social mores but intellectual culture. He makes the observation in his poem, which he puts more pointedly in the parody of the statute at the beginning of the 'Analogies' essay, that the corollary of such fussy procedural pedantry is intellectual dullness:

> The Lecture Room no more, John,
>   Shall hear thy drowsy tone,
> No more shall men in Chapel
>   Bow down before thy throne.[13]

The Peterhouse rule announced at the start of the 'Analogies' essay draws attention to scholarly protocols, a more subtle form of university regulation: 'But since we are still in doubt as to the legitimacy of reasoning by analogy, and as reasoning even by paradox has been pronounced less heinous than reasoning by puns, we must adopt the direct method with respect to analogy, and then, if necessary, deduce by reciprocation the theory of puns.'[14] The vogue amongst the early to mid-Victorians for puns, amply apparent from such books as Paris's *Philosophy in Sport* and Percival Leigh's *The Comic English Grammar* (1840) and such popular magazines as *Punch*, *The Comic Almanack* and *Hood's Magazine*, would not have helped their case for recognition as a logical form. The prohibition upon 'reasoning by puns' is implicit in the canons of scholarship. By casting it in the banal but overt form of college rules, paralleling it to what are depicted in the 'Valedictory Address' as arbitrary and petty codes of behaviour, Maxwell draws attention to the ways that conservative institutional forms of knowledge can quietly discipline thought and restrict its creativity.

Lecercle argues that 'nonsense as a genre is a by-product of the development of the institution of the school ... where not only rules of grammar, but also maxims of good behaviour, linguistic and otherwise, are learnt'.[15] Identifying the university with the propagation of such arbitrary social and linguistic manners, Maxwell affronts this model of sense with his epistemology of puns, which like the nonsense of Lear's limericks harks back to the play of children not yet socialised by formal education. Scholarly protocols are identified as arbitrary manners, akin to the rules of etiquette around mutton and plum pudding in Carroll's *Through the Looking-Glass*, and the normative code that Lear's 'they' police so cruelly in his limericks.

A poem from 1853, which bears the unwieldy title '*Lines written under the conviction that it is not wise to read Mathematics in November after one's fire is out'*, depicts the university's rules as proudly restricting the play of intelligence in order to produce pedants:

> Then, I said, 'These haughty Schools
> Boast that by their formal rules
> They produce more learned fools
>   Than could be well expected.'[16]

Maxwell characterises pedants as sober but not sensible in a poem from 1848, 'Song of the Edinburgh Academician', in which, a year after matriculating to Edinburgh University, he looks back to his schooldays:

> Let Pedants seek for scraps of Greek,
>   Their lingo to Macadamize;
> Gie me the sense, without pretence,
>   That comes o' Scots Academies.[17]

After an unsuccessful period of home schooling, Maxwell was sent at the age of ten to the gentlemanly Edinburgh Academy, where he was in the same year as Campbell and a year ahead of Tait and Fleeming Jenkin, another friend and future scientific ally. This early education evidently left Maxwell with little tolerance for the bad faith of the pedant, who stands accused in the 'Song of the Edinburgh Academician' of using language as a vehicle not of 'sense' but of 'pretence'. Akin to the college rules and Frere's enforcement of them in the 'Address', pedantry is seen to be unyielding, as it renounces the fluency of commonsense usages for its own language, which is levelled and compacted, made impenetrable, through its use of archaeological fragments of Greek, like the evenly sized stones applied in layers in J. L. McAdam's method of making and repairing roads. This even surface is likely to support only a predictable and pretentious traffic of ideas, rule-bound thought that runs along well-established trajectories.

   There is no evidence that Maxwell wrote any poems lampooning Edinburgh University. He was well disposed to the University, and it to him, several years before he came to study there. The young Maxwell had devised an elegant and indeed definitive mechanical means for drawing a perfect oval, which he elaborated in February 1846 in his paper 'Observations on Circumscribed Figures having a plurality of Foci, and Radii of Various Proportions'. Maxwell's father John showed the paper to Forbes at Edinburgh University, who passed it to the professor of mathematics Philip Kelland. The paper greatly impressed both of his future lecturers, with Forbes reporting back to John Clerk Maxwell that '[Kelland's] opinion of your son's paper agrees with mine; namely, that it is most ingenious, most creditable to him, and, we believe, a new way of considering higher curves with reference to foci.'[18] Forbes presented an account of the paper on the schoolboy's behalf to the Edinburgh Royal Society in April 1846, in which form it became Maxwell's first scientific publication.[19] He arrived at Edinburgh in autumn 1847, and for the following three years of his degree was granted privileged access to the university laboratories by Forbes and his colleagues.

Maxwell found the Scottish system deficient in the pedantry and cynicism that provoked him into verse at Cambridge. The Scots system, with its five universities, and much smaller population than England with its three universities, was characterised by, as George Davie puts it, a 'democratic intellectualism' that eschewed the class-based privilege of the Oxbridge system. While Edinburgh taught natural philosophy through experiments, an activity that was still widely identified with manual labour, Cambridge considered 'analysis' to be a more appropriate pursuit for gentlemen taking its Mathematical Tripos. Tait recalls that 'Clerk-Maxwell spent the years 1847–50 at the University of Edinburgh, without keeping the regular course for a degree. He was allowed to work during this period, without assistance or supervision, in the Laboratories of Natural Philosophy and of Chemistry: and he thus experimentally taught himself much which other men have to learn with great difficulty from lectures or books.'[20]

Maxwell also attended Sir William Hamilton's lectures and classes on metaphysics and logic at Edinburgh from 1847 to 1849. An early letter to Campbell alludes to the curious way in which the class sessions were conducted: 'In Logic we sit in seats lettered according to name, and Sir W. takes and puts his hand into a jam pig full of metal letters (very classical), and pulls one out and examines the bench of the letter.'[21] Eneas Sweetland Dallas, a student at Edinburgh three years Maxwell's senior, writes that 'Sir William Hamilton's class was perhaps the most marvellously conducted class in any university', and in justifying his claim provides a gloss for Maxwell's intriguing reference:

About 150 students were ranged on seats before the professor, who lectured three days in the week, and on two days held a sort of open conference with his pupils, which was conducted in this wise: – Sir William dipped his hand into an urn and took out a letter of the alphabet – say M. Any student whose name began with M was then at liberty to stand up and comment on the professor's lectures – attack them – illustrate them – report them – say almost anything, however far-fetched, which had any relation to them. A couple of Macs get up at once. The first merely raises a laugh by topping one of his William's philosophical anecdotes with another which he fancies to be still better. The second gets up, and has a regular tussle with his master about the action of the mind in sleep, and in a state of semi-consciousness. It is all over in five minutes, the student at length sitting down in a state of profuse perspiration, highly complimented by Sir William for his ingenuity, and feeling that he has done a plucky thing which thoroughly deserves the cheers of 149 fellow-students. These exhibitions are quite voluntary, and it appears that among the M's there is no more heart to get up and speak. The letter C is therefore next taken out of the urn ... So the hour passes, each letter of the alphabet being presented in turn, and all the students who desire it, having a chance of speaking. Sometimes the exercise was varied by essays being read, or by Sir William Hamilton suddenly propounding a difficult question as to the use of a term ... Never was there a class in which so much enthusiasm manifested itself.[22]

Hamilton's classes would have not only galvanised Maxwell's interest in philosophy, but informed the approach he took in pursuing it as a Cambridge Apostle. They evidently encouraged the humour, sociability and free play of ideas noted earlier of the 'Analogies' essay.

It is unsurprising, given the freedom and independence Edinburgh allowed him, that Maxwell should find Cambridge rule-bound, authoritarian and oppressive. He missed studying philosophy and objected to studying natural philosophy through repetitive mathematical problems rather than physical experiments (an element that he would introduce to the Tripos in the 1870s as the inaugural Cavendish Professor). The Scots system did not share the Newtonian confidence of the Cambridge Tripos that reality could be apprehended by pure reason, but in accordance with Hume's empiricism emphasised the role of experiment, a sceptical perspective that is fundamental to Maxwell's science. Hume's university, Edinburgh, one of the principal universities of the Enlightenment, furnished Maxwell with certain ideals, with Forbes and Hamilton, in particular, emphasising attention to intellectual integrity and epistemological first principles. Maxwell would judge the Cambridge system by such criteria during his original tenure at the university from 1850 to 1856, as a student for four years, then as a scholar and, briefly, a fellow, before leaving to take the Chair in natural philosophy at Marischall College, Aberdeen.

While at Edinburgh, Maxwell would have been aware of its radical differences with Cambridge through a series of public controversies that Hamilton had been having with Whewell, the master of Maxwell's future college, Trinity. From the 1830s, Hamilton had been publicly defending the Scots tradition, in which mathematics was taught through Euclidean geometry as integral to classics and philosophy, against the modern Tripos's more specialist concerns. Whewell maintained that the best means of developing students' capacities for logic was not the formal philosophical study of the subject, as Hamilton argued, but the mathematical focus of the Tripos.[23] Hamilton's edition of Thomas Reid, the dissertations to which Maxwell was reading in July 1848, states his position bluntly: 'A man is made "to reason justly in mathematics", in the same manner in which a man is made to walk straight in a ditch.'[24] It was also largely through Whewell's efforts that continental analysis had been placed at the heart of the Tripos, while the Newtonian calculus or fluxions was retained in 'the place of honour in the syllabus' of the Scots system.[25] This was the subject of another public debate between Whewell and Hamilton, whose emphatic opinion 'On the pernicious influence of the modern analysis, in an educational point of view',[26] as he puts it in his edition of Reid, would have informed Maxwell's expectations of Cambridge. In an 1874 letter to *Nature*, Tait makes a veiled reference to Maxwell's partisanship in this matter: 'One of my most intimate friends in Cambridge, who had been an ardent disciple of the late Sir W. Hamilton, Bart., and had adopted the preposterous notions about mathematics inculcated by that master, was consequently in great danger of being plucked [i.e., of failing his examination].' His college tutor gave him additional lessons in algebra, but, Tait reports, 'my friend's early mental bias [towards Hamiltonian scepticism, and drollness] too soon recovered its sway, and he cried out in an agony of doubt and despair, "But what if x should turn out, after all, *not* to be the unknown quantity?"'[27] In his definitive study of nineteenth-century Scots universities, Davie writes that 'Maxwell was perhaps the last great representative of the line' shaped by this distinctive system, of men whose 'scientific work was informed by a certain metaphysical spirit'.[28] Maxwell also came to Cambridge with a precocious back-catalogue of scientific publications, which would have allowed him some licence to pass judgement on the Mathematical Tripos. As one of his peers from Trinity College recalls, 'We understood even then that, though barely of age, he was in his own line of inquiry not a beginner, but a master. His name was already a familiar name to men of science.'[29]

THE CAMBRIDGE MATHEMATICAL TRIPOS

The final examinations for the Mathematical Tripos were by mid century, when Maxwell began studying for them, pitched at a shrill level of difficulty and competitive fervour. The 1848 examinations were twice as long as those of twenty years earlier, when the modern form of the Tripos was consolidated. Some regulations that came into effect that year had extended the examination period from six to eight days, a total of forty-four and a half hours. Twelve hours were devoted to new examples of mathematical problems to be solved, while the remainder tested 'bookwork', knowledge that could be memorised from books and lectures, such as definitions, laws, theorems and proofs.[30] This long season of examinations, which Maxwell sat in January 1854, contrasts with the five hours of formal examinations he took at Edinburgh in 1847–8, and the six of the following year.[31] The Tripos began with a three-day examination on elementary mathematics, the results of which determined the list of Honours candidates that the moderators announced seven days later. These students could take the five-day examination on higher mathematics, which began two days later.[32] After the Honours examination the best students could take a further three days of even more difficult examinations for the two Smith's prizes. Maxwell sat these additional papers, to emerge second wrangler (that is, second in the first class) and joint winner of the prize.

Some important senate regulations that were passed in 1827 and came into effect a year later finally deprived examiners of the power to give *viva-voce* questions and prescribed that the written examination questions be printed, rather than dictated to the candidates. The change from oral to written examinations had, as Andrew Warwick details, revolutionary consequences for the Tripos.[33] Accompanying the regularisation of the examination procedures, the marking of papers also became progressively more even and fair during the 1830s and 1840s. Walter William Rouse Ball writes in 1889 that the 1836 Tripos 'is said to have been the earliest one in which all the papers were marked', and explains that before this time examiners had 'partly relied on their impressions of the answers given', a practice carried over from the old 'Acts' system, in which evaluations of the *viva voce* were necessarily impressionistic and their ranking dependent upon the examiners' memories: 'The judgement formed of oral performances', Whewell concedes in 1845, 'must, necessarily, be rapid, and may easily be conceived to be hasty and inaccurate.'[34] In contrast to the open-ended essay reading and debating sessions of the 'Acts', which were judged in a discretionary manner by criteria that included civic virtues and gentlemanly niceties,[35] the set examination paper meant that all candidates were judged and compared by the same technical criteria, the same novel questions answered within a fixed time, each of which could be marked according to an examiner's schema. Such scrupulous protocols could be trusted to accurately grade candidates, and by giving greater credibility and hence prestige to the final rankings, exacerbated the competition amongst increasing numbers of students for the higher places in the Tripos. Indeed, as Smith observes, during the peak of the Cambridge competitive examination system from 1830 to 1870, the examiner's marking systems were finely tuned not only to calibrate minute differences between the performances of peers but also to allow comparison between men from different years.[36] By raising and enforcing the standards of educational qualifications, such reforms provided a crucial foundation for the professional practice of the mathematical and physical sciences in Victorian Britain.

Maxwell and his peers use such terms as 'drilling', 'grinding' and 'milling' to describe what they evidently regarded as the gruelling mechanistic tasks of studying for the Tripos, as they learned, often by rote, for the 'bookwork' sections and repetitively attacked mathematical exercises to prepare for the problems and examples sections. The pattern that this gives to the student's life is rendered in the monotonous rhythm of the following lines from Maxwell's 1852 poem, 'A Vision: *Of a Wrangler, of a University, of Pedantry, and of Philosophy*':

> Late to bed and early rising,
> Ever luxury despising,
> Ever training, never 'sizing',
>   I have suffered with the rest.
> Yellow cheek and forehead ruddy,
> Memory confused and muddy,
> These are the effects of study
>   Of a subject so unblest.

The brisk regular rhythm of the first tercet, with its short vowel sounds cast into scrupulous trochaic tetrameter and flurries of feminine rhyme, is attuned to a set of stoic practices, the sacrifices that Tripos students make, apparently gladly, for their studies. This rallying start, as if to a college song, is of course cut short, disrupted and deflated by the conclusion to its sentence, the plain and plaintive truth that the poet has 'suffered with the rest' of his peers. The second tercet answers the first by describing and evoking this oppression in viscous trochaic tetrameter and rhymes weighted down with dull consonants. The eccentric lines that prevail over each of the tercets, and cut through their music with their masculine rhyme, conjoin the students' suffering and its cause, Tripos mathematics, 'a subject so unblest'.

'A Vision' describes a dream that its student persona has about the Cambridge system. It begins with a procession of the various academic staff responsible for preparing the Tripos examination itself and the students who will take it. After this the student meets various university types, principally the mercenary Wrangler. He is 'the learned fool' who, silencing 'restive reason', rejects 'Every scruple' in favour of a comfortable 'place and pension', the 'lofty station' of a college fellowship. This vision of the university effectively dissolves into its essence, the spirit of Pedantry. Made of her tools in trade – 'Hair of pens and skin of paper; / Breath, not breath but chemic vapour;' 'Eyes of glass, with optic axes' – Pedantry embodies a purely instrumentalist approach to intellectual endeavours.

She belongs to a curious species, visual puns in which human faces and figures are composed of mechanical, industrial and consumerist artefacts. A baroque conceit, best known from the predominantly organic composites of Giuseppe Archimboldo's portraits, it can be traced back to Giovanni Bracelli's *Capricci* (1624), Nicolas de Larmessin's *Habits de métiers* from the latter half of the century, and Hogarth's mid-eighteenth-century satirical engraving 'Some of the Principal Inhabitants of the Moon' (1724), where the figures consist of coins, mallets, teapots, mills and other mechanical parts. Nineteenth-century examples include Thomas Rowlandson's 'Twelfth-Night Characters' (1832), which are made of kettles, bottles, tankards, pipes and glasses, and works by J.J. Grandville and George Cruikshank in the 1840s and 1850s.[37] 'Hair of pens and skin of paper' describe Pedantry's broad form and main substance. She is constituted by what Maxwell refers to in his 1871 inaugural Cavendish lecture as the 'familiar apparatus of pen, ink, and paper' and recognises as the emblematic tools of change in the Tripos during the first half of the nineteenth century, as it moved from oral to written examinations.[38]

No more than the sum of her parts, a perfunctory aggregate of materials for scientific study that neither cohere organically nor transcend themselves in the creation of new knowledge, Pedantry embodies a dismally reductive vision:

> ... those dull, unmeaning eyes.
>
> Such the eyes, through which all Nature
> Seems reduced to meaner stature.
> If you had them you would hate your
>   Symbolising sense of sight.
> Seeing planets in their courses
> Thick beset with arrowy "forces",
> While the common eye no more sees
>   Than their mild and quiet light.[39]

The 'planets in their courses' are represented crudely, 'Thick beset with arrowy "forces"', as Pedantry's soulless eyes impose upon them her mechanistic schemata. Rather as Rowan Hamilton does the earth and the moon in his 'Ode to the Moon under Total Eclipse', Maxwell indicates that the planets could also be appreciated imaginatively, through a free 'Symbolising sense of sight', for example in their Pythagorean significance, as he does six months later in his poem 'A Student's Evening Hymn', where 'the stars most musically / Move in endless rounds of praise'.[40] This harks back to the Hellenising philosophical practice of geometry that Sir William Hamilton champions, and opposes to 'the mechanical process of the algebric calculus',[41] the analysis favoured by the Cambridge Tripos. By treating knowledge instrumentally, as a means to an end, Pedantry is seen to denude the planets of their mystery, the subtle lyrical possibilities that 'the common eye' apprehends in 'their mild and quiet light'. Twenty-four years later, in his poem 'Report on Tait's Lecture on Force:—B.A., 1876', Maxwell congratulates his friend for banishing Pedantry's 'arrowy "forces"' from the universe; 'No more the arrows of the Wrangler race.'[42]

PUNNING MACHINES

The frustrations and pressures of the Tripos that Maxwell sketches in such poems as 'A Vision' led to a breakdown after his third-year college examinations in June 1853, while he was staying with Reverend Tayler and his family in their home at Suffolk. 'The long continuous strain of the past months', Campbell writes, 'had been too much for him.' Tayler described the illness 'as a sort of brain fever', from which he and his family nursed Maxwell back to health for 'more than a month'. Immediately after his recovery, having returned to Trinity, Maxwell wrote the letter to Tayler cited in Chapter 1, in which he reports meeting with Macmillan, who 'had a good report of "Philosophy in Sport made Science in Earnest", which I spoke of with you'. It is not surprising that in recovering from his breakdown, Maxwell should recommend to his host a book he probably knew from childhood that affirms the playful experimental ethos he finds baulked by his Tripos studies. Indeed Paris's book presents a burlesque allegory of this conflict of values, with Reverend Twaddleton, the Cambridge man, embodying dull pedantry and Seymour representing the creative pedagogy of play and experiment, which, as with Maxwell, has its anarchic expression and emblem in the pun. Seymour is devoted to punning, which Twaddleton recognises as an affront to pedantry: 'That you should compare the vile practice of punning with the elegant and refined habit of conveying our ideas by classic symbols, does indeed surprise and disturb me.'[43]

Seymour is most clearly his author's proxy, and indeed commercial agent, when he announces the invention of the 'thaumatrope', 'a small machine... which is well calculated to furnish us with some capital puns and well-pointed epigrams', as the following brief advertisement for Paris's toy demonstrates:

> The Thaumatrope;
> being
> *Rounds* of Amusement,
> or
> How to please and surprise
> by *Turns*.

The reference to the toy as 'a small machine' is made to yield a further dimension of punning, as the thaumatrope not only furnishes pretexts for making puns, it also literally manufactures them: 'the "Quarterly Review" has asserted, that a certain English poem was fabricated in Paris, by the powers of a steam-engine; but the author of the present invention claims for himself the exclusive merit of having first constructed a hand-mill, by which puns and epigrams may be *turned* with as much ease as tunes are played on the hand-organ, and old jokes so *rounded* and changed, as to assume all the airs of originality'.[44] The punning effect of the thaumatrope, as two pictures are combined dynamically into one, was often enhanced textually by distributing short riddles or epigrams along either the circumference or diameter of the disc's sides. So, in an early example, the head of a droll-looking bald man appears on one side, with the question running around the top of the disc, 'Why does this man appear over head and ears in debt[?]' The other side depicts a carefully positioned peruke, with the answer placed underneath: 'Because he has not paid for his wig.'[45] Paris has Seymour furnish a verse epigram for a thaumatrope of Orpheus and Eurydice, giving 'to it a classical *turn*', *ut pictura poesis*:

> By *turning round*, 't is said, that Orpheus lost his wife;
> Let him *turn round* again, and she'll *return* to life.

Twaddleton responds predictably to this example: 'I should have preferred a quotation from the fourth Georgic, so beautifully descriptive of the fable.'[46]

Jonathan Crary observes that whereas in the seventeenth and eighteenth centuries the camera obscura provided a model for human vision that enforces a clear distinction between an isolated perceiving subject and its object world, such early nineteenth-century optical devices as the thaumatrope make the subject the 'active producer of optical experience',[47] as the mind melds and makes sense of its after-images. Paris had first used the thaumatrope to demonstrate this phenomenon of the persistence of vision in a lecture he gave to the Royal College of Physicians in 1824. By disclosing the most fundamental acts of perception to be active and synthetic, such devices offer an experimental corroboration of Coleridgean principles of mind. Paris's conception of the thaumatrope as a visual pun makes perception consistent with the complex functions of mind required to make and interpret the verbal pun, creative capacities that, like Maxwell after him, he recognises as antithetical to stolid and unyielding pedantry.

Seymour's pedantry-punishing puns have a counterpart in Maxwell's humorously equivocal verse versions of Tripos problems. While Maxwell proved his capacity for solving such problems definitively in the final examinations, a few weeks after completing the Tripos, on 19 February 1854, he also demonstrated a rare ability to state and solve them in poetry:

> An inextensible heavy chain
> Lies on a smooth horizontal plane,
> An impulsive force is applied at A,
> Required the initial motion of K.

'A Problem in Dynamics' continues at some length in reaching its solution, apparently exorcising such exercises, which Maxwell describes grumpily in 'A Vision' as 'Problems made express to bore me.'[48]

While the student protagonists of such poems as 'A Vision' describe their subjection to the Tripos system and its mechanistic exercises, a Rigid Body, a staple of the problems that so oppressed these personae, similarly speaks for itself in the poem that Campbell publishes under the title 'In Memory of Edward Wilson, *Who repented of what was in his mind to write after section.*' Verse itself is presented as a playful liberating form, directly opposed to the constraining genre of the Tripos problem, as, something of a poet-seer, the '*Rigid Body . . . sings*'. While being itself subject matter, material that, akin to the students, is subject to the mixed mathematics curriculum, it reveals that it has little interest in the constructions that are placed upon it by the course, and accordingly expresses its freely swinging nature in song rather than 'By analytics high'. Defying differential calculus, the 'analysis' by which varying rates of change and curvature are measured, the Rigid Body represents a mischievous break with the Tripos curriculum, while the title of its poem signals a corresponding lapse in its dedicatee's mental focus.

Lawson recalls 'Maxwell coming to me one morning with a copy of verses beginning – "Gin a body meet a body Going through the air", in which he had twisted the well-known song into a description of the laws of impact of solid bodies.'[49] Evidently first drafted in the early 1850s, 'In Memory of Edward Wilson', a title it received twenty years later, is a parody of Robert Burns's song 'Comin' thro the Rye'. Maxwell based it upon the second, unsigned, setting of the Scots folk song, which as Burns's modern editor James Kinsley notes was 'the more widely current version in Scotland'.[50] The first of this version's three verses establishes a formula that is varied only slightly by its successors:

> Gin a body meet a body, comin thro' the rye,
> Gin a body kiss a body, need a body cry;
> Ilka body has a body, ne'er a ane hae I;
> But a' the lads they loe me, and what the waur am I.

Further meetings 'comin frae the well' and 'frae the town' in the second and third verses respectively question the need for a body to 'tell' and to 'gloom' after it has kissed another. The only other departure from the first verse occurs in the penultimate line of the third, which renders more definite the sexual suggestion of the earlier references to every body, apart from the persona, having somebody; 'Ilka Jenny has her jockey, ne'er a ane hae I.'[51]

'In Memory of Edward Wilson' displaces the sly sexual innuendo of its model, which presumably represents the type of thoughts that Wilson '*repented of*', with verses that focus upon the activities of those inorganic bodies that are more properly thought about by young students of physical science. Maxwell's parody springs from its quibbling use of the word 'body', which mobilises and juxtaposes the starkly contrastive connotations that each of its two referents, the human body and the inorganic body studied by mechanics, acquire when they make physical contact with their own type. Not quite coquettish, the '*Rigid Body*' that '*sings*' here is nonetheless depicted as the flighty object of young men's attentions:

> Gin a body meet a body
>   Flyin' through the air
> Gin a body hit a body,
>   Will it fly? And where?
> Ilka impact has its measure,
>   Ne'er a ane hae I,
> Yet a' the lads they measure me,
>   Or, at least, they try.
> Gin a body meet a body
>   Altogether free,
> How they travel afterwards
>   We do not always see.
> Ilka problem has its method
>   By analytics high;
> For me, I ken na ane o' them,
>   But what the waur am I?

The Rigid Body teases its young men with Tripos problems cast as riddles. Indeed the first of these appears to be insoluble, for while every 'impact has its measure, / Ne'er a ane hae I'. This suggests that this body is merely heuristic, an ideal model akin to the object and protagonist of 'A Problem in Dynamics', the 'heavy chain' that is 'endowed with a property incomprehensible', that of being 'inextensible'.[52] The Rigid Body is like the protagonists of Burns's song, or indeed of Lear's limericks, who defy others to take the measure of them by the normative standards of society, in this case conventional mathematical measurement.

Maxwell's ostensibly repentant re-writing '*of what was in* [Wilson's] *mind*' works ironically to give credence and authority to Burns's guiltless naturalistic account of amorous relations between human bodies by making it analogous to the respectable, morally neutral, model of mechanistic relations between inorganic bodies sanctioned by Tripos mathematics. The visiting American student Charles Bristed records that a sexually licentious ethos prevailed amongst Cambridge undergraduates during the 1840s and 1850s, but notes that the aspiring wrangler remained abstinent as a means of 'training with reference to the physical consequences alone'.[53] He restricted his attentions to the mechanical bodies on the curriculum. Through the parody of 'Comin' thro the Rye', the Rigid Body comes to represent a lyrical and libidinal freedom from intellectual oppression and physical repression, a release from the frustration with the Tripos that is expressed bitterly in '*Lines written under the conviction*':

> Why should wretched Man employ
> Years which Nature meant for joy,
> Striving vainly to destroy
>   Freedom of thought and feeling?[54]

The parody's preposterous textbook creature has its *raison d'être* in such freedom.

ANALOGIES

Displaying a lyric and capricious nature that defies quantification and knows of no method, the Rigid Body of 'In Memory of Edward Wilson' represents a creative freedom of play that Maxwell exercises at Cambridge in his extra-curricular activities of experiments, poetry and essays. Forbes writes to his friend and ally Whewell in 1852 of Maxwell's 'uncouthness', saying that he 'thought the Society and Drill of Cambridge the only chance of taming him & much advised his going'. Tait similarly recalls that Maxwell brought to Cambridge 'a mass of knowledge which was really immense for so young a man, but in a state of disorder appalling to his methodical private tutor ... William Hopkins'.[55] This disorganised mass was accompanied by what Maxwell and his father referred to as his 'dirt', pieces of unannealed glass and other materials for making experiments that he brought from the family home at Glenlair House. Such experiments presented themselves to Maxwell, rather as 'Science' had for Rowan Hamilton earlier in the century, as a distraction from his examinations. He writes to his aunt in the midst of his Tripos finals, in January 1854, that he and his friends were relaxing 'at-night working with gutta-percha, magnets, etc. It is much better than reading novels or talking after $5\frac{1}{2}$ hours' hard writing.'[56] Maxwell found Cambridge science too abstract and *a priori* in its disregard for experiment, but also insufficiently so in its failure to engage with metaphysics. He accordingly compensated for the lack of philosophical studies with his Apostles essays, while in poetry he found a medium for both counterfactual experiment and metaphysical speculation.

Pedantry's antipathy to poetry in 'A Vision' provides an index of the instrumentalist ethos of the Mathematical Tripos:

> As for Poetry, inter it
>   With the myths of other days.
>   'Cut the thing entirely, lest yon
>   College Don should put the question,
>   Why not stick to what you're best on?
> Mathematics always pays.'

'Grave and hard-reading students shook their heads at [Maxwell's] discursive talk and reading', 'Freshman Tayler' recalls, 'and hinted that this kind of pursuits would never *pay* in the long run in the Mathematical Tripos.'[57] Poetry is an embarrassingly impecunious pursuit in a mercantile age, while, as the College Don demonstrates and observes, mathematics offers the financial security of a college fellowship. By being identified with 'the myths of other days', banished to a superstitious past, Poetry is positioned in 'A Vision' as the antithesis not of scientific enlightenment, as Pedantry implies here, but of a modern acquisitiveness. Poetry has become culturally marginal next to the instrumentalist applications of science it affronts here. Rejecting such mercenary practices, the young Maxwell is concerned rather with establishing epistemological foundations for scientific discovery. In pursuing this quest he focuses upon the nature and authority of analogy, as a means of both grasping physical phenomena and making audacious connections between them that can generate new knowledge. Poetry, the art that is defined by formal and semantic parallellisms, is accordingly integral to Maxwell's exploration of analogy and its epistemological possibilities during the 1850s.

Analogy was an important principle for physics around mid century as it offered the means of discerning and representing such suppositious physical entities as atoms, molecules, ethers, energy and electromagnetic radiation. At the time he was writing his Apostles essay on 'Analogies in Nature', Maxwell was developing what he distinguished as a 'physical analogy' of a 'purely imaginary fluid' to represent electromagnetic phenomena in his paper 'On Faraday's Lines of Force'.[58] The first part was presented to the Cambridge Philosophical Society in December 1855 and the second in February 1856, the date that Campbell gives for the essay on 'Analogies', although Tait writes that '[Maxwell] showed me the MS. of the greater part of it in 1853',[59] a year in which, as Chapter 3 will show, his interest in analogy was particularly focused and urgent. He outlines the form of 'physical analogy' in a draft abstract for the paper in December 1855:

> There is, however, one method which combines the advantages, while it gets rid of the disadvantages both of premature physical theories and technical mathematical formulae. I mean the method of Physical Analogy. Of this we have instances in the substitution of numbers for quantities in all calculations, in the use of lines in mechanics to represent forces and velocities, in the partial analogy between the motion of light and that of a particle, and the more complete analogy between the motion of light and that of a vibration in an elastic medium.[60]

Apparently drawing upon his teacher Whewell's principle of consilience, Maxwell writes that 'By a physical analogy I mean that partial similarity between the laws of one science and those of another which makes each of them illustrate the other.'[61] So, for instance, the dynamics of elastic spheres provides a physical analogy for the behaviours of gas molecules. In another example, the mechanical analogy of an incompressible fluid, which Maxwell uses to illustrate various phenomena of electromagnetism in the successive parts of his paper 'On Faraday's Lines', is seen to furnish the means of specifying and integrating the present understandings of such phenomena without rushing to 'premature physical theories' that could restrict further understandings. A much more cautious principle than Whewell's consilience, physical analogy shows Maxwell wary of too readily imposing our forms of thought upon the phenomena of nature.

Illustrating the purely heuristic application of 'physical analogy', the motions of a hypothetical fluid in 'On Faraday's Lines of Force' describe a 'resemblance in mathematical form'[62] to various phenomena of electric currents and magnetic forces, without any pretence of offering a physical explanation of them. Similarly, and more audaciously, such a mathematical resemblance furnishes the formal ground for the correspondence Maxwell observes at the start of his essay on 'Analogies in Nature' between the analogy and the pun. With this essay he explores the possibility of using analogy more boldly as a means of not only describing scientific knowledge, as in 'physical analogy', but of discovering it. Beginning by noting the institutional taboo on reasoning by puns, and the more prevalent but nonetheless cautious tolerance for reasoning by analogy, Maxwell's Apostles essay deconstructs these conventions mathematically, by demonstrating their equivalence through reciprocation:

> Now, as in a pun two truths lie hid under one expression, so in an analogy one truth is discovered under two expressions. Every question concerning analogies is

therefore the reciprocal of a question concerning puns, and the solutions can be transposed by reciprocation.

The mathematical equation, itself an apodeictic form of correspondence and a bearer of multiple applications, is wittily applied to the relation between the analogy and the pun; the latter being described as a fraction of one expression over two truths, the former reciprocally of two expressions over one truth, the product of both being one (that is, the two terms are seen as mathematically equivalent). These terms are mirror images of one another, a pattern of equivalence that is also illustrated both imagistically and formally by Maxwell's 1853 poem, 'Reflex Musings: Reflection from Various Surfaces'. The title depicts the analogy, originally made by John Locke, between the mind's intrinsic capacity to focus upon its own operations and the physical reflections of light and sound as they strike the surfaces of the world (and so discloses them to our senses).

Locke argues in *An Essay Concerning Human Understanding* that all our ideas come from either reflection, 'the internal operations of our minds, perceived and reflected on by ourselves', or sensation, the mind's observation of the impressions that external objects make upon it through the senses.[63] Maxwell's professors, Hamilton at Edinburgh and Whewell at Cambridge, both made available to him a range of philosophical understandings of mental reflexion, albeit ones in which their respective idealist principles prevail, with Hamilton in particular emphasising Locke and recognising his original metaphoric extension of the term.[64] The juxtaposition of terms made by the title of Maxwell's poem suggests a formal, epistemologically potent, analogy between our mental thought and phenomenal intuitions. Exemplifying the formal pattern that he later describes mathematically in his essay on 'Analogies', it can be read as a single truth of reflection offered under two expressions, one being mental, the other physical, or, alternatively and equally, as a pun on the word 'reflection' that incorporates the respective ideas of mental reflexion and physical reflection that Maxwell distinguishes with the variant spellings. The metaphysical breadth of his thought here encompasses and interfuses the literary and the scientific easily and drolly.

CHAPTER 3

Knowing more than you think: James Clerk Maxwell on puns, analogies and dreams

Maxwell's poem, 'Reflex Musings: Reflection from Various Surfaces', poses the question of knowledge in modern terms of the relation between subjective mind and its objects, terms that its poised title balances nicely about the fulcrum of the colon. If as a young man he been asked 'What is your name?', he recalls in his late essay 'Psychophysik', 'The instructors of my youth would have expected me to answer – My name is the Conscious Ego, one and indivisible, the Subject, in relation to whom all other beings, material, human, or divine, are mere Objects.'[1] He outlines this relation in the notes he took from the second of Sir William Hamilton's 1847 lecture series on logic:

Thought is the product of the discursive faculty, or the faculty of relations or comparison, and there are 3 things to be considered in thought,
 1st the thinking subject which exerts the faculty,
 2nd the object thought of, or the matter of thought.
 3rd the relation between the subject and object, or the form of thought.[2]

John Hendry observes that 'Hamilton's contribution to science consisted solely of his advocacy and analysis of the analogical method, and was effectively communicated only to his students.'[3] Dallas recalls that 'An immense interest was excited in [Hamilton's] lectures',[4] while Maxwell reports to Campbell in 1847 that of all the lectures he had attended at Edinburgh, 'The Logic lectures are far the most solid and take most notes.'[5] Campbell testifies that throughout his friend's life 'the ideas received from Sir William Hamilton were his habitual vantage-ground'.[6]

Like Reid and his other forebears in the Scottish Commonsense tradition, Hamilton develops his philosophy in response to Hume's sceptical epistemology, which regards knowledge as the mind's intuitive but ungrounded consecration of contingent associations amongst its sense perceptions. Reid argues that the mind admits direct sensory perceptions that, together with intuitively known general principles, are able to produce real knowledge of the object world. These mental principles furnished an opening for idealism to enter, a breach that Hamilton makes good use of through his reading of Immanuel Kant. He critiques and develops them and their complement, direct sensory perception, in his doctrine of 'the Duality of Consciousness':

> When I concentrate my attention in the simplest act of perception, I return from my observation with the most irresistible conviction of two facts, or rather two branches of the same fact; – that I am, – and that something different from me exists. In this act, I am conscious of myself as the perceiving subject, and of an external reality as the object perceived; and I am conscious of both existences in the same indivisible moment of intuition.[7]

The principle of the simultaneous interaction of the subject and object by which Hamilton explains perception is put pithily by Maxwell in his Apostles essay 'What is the Nature of Evidence of Design?', which, like 'Reflex Musings', dates from spring 1853: 'Perception is the ultimate consciousness of self and thing together.'[8]

Hamilton intends his doctrine of 'the Duality of Consciousness' to counter 'the pollution of cosmothetic idealism', as Maxwell refers to it in his 'Design' essay, the sort of hypothetical realism, seen to pervade modern philosophy, that affirms the existence of an object world, but denies that we can have an immediate sense perception of it. Maxwell is, however, critical of Hamilton's doctrine. Campbell notes that the manuscript essay on 'Design', which has since been lost, includes 'incidentally a statement of the Hamiltonian doctrine of Perception', a statement that comes, as his editor notes, 'with the following significant corollary':[9]

> If we admit, as we must, that this *ultimate* phenomenon is incapable of further analysis, and that subject and object alone are immediately concerned in it, it follows that the fact is strictly private and incommunicable. One only can know it, therefore two cannot agree in a name for it. And since the fact is simple it cannot be thought of by itself nor *compared alone* with any other *equally simple fact*. We may therefore dismiss all questions about the absolute nature of perception, and all theories of their resemblances and differences. We may next refuse to turn our attention to perception in general, as all perceptions are particular.[10]

The doctrine of 'the Duality of Consciousness' is seen to radically diminish the possibility of knowledge, rendering it particular, private and incommunicable. Maxwell sees it to confuse fatally what Hamilton calls 'the great problem of philosophy', that is, of how 'to distinguish what elements are contributed by the knowing subject, what elements by the object known'.[11]

James Clerk Maxwell on puns, analogies and dreams

Maxwell works independently to disentangle 'the Duality of Consciousness', as he poses and addresses Hamilton's 'great problem' afresh, not only in his essays, but in 'Reflex Musings: Reflection from Various Surfaces':

> In the dense entangled street,
>   Where the web of Trade is weaving,
> Forms unknown in crowds I meet
>   Much of each and all believing;
>   Each his small designs achieving
> Hurries on with restless feet,
>   While, through Fancy's power deceiving,
> *Self* in every form I greet.
>
> Oft in yonder rocky dell
>   Neath the birches' shadow seated,
> I have watched the darksome well,
>   Where my stooping form, repeated,
>   Now advanced and now retreated
> With the spring's alternate swell,
>   Till destroyed before completed
> As the big drops grew and fell.
>
> By the hollow mountain-side
>   Questions strange I shout for ever,
> While the echoes far and wide
>   Seem to mock my vain endeavour;
>   Still I shout, for though they never
> Cast my borrowed voice aside,
>   Words from empty words they sever –
> Words of Truth from words of Pride.
>
> Yes, the faces in the crowd,
>   And the wakened echoes, glancing
> From the mountain, rocky browed,
>   And the lights in water dancing –
>   Each, my wandering sense entrancing,
> Tells me back my thoughts aloud,
>   All the joys of Truth enhancing
> Crushing all that makes me proud.[12]

An allegorical quest to find a ground for knowledge, the poem begins sceptically. The first stanza presents a correspondence between subject and object that instances not analogy, but solipsism. A quibble early in the stanza on 'Forms unknown' invokes both ideal and phenomenal characterisations of knowledge, as the phrase immediately suggests principles of ideal Form, while, as the conventional entitlement of the first word in a line of poetry, its capitalisation can also be discounted to yield the phenomenal sense of forms as shadows or shades.

'Reflex Musings' locates its instance of perceptual solipsism in the city, itself a monument to human wilfulness and the venue for its further exercise and expression. While the ethereal 'web of Trade' appears to give order to the concrete chaos of the 'dense entangled street', such imagery also describes mercantile city life as inauthentic, a shadowy form, evanescent in the semi-transparency, fragility and silvery greyness of the spider's web. The city is introduced in terms that recall Walter Scott's observation, since passed into proverb, 'O, what a tangled web we weave, / When first we practise to deceive!'[13] In the final lines, as the persona's 'deceiving' chimes in with the earlier 'weaving', the wilful 'Fancy' that informs his perceptions of his peers also describes his affinity with them, as the further rhyme underlines: 'Each his small designs achieving.' The ambiguity of the 'Forms unknown' is resolved arbitrarily, 'through Fancy's power deceiving', as the ideal '*Self*' is found 'in every form', ostensibly extracted from the phenomenal shadows. The persona is the knowing urban counterpart to the 'enamoured rustic' of Coleridge's poem 'Constancy to an Ideal Object', who perceives in his own shadow cast upon mountain mists another bright figure, the phenomenon of the Brocken spectre, 'Nor knows he makes the shadow, he pursues!'[14]

The ideal and phenomenal forms that are confused early in 'Reflex Musings' coincide nicely in the natural theological vision of its companion poem, 'A Student's Evening Hymn':

> Through the creatures Thou hast made
>   Show the brightness of Thy glory,
> Be eternal Truth displayed
>   In their substance transitory,
>   Till green Earth and Ocean hoary,
> Massy rock and tender blade
>   Tell the same unending story –
> "We are Truth in Form arrayed."

Dated a week apart, at 18 April and 25 April respectively, 'Reflex Musings' and the 'Hymn' consist of octaves with an *ababbaba* rhyme scheme, a pattern of reversal that is sustained by its trochaic tetrameters, where those finishing on the *a* rhyme take the catalectic form of the line. Both the student persona of the 'Hymn' and the reflective persona of the slightly earlier poem invite identification with the author, who was, of course, an undergraduate at the time he wrote them. The singing of hymns and metrical psalms, rather than the saying of prayers, was the principal form of worship for Scots Presbyterians such as Maxwell. As a song of praise, the student's 'Hymn' is presented as a humble but harmonious parallel to the 'songs' that the evening sky offers to the Creator:

> While the world is growing dim,
>   And the Sun is slow descending
> Past the far horizon's rim,
>   Earth's low sky to heaven extending,
>   Let my feeble earth-notes, blending
> With the songs of cherubim,
>   Through the same expanse ascending,
> Thus renew my evening hymn.

As a figure for divine Design the hymn is almost over-determined, being both literary, a version of the natural theological trope of the Book of Nature, and musical, Pythagorean in its inspiration and application here: 'And the stars most musically / Move in endless rounds of praise.'[15] The Design hypothesis answers Maxwell's question 'Are there Real Analogies in Nature?' unequivocally, as the harmonious relations amongst its parts, including the perceiving mind, follows from their shared Creator. This answer, however, defers the question of knowledge to one of the doctrine's own authority, which Maxwell questions directly in 'What is the Nature of Evidence of Design?'

THE MAGAZINE OF NATURE

Associating Design with the Pythagorean cosmology, much as he does in the 'Hymn', Maxwell's essay on 'Design' also observes that the hypothesis may covertly inform contemporary science:

> Why should not the Original Creator have shared the pleasure of His work and His creatures and made the morning stars sing together, etc.?
> I suspect that such a hope has prompted many speculations of natural historians, who would be ashamed to put it into words.[16]

While modern professional science is reticent about making such admissions, poetry allows Maxwell to put 'such a hope' unashamedly 'into words'.

As the contemporary natural historian's shy secret, the Design hypothesis belongs to an earlier age. Such Anglican 'gentlemen of science' as Herschel and Rowan Hamilton embraced the Book of Nature as an epistemologically enabling trope and hence a ground for their science, as it asserts the pre-adaption of human perception to the phenomenal world. Design endows the various parts of the Creation with structural and stylistic affinities, like a book, an idea that suggests Whewell's theory of consilience, which finds in formal correspondences between distinct scientific fields the criterion by which their truth is mutually confirmed.

Maxwell's variant use of the trope of the Book of Nature in his 'Hymn' can be usefully compared with Herschel's 1851 poem 'Man the Interpreter of Nature',[17] which makes the epistemological principle of Design axiomatic and its application a moral duty. While the poem takes its title from Francis Bacon, '*Homo, naturæ minister et interpres*' ('Man, the servant and interpreter of nature'), it makes of this humanist prescription for modern scientific practice a text for a sermon. Delivered as a catechism, its opening lines strike the keynote in an imperative mode: 'Say! when the world was new and fresh from the hand of its Maker / Ere the first modelled frame thrilled with the tremors of life.' The Creation is introduced synecdochically with the first man coming to life, while the later part of the poem, maintaining the catechetical mode, inquires about the end he was created to realise:

> Say! Was the WORK wrought out! Say, was the GLORY complete?
> What could reflect, though dimly and faint, the INEFFABLE PURPOSE
> Which from chaotic powers, Order and Harmony drew?
> What but the reasoning spirit, the thought and the faith and the feeling?
> What, but the grateful sense, conscious of love and design?
> Man sprang forth at the final behest. His intelligent worship
> Filled up the void that was left. Nature at length had a soul.

Man is 'the reasoning spirit' in which 'the thought and the faith and the feeling' draw together to both 'reflect' nature and reflect upon it, bring the analogous synthesis of Creation, the 'Order and Harmony', to conscious recognition.

Maxwell's 'Hymn' similarly enacts a process in which Nature is brought to consciousness. However, in contrast to Herschel's encompassing principle of 'intelligent worship', it yields only a particular register of truth:

> Thou that fill'st our waiting eyes
>   With the food of contemplation,
> Setting in thy darkened skies
>   Signs of infinite creation,
>   Grant to nightly meditation
> What the toilsome day denies –
>   Teach me in this earthly station
> Heavenly truth to realise.

Nature is indexical here, characterised by 'Signs of infinite creation' that, in a rather mannerist gesture, point to the transcendent 'Heavenly truth'. The hermeneutic exercise directs us from the evening heavens to Heaven itself, from this world to the next:

> Teach me so Thy works to read
> That my faith, – new strength accruing, –
> May from world to world proceed.[18]

The contemplative faith of the 'Hymn' contrasts with the insistent didacticism of Herschel's poem. The early parts of 'Man the Interpreter of Nature' list some easily perceptible regularities in nature, simple phenomena of cause and effect, that the reader is asked rhetorically to affirm date back to the time of the Creation; 'Heaved not ocean, as now, to the moon's mysterious impulse? / Lashed by the tempest's scourge, rose not its billows in wrath?' It describes the 'glorious forms of Creation' in terms that register its immutable regularities, and so represent the scientific principle of law, rendering them in appreciative layers of creamy poeticism: 'Roseate morn, and fervid noon, and the purple of evening – / Night with her starry robe solemnly sweeping the sky.' Science and poetry are homogenised in Herschel's poem, subordinated to the overarching imperative of 'intelligent worship' to interpret the Book of Nature, whereas Maxwell's poem makes this function the exclusive province of poetry and personal faith. 'A Student's Evening Hymn' derives from the trope of the Book of Nature a lyrical conceit that recognises the sublime totality of the Creation, not a scientific means of discovering knowledge of its aspects and its Creator.

'What is the Nature of Evidence of Design?' endeavours to theorise the grounds of the faith dramatised in the 'Hymn' with the following dictum: 'The belief in design is a necessary consequence of the Laws of Thought acting on the phenomena of perception.'[19] The 'Laws of Thought' evidently function in a regulative capacity akin to the apparatus of the Kantian faculty psychology, idiosyncratic versions of which were available to Maxwell from both Hamilton and Whewell. The phrase, however, belongs to Hamilton, who defines Logic in his lectures as 'the science which is conversant about the Laws of Thought', that is, the formal operations of 'Thought Proper simply and in itself'.[20] Hamilton defines 'Philosophical knowledge, in the widest acceptation of the term, and as synonymous with science', in canonical Newtonian terms as 'the knowledge of effects as dependent on their causes',[21] while Maxwell's essay similarly specifies that 'It is the business of science to investigate these causal chains.'[22]

Hamilton sees causality to proceed from two Laws of Thought, Existence and Time, which are for him necessary conditions for thinking any object. As Hamilton puts it, 'the application of the law of the conditioned [by which he sees the conceivable to lie 'between two inconceivable extremes' of the infinite and the absolute] to any object, thought as existent, and thought as in time, will give us at once the phænomenon of causality'. These Laws make it impossible to imagine an existence coming into being *ex nihilo*, so that a new phenomenon must always be thought of within time as an altered form of something else. God is accordingly established as the original instigator of all forms of being: 'creation ... is conceived, and is by us conceivable, merely as the evolution of a new form of existence, by the fiat of the Deity'. The causal chains that our 'Laws of Thought' discriminate accordingly yield faith in God but no knowledge of Him, no attributes other than ontological primacy. The Design hypothesis asserts the fallacy that, as Hamilton puts it in his lectures, 'the infinite can be known, but only known as finite'.[23]

In his essay on 'Design', Maxwell considers the Deist analogy of the watchmaker, which was established for nineteenth-century Britain by William Paley in his much reprinted *Natural Theology or, Evidences of the Existence and Attributes of the Deity, Collected from the Appearances of Nature* (1802).[24] This notorious argument reasons by analogy that, just as in chancing upon a watch we would naturally assume a maker behind it, so too any natural object found in the world must also have its Maker. Maxwell considers a version of this analogy that focuses upon the discernible functions of objects as an index of intention and Design. He returns to this argument three years later in his paper on 'Analogies in Nature'. Restating the 'analogy ... between the principle, law, or plan according to which all things are made suitably to what they have to do, and the intention which a man has of making machines that will work', he argues that the application of this analogy is purely heuristic and limited to knowledge of nature, it does not extend to knowledge of God's attributes. He is accordingly able to both affirm Hamilton's teaching and vindicate the 'many speculations of natural historians' that in his earlier essay he saw the Design hypothesis to prompt: 'The doctrine of final causes, although productive of barrenness in its exclusive form, has certainly been a great help to enquirers into nature; and if we only maintain the existence of the analogy, and allow observation to determine its form, we cannot be led far from the truth.'[25]

Maxwell's 'Hymn' affirms God as the transcendent First Principle, who can be known poetically through the sublime, Kant's *'starry heaven above'*,[26] not substantively and scientifically from the human eye or the insect's wing, in the manner of natural theology. The poeticising drive to transcend the particular conditions of nature through faith is demonstrated neatly in the final stanza of the 'Hymn', where, allowing that 'led by shadows fair' he could 'have uttered words of folly', the persona appeals to the physical medium of the air that propagates sound to cancel them: 'Let the kind absorbing air / Stifle every sound unholy.'[27] In being assimilated to the poem's cosmology, the persona's scientific understanding of the physical world is construed fancifully here, as the air is anthropomorphised and requested to behave as an occult power. By recognising the expansive vision of the 'Hymn' as a distinct mode of acknowledging Creation, Maxwell quarantines the Design hypothesis from his natural philosophy. When it comes to devising particular models for physical phenomena, Maxwell is, as he puts it in 'On Faraday's Lines' (and his cautious use of analogy in this essay demonstrates), wary of 'rashness in assumption'.[28]

Rather than cohering as a book, the relations between the parts of nature could, as Maxwell observes in his essay on 'Analogies in Nature', correspond to another, peculiarly modern, type of publication:

> Perhaps the 'book', as it has been called, of nature is regularly paged; if so, no doubt the introductory parts will explain those that follow, and the methods taught in the first chapters will be taken for granted and used as illustrations in the more advanced parts of the course; but if it is not a 'book' at all, but a *magazine*, nothing is more foolish to suppose [than] that one part can throw light on another.[29]

There may be no 'Analogies in Nature' as such. The attribution of a particular genre to the author of the world could be, as 'A Student's Evening Hymn' implies, a conceit that furnishes an objective correlative for the mystery of faith, but not a means for the inductive scientist to discover new knowledge. As a hypothesis applied to the phenomena of nature, Design highlights inductivism's attendant risk of tautology, of simply finding in nature only what it has already placed there (i.e., a beneficent Creator). It is the problem of solipsism writ large, larger than the Brocken spectre, of finding '*Self* in every form I greet', as man describes nature as he conceives of himself, to be made in God's image.

CAUSES AND FORCES

Maxwell approaches the problem of scientific knowledge in his essay on 'Analogies in Nature', much as he does in 'Reflex Musings', by acknowledging the pervasive risk of solipsism and the scepticism it licenses: 'For, not to mention all the things in external nature which men have seen as the projections of things in their own minds, the whole framework of science, up to the very pinnacle of philosophy, seems sometimes a dissected model of nature, and sometimes a natural growth on the inner surface of the mind.'[30]

This idea of projection is fundamental to Hume's thought, which is similarly wary of the mind's capacity for 'guilding or staining all natural Objects'.[31] Maxwell suggests that our ideas about external nature could be like the pearl that forms in the oyster, as the irritant empirical grit of the object world is transformed aesthetically but unrecognisably by the inherent nature of the mind. The Pythagorean cosmology alluded to in the 'Hymn' and indeed modern mathematical physics could be extravagant demonstrations of what Nietzsche describes as our 'constant falsification of the world through numbers'.[32]

The solipsistic tendency introduced in the first stanza of 'Reflex Musings' is developed in the second with an image that recalls the myth of Narcissus, as the persona, like his romantic kindred in such poems as Wordsworth's 'Tintern Abbey', takes leave of the city for the country and the contemplation of nature:

> Oft in yonder rocky dell
>   Neath the birches' shadow seated,
> I have watched the darksome well,
>   Where my stooping form, repeated,
>   Now advanced and now retreated
> With the spring's alternate swell,
>   Till destroyed before completed
> As the big drops grew and fell.

External nature is presented as an independent system that functions through its own principles and laws. Rather than obliging the persona with a faithful reflection in a perfectly mirroring surface, nature acknowledges him on its own mechanistic terms of hydrostatics and optics. It exemplifies the ontology of matter in motion that Forbes and Hamilton each propagated amongst their students at Edinburgh, and upon which Maxwell based his current researches on light and elastic solids. The first quatrain of the stanza establishes the poet's reflection, his image 'repeated', while the second presides over its destruction. Nature offers him an unstable image of himself, a reflection that draws alternately towards and away from his stationary form as the spring causes the level of the well to rise and subside, before the surface is obliterated by falling drops of water. While the first stanza of the poem documents the mind's propensity for projection, for finding its own images in the phenomenal world, the second demonstrates the reluctance of nature to cooperate with such wilfulness. It suggests criteria by which subject and object could be extricated from what Maxwell sees as their dissolution in 'the Duality of Consciousness'.

The allegory of 'Reflex Musings' suggests a romantic narrative of growth and development, a modest counterpart to those of Lorenz Oken and G. W. F. Hegel. In the first stanza of the poem the persona is like the infant who confuses himself with the objects around him, an original narcissistic tendency that is subsequently frustrated as this external world asserts itself in the second stanza. With this particular mirror stage comes resistance from the laws of nature, rather as the Law of the Father supervenes in Jacques Lacan's developmental psychology. Indeed, the disjuncture between the self and the object world, the moral of this episode, necessitates his use of language in the third stanza, as his alienation is registered in the 'Questions strange' he directs at nature. 'In order to advance', Maxwell writes to Campbell late in 1851, 'the soul must converse with things external to itself':[33]

> By the hollow mountain-side
>   Questions strange I shout for ever,
> While the echoes far and wide
>   Seem to mock my vain endeavour;
>   Still I shout, for though they never
> Cast my borrowed voice aside,
>   Words from empty words they sever –
> Words of Truth from words of Pride.

Addressing appropriate questions to nature, whether theoretically or practically by framing experiments, is a defining activity of the scientist. While the 'Questions strange' that the persona hurls at the object world 'for ever' are both loud and voluble, they are greeted coolly by nature. Maxwell observes in his 1860 inaugural lecture to King's College that 'when we meet them out of doors' 'physical facts' present themselves in a '*natural retiring* form'.[34]

The protagonist's obsessive interrogation of natural phenomena can be assumed to proceed from what the 'Design' essay describes as the scientist's definitive preoccupation with causality. It suggests a 'restless questioning about the *why* of them all'. Nature, however, recognises the protagonist's 'Questions strange' only as utterance, instances of physical force that accordingly elicit its echoes. Maxwell clarifies this distinction between cause and force in the essay on 'Analogies in Nature':

Cause is a metaphysical word implying something unchangeable and always producing its effect. Force on the other hand is a scientific word, signifying something which always meets with opposition, and often with successful opposition, but yet never fails to do what it can in its own favour. Such are the physical forces with which science deals, and their maxim is that might is right, and they call themselves laws of nature.[35]

Maxwell was familiar with Humean critiques of causation from Hamilton's lectures. That objective causes cannot, as Hamilton observes, be directly intuited empirically encourages scepticism about them, justifying the 'many philosophers who surrender the external perception, and maintain our internal consciousness, of causation or power'.[36] Maxwell follows his teacher in rejecting the French philosopher Maine de Biran's thesis that cause is a reification of subjective will that we anthropomorphically attribute to phenomena;[37] 'Some had supposed that in will they had found the only true cause, and that all physical causes are only apparent', Maxwell writes in his 'Analogies' essay, 'I need not say that this doctrine is exploded.'[38] Cautious of the 'metaphysical word', Maxwell draws upon Hamilton's philosophy in the second and third stanzas of 'Reflex Musings' to describe relations between the subject and its objects naturalistically with the principle of force.

Hamilton explains perception as the consequence of direct and unmediated physical forces that press upon our sense organs. The paradigmatic sense in his theory is accordingly touch: 'all our senses are only modifications of touch; in other words ... the external object of perception is always in contact with the organ of sense'.[39] In an exercise 'On the Properties of Matter' (1848–9) that he wrote for Hamilton, and which so impressed his teacher that he kept it,[40] Maxwell describes sense perception in this way as the mechanical effect of force: 'Now the only thing which can be directly perceived by the senses is Force, to which may be reduced heat, light, electricity, sound, and all the things which can be perceived by any sense.' He defines each of the senses by this principle, so that, taking the case that applies to the mountain's echoes, 'By the sense of hearing we perceive the intensity, rapidity, and quality of the vibrations of the surrounding medium.'[41]

Hamilton's theory finds objects transmitted to our senses through space, a medium he considers to be both metaphysical and physical. He understands space to exist and be directly accessible to us, both in the Kantian manner, as an *a priori* Form of Sense, and also, contra Kant, objectively, as entrenched in external reality.[42] Maxwell endorses Hamilton's understanding of space in 'On the Properties of Matter', where he writes that 'length, breadth and thickness' do not 'belong exclusively to matter ... for they belong also to geometric figures, which are forms of thought and not of matter'.[43] As Campbell observes of his friend, 'his geometrical imagination predisposed him to accept the doctrine of "natural realism"'.[44] Maxwell's adoption of Hamilton's principle of space is clear from 'Analogies in Nature', where he draws from it the conviction that 'we have a *real* analogy between the constitution of the intellect and that of the external world'.[45] Indeed, this common medium of space allows Maxwell to renew the Lockean metaphor of mental reflection, specifying phenomena concretely through the mechanistic ontology of matter and motion as 'Reflection from Various Surfaces', whilst reciprocally seeing thought to operate in analogous configurations as 'Reflex Musings'. His fundamentally optical metaphor of reflection conversely suggests that space, as the formal substrate common to mind and the object world, is like an ether, able to propagate distinct but transformable modes of energy, as occurs, for instance, with the identification Maxwell famously makes of light with transverse waves in his model of the electromagnetic ether in the third part of 'On Physical Lines of Force' (1862).[46]

In his *Lectures on Metaphysics*, Hamilton speaks of perceptions being shaped by the nature of our mind and the quality of its interactions with matter: 'in different perceptions, one term of the relation may predominate, or the other'. He accordingly distinguishes three 'qualities of matter', three categories that describe the possibilities and limitations of our knowledge, which he names the Primary, Secondary and Secundo-primary. Maxwell would have been aware of these qualities from attending Hamilton's lectures, and more particularly from his reading in July 1848 of Hamilton's *Dissertations* to his edition of Reid, where in Note 'D' they receive their fullest explication.[47]

Hamilton's Primary qualities of matter are apprehended through the *a priori* form of space, as we recognise its reality through experience of our bodily mass and senses: 'In the apprehension of the Primary qualities the mind is primarily and principally active,' he explains in his *Dissertations*, 'it feels only as it knows.' The Secondary qualities are correspondingly experienced directly as sensory affections by physical forces, so that 'the mind is primarily and principally passive; it knows only as it feels'. As the third term suggests, the Secundo-primary qualities describe the coincidence of the other qualities: 'In that of the Secundo-primary the mind is equally and at once active and passive; in one respect, it feels as it knows, in another, it knows as it feels ... Perception and Sensation, Activity and Passivity, are in equipoise.'[48] Scientific knowledge is theorised in Hamilton's dynamic physicalist model of perception as the consequence of pressure and resistance, forces exerted mutually between material objects and embodied minds: 'On their Primary or objective phasis they manifest themselves as *degrees* of resistance opposed to our locomotive energy; on their secondary or subjective phasis, as *modes* of resistance or pressure affecting our sentient organism.'[49] The situations described by 'Reflex Musings' dramatise its title's ambiguous juxtaposition of terms as a series of Hamiltonian tussles between subject and object, as each confronts the other with forces of projection and resistance.

'THE BLACK ROCKS OF ONTOLOGY'

While the second stanza of the poem demonstrates that the phenomenon of reflection hinges upon the principle of resistance, the resistance that 'Various Surfaces' offer to light, sound and other such physical forces, the third exemplifies it with what Maxwell describes in his 1854 essay 'Has Everything Beautiful in Art its Original in Nature?' as the definitive instance of such a surface:

> Rivers and mountains have not even an organic symmetry; the pleasure we derive from their forms is not that of comprehension, but of apprehension of their fitness as the forms of flowing and withstanding matter.[50]

The 'Questions strange' that the persona of 'Reflex Musings' shouts at the mountain can be understood literally as his vocal projections upon its surfaces, and metaphorically, in the Humean phrase from the 'Analogies' essay, as the 'projections of our mental machinery on the surface of external things'. As the resultant of the force of the poet's shouted words and that offered by the resistant surface of the mountain, the secondary acoustic image of the echo effectively dramatises Hamilton's principle of the secundo-primary quality of perception, a mediate status that licenses Maxwell's conceit of personifying them. Indeed, as the Hegelian analogy of the resultant suggests, it promises a resolution of the dialectic between subject and object in a concrete idea, an instance of knowledge.

The mountain-side's telling echo returns the question to its author as an objective phenomenon. By treating it as utterance, the mountain empties the question of its intended meaning (so that it 'Seem[s] to mock my vain endeavour'), furnishing an acoustic reflection from its surface that facilitates the persona's 'Reflex Musings', the self-conscious severance of 'Words from empty words'. Through its integrity and consistency the sphinx-like mountain restates the persona's question as a riddle or puzzle. Inflected by the surface of the mountain that they strike and accordingly articulate, the poet's utterances are transformed into echoes that not only deliver them back to their speaker in an objective form for him to consider, but also precisely register the nature of these surfaces. The fourth and final stanza of the poem points the significance of this episode figuratively, as nature's coming to consciousness through its 'Reflection from Various Surfaces':

> And the wakened echoes, glancing
> From the mountain, rocky browed.

The 'wakened echoes' suggest eyes that have opened, and following upon this image, the metaphor of acoustic reflection as a knowing ocular 'glancing' from beneath emphatic brows.

The poem's anthropomorphic image of the stolidly resistant mountain, a giant 'wakened' by the sound of the poet's shouting, itself has an echo in Maxwell's contemporaneous essay on 'Design'. As if offering a rejoinder to the confident catechism of Herschel's 'Man the Interpreter of Nature', Maxwell's essay opens satirically, in the manner of a fiery sermon, on a clamorous note:

> Design! The very word . . . disturbs our quiet discussions about *how* things happen with restless questioning about the *why* of them all. We seem to have recklessly abandoned the railroad of phenomonology [*sic*], and the black rocks of Ontology stiffen their serried brows and frown inevitable destruction.

The Design hypothesis is presented as luring science away from the modern study of phenomena to its doom on the rocks of the *a priori* science of Being, which is figured here like Michelangelo's sculpture of Moses, as ancient, monumental and heavily browed and frowning. Implicitly identified with the railway train, the great Victorian icon of progress, modernity and the utilitarian application of scientific knowledge, science threatens to go off the rails, the empiricist principles that are seen to ground and direct it ever forward in a simple linear trajectory. Maxwell, however, objects to thought being marshalled along any such established paths, as he makes clear in an 1873 letter to Herbert Spencer: 'Mathematicians, by guiding their thoughts always along the same tracks, have converted the field of thought into a kind of railway system and are apt to neglect cross-country speculations.'[51] The metaphor may have been drawn from Hamilton, who likens 'The mathematical process in the symbolical [i.e., algebraic] method' to 'running a rail-road through a tunnelled mountain . . . a short and easy transit, to our destined point, but in miasma, darkness and torpidity, whereas the [Hellenising geometrical method] allows us to reach it only after time and trouble, but feasting us at each turn with glances of the earth and of the heavens'.[52] By limiting the range and free play of thought both positivist phenomenalism and *a priori* systems of mathematics threaten to preordain further scientific knowledge. Written two decades apart and from complementary perspectives, Maxwell's remarks indicate his sustained striving for the means by which hitherto unknown truths of nature may be apprehended.

The peculiar parable of Echo and Narcissus that Maxwell elaborates in 'Reflex Musings' eschews any easy communications and correspondences between the subject and its objects. Nature is insistently autonomous, leaving the persona, like Moses, 'a stranger in a strange land'.[53] As it finds its voice in the poem, 'rocky browed' and forbidding, nature suggests not the idealised anthropomorphic paradigm of Christ to which natural theology implicitly parallels it, but an Old Testament prophet who speaks obliquely of 'a god that hidest thyself'.[54] It also casts aside 'words of Pride', words that by degrees reproduce Lucifer's sin, his selfish and narcissistic *non serviam*, and express the wilfulness demonstrated by the protagonist and his urban peers in the opening stanza of the poem. Like Moses, the echoes in Maxwell's poem bring the truth, certain laws of nature, down from the mountain-side.

The stern imagery deployed early in the 1853 Apostles essay slyly defends a principle of metaphysical idealism it finds at the core of the Design hypothesis, a mysterious and formidable principle of Being, as 'the black rocks of Ontology stiffen their serried brows' against a prevalent and predictable phenomenalism. Metaphysics is similarly menacing in Maxwell's presidential address to Section A at the 1870 Liverpool meeting of the BAAS, where, searching for the 'hidden and dimmer region where Thought weds Fact', he asks: 'Does not the way to it pass through the very den of the metaphysician, strewed with the remains of former explorers, and abhorred by every man of science?'[55] The cave-like habitat of the metaphysician suggests 'the hollow mountain-side' in the third stanza of 'Reflex Musings', and indeed 'the darksome well' in the second stanza. The reflections that each of these yield are depicted in the poem as proceeding not simply from superficies, in the manner of positivist 'phenomonology [sic]', but resonantly and austerely through depths, as forces that are inflected significantly and expressively with telling details of autonomous nature. Being is pictured in 'the wakened echoes, glancing / From the mountain, rocky browed' as rumbling into consciousness.

The process of realising knowledge of nature is described in the 'Hymn' with the biblical and neoplatonic metaphor of harvesting, 'to reap / All that Shadows, world-concealing, / For the bold enquirer keep'.[56] 'Reflex Musings', however, recognises no phenomenal husks to cast aside. Its persona comes to appreciate phenomena not as obfuscating appearances, but as the integral expression of ontological depth. 'Reflex Musings' argues that it is not nature that provides obstacles to knowledge but its percipient, who accordingly needs to perform what Maxwell refers to in the 'Analogies' essay as an 'act of self-excenteration',[57] to pass through a personal

'EMPTY BUBBLES, FLOATING UPWARDS THROUGH THE CURRENT OF THE MIND'

Copernican revolution, in order to be able to grasp and interpret the truths offered by the object world.

In a curious counterpoint to 'Reflex Musings', Maxwell describes his selfhood by analogy with the poem's mechanistic model of nature in a slightly later poem, 'Recollections of Dreamland', which, like the 'Analogies' essay, dates from 1856. Indeed, this poem completes by reciprocation the analogy between mind and phenomena that is introduced by the title of 'Reflex Musings' and explored in the poem itself. The relationship between the unconscious and conscious mind, upon which the anthropomorphic analogy of the 'wakened' mountain-side pivots in 'Reflex Musings', becomes the tenor of the analogy in 'Recollections'. Its vehicle is furnished conversely by the mechanistic ontology of matter in motion, with a hydrodynamic model akin to the spring-fed 'darksome well' of the earlier poem:

> What though Dreams be wandering fancies, by some lawless force entwined,
> Empty bubbles, floating upwards through the current of the mind?
> There are powers and thoughts within us, that we know not, till they rise
> Through the stream of conscious action from where Self in secret lies.
> But when Will and Sense are silent, by the thoughts that come and go,
> We may trace the rocks and eddies in the hidden depths below.[58]

In this summary account of the analogy from the concluding part of the poem, dreams suggestively 'trace the rocks and eddies', the peculiar personal forms and forces that compose the mind's subconscious ground. Parallel to the mountain from which the 'wakened echoes' proceed expressively in 'Reflex Musings', the rocks in 'Recollections of Dreamland' entail certain dynamic consequences, as they inflect 'the currents of the mind' and so become apparent 'Through the stream of conscious action' and in 'the thoughts that come and go'. Having established the terms of his metaphor with his description of 'the black rocks of Ontology' in the 'Design' essay, Maxwell represents this foundational metaphysical principle as underpinning both the object world in the 'rocky browed' mountain of 'Reflex Musings' and, in the form of its rocky substrate, subjective mind in 'Recollections'. Each is figured as an unconscious depth that is continuous with conscious mind and hence amenable to understanding.

'Recollections of Dreamland' appears to have been prompted by dreams that Maxwell had about his father, who died in early April 1856, as it is fascinated by the power of dreams to bring the past to mind as a vivid presence. Its title figures the dreaming mind as another country, distinct from the waking state in which it is recollected. Developing this conceit, the poem opens with a quibble drawn from dream scepticism. Maxwell would have known this argument from reading Descartes. Consistent with Maxwell's early propensity for exploring forms of scepticism, as he worked to establish grounds for scientific knowledge and discovery, Descartes was, according to Campbell, one of the 'metaphysical writers who had received most of his attention before going to Cambridge'.[59] Descartes reasons that because the same conviction and certainty can accompany our immediate experiences in dreaming as in waking life, 'there are never any sure signs by means of which being awake can be distinguished from being asleep'.[60] Maxwell addresses his readers as those who are sunk in the impoverished slumber of waking life:

> Rouse ye! torpid daylight-dreamers, cast your carking cares away!
> As calm air to troubled water, so my night is to your day;
> All the dreary day you labour, groping after common sense,
> And your eyes ye will not open on the night's magnificence.
> Ye would scoff were I to tell you how a guiding radiance gleams
> On the outer world of action from my inner world of dreams.
> When, with mind released from study, late I lay me down to sleep,
> From the midst of facts and figures, into boundless space I leap;
> For the inner world grows wider as the outer disappears,
> And the soul, retiring inward, finds itself beyond the spheres.
> Then, to this unbroken sameness, some fantastic dream succeeds,
> Vague emotions rise and ripen into thoughts and words and deeds.
> Old impressions, long forgotten, range themselves in Time and Space,
> Till I recollect the features of some once familiar place.[61]

Like Carroll's Alice, who having passed through the Looking-Glass 'noticed that what could be seen from the old room was quite common and uninteresting, but that all the rest was as different as possible',[62] the first stanza of Maxwell's poem depicts waking life as dull and commonsensical next to the reflex reality of 'Dreamland' described by the second. Elemental and dynamic in their difference from one another, 'As calm air to troubled water, so my night is to your day', the states of sleeping and waking are, as these balanced antitheses indicate, nonetheless attributed with an ontological parity.

The conception of the mind introduced in these lines, as always active, even in sleep, was available to Maxwell from Hamilton. Dallas's account of Hamilton's classes, cited in Chapter 2, includes a student who 'gets up, and has a regular tussle with his master about the action of the mind in sleep, and in a state of semi-consciousness'. Hamilton's Lecture XVII argues that the mind is nonetheless awake while the body and the senses are 'torpid'[63] in sleep, an adjective that the poem applies conversely in its description of the 'torpid daylight-dreamers'. 'In our waking and our sleeping states, we are', he writes, 'placed in two worlds of thought', a characterisation that Maxwell carries over to his poem as 'the outer world of action' and the 'inner world of dreams'. Hamilton's lecture draws heavily upon the work of the French translator and champion of Scottish philosophy, Théodore-Simon Jouffroy, with ten pages, over a third of the lecture, given to quoting a passage in which he argues that an extended 'analogy' exists between waking and sleeping states of mind, as 'When we dream, we are assuredly asleep, and assuredly also our mind is not asleep, because it thinks.'[64]

As Maxwell would have known from his reading of the philosopher, Descartes famously traces his decision to reform philosophy to the inspiration of three dreams he had on the night of 10 November 1619. Sharing this idea of dreaming as creative thought, Maxwell writes to Campbell in July 1848 that he is able to solve certain problems, such as an impediment to a machine working and the loss of a key, only once 'I had dreamt over it properly, which I consider the best mode of resolving difficulties of a particular kind, which may be found out by thought, or especially by the laws of association'.[65] In a suggestive anticipation of Freud, Maxwell not only sees dreams to work by a logic of association, but also to summon the mind's residue of waking experiences,[66] as 'Old impressions, long forgotten, range themselves in Time and Space'. The capital letters given to these coordinates indicate that the 'Old impressions' are arranged in accordance with the *a priori* Forms of Sense that Kant attributes to conscious mind, versions of which Maxwell knew not only from Hamilton's lectures but more recently from Whewell's doctrine of the 'Fundamental Ideas'.[67] These shared Forms underpin the complementary but continuous relation between the dreaming and the waking mind in Maxwell's poem. Dreaming suggests a sequel to the Pythagorean apprehension of the evening sky in 'A Student's Evening Hymn', as 'the soul, retiring inward, finds itself beyond the spheres'. Space is unleashed from both our finite waking experience of it and its *a priori* mathematical representations, as the persona reports that, forsaking the conventional passive spatial metaphor of *falling* asleep, 'From the midst of facts and figures, into boundless space I leap.'

The form of space that furnishes the condition for the complementary phenomenal 'facts' and abstract 'figures' that compose the student's apprehension of the world in waking life is described reflexively as unconditioned in sleep, a 'boundless' realm of possibility, 'For the inner world grows wider as the outer disappears.' Latent moral and emotional dimensions to the subconscious mind are expressed in dreams, as 'Vague emotions rise and ripen into thoughts and words and deeds.' Indeed this ground is seen to furnish promising and unexpected capacities for knowledge, 'powers and thoughts within us, that we know not, till they rise'. He reiterates this idea with another liquid analogy a year later, in a letter to his friend Richard Litchfield from May 1857: 'But I believe there is a department of mind conducted independent of consciousness, where things are fermented and decocted, so that when they are run off they come clear.'[68] The idea is referred to in similar terms in the final stanza of the poem: 'Let me dream my dream till morning; let my mind run slow and clear.' This principle of the subconscious correlates with Hamilton's lectures, which maintain that 'the sphere of our conscious modifications is only a small circle in the centre of a far wider sphere of action and passion, of which we are only conscious through its effects'. Maxwell's image of 'treasure-caves in Dreamland', presumably hewn in the mind's rocky substrate, recalls Hamilton's observation that 'the infinitely greater part of our spiritual treasures, lies always beyond the sphere of consciousness, hid in the obscure recesses of the mind'.[69] The mind is seen to have an innate capacity for the 'self-excenteration' that 'Reflex Musings' opposes to solipsism and makes a prerequisite for discovering new knowledge, as through its emissaries the subconscious mind becomes immanent to consciousness. Subconscious mind is presented as an active principle that not only surfaces consciously as knowledge of the self, but also facilitates new knowledge of the object world.

Building upon his proposition in lecture XVII that the dreaming mind is awake as the body sleeps, Hamilton devotes the subsequent lecture to renewing the Platonic doctrine of anamnesis, as he argues for the subconscious origins of knowledge.[70] In it he offers an anecdote from Lord Monboddo about a Comtesse who could speak Breton only in her sleep, and cites a complementary case from Coleridge's *Biographia Literaria*, of an illiterate young German woman who, struck with 'a nervous fever ... continued incessantly talking Latin, Greek, and Hebrew, in very pompous tones, and with most distinct enunciation'.[71] Hamilton then argues again from the impossibility of absolute change, of something coming from nothing, that any mental 'modification must be present, before we can have a consciousness of the modification', so that consequently 'we can have no consciousness of its rise or awakening; for its rise or awakening is also the rise or awakening of consciousness'. This leads him to make the following bold declaration: 'I do not hesitate to maintain that what we are conscious of is constructed out of what we are not conscious of, – that our whole knowledge, in fact, is made up of the unknown and the incognisable.'[72]

Campbell lists this doctrine of 'Unconscious Mental Modifications' as one of four recurrent 'Hamiltonian notions' he finds in Maxwell's letters and occasional writings.[73]

While the 'questions strange' in 'Reflex Musings' bespeak tensions between nature and the terms that the persona applies to it, by eliciting its echoes, wakening its cavernous depths, they can in turn call out analogously deep resources within the mind, which are correspondingly imaged in 'Recollections' as stretching back to its own rocky substrate, and by such reflexive means facilitate the separation of 'Words of Truth from words of Pride'. Just as the persona's reverie over various phenomena occurs in 'Reflex Musings' with the abeyance of his wilfulness, so correspondingly in the *a priori* environs of 'Dreamland' the subtle 'hidden depths below', 'where Self in secret lies', are disclosed 'when Will and Sense are silent'. Nature is described similarly in 'a "kind of allegory"' Maxwell wrote in January 1858, as 'a spirit of melody [that] found me, / And taught me in visions and dreams', but which human wilfulness is destroying; 'We are drowning in wilful confusion / The notes of that wonderful song.'[74]

Dreams are seen in 'Recollections' to instance the mind's free exercise: 'Empty bubbles, floating upwards through the currents of the mind.' Instances of apparent emptiness, mere curvilinear forms, bubbles nonetheless assert their extension, a delicate but insistent materiality, through the medium of the mind. The freely meandering ascension of the bubbles emblematises the oneiric 'wandering fancies' as they are drawn together through a principle of free association, 'by some lawless force entwined'. Invoking an appropriately submerged pun on *pneuma*, being portions of air like breaths,[75] the 'Empty bubbles' are little figures, both moral and intellectual, of aspiration and inspiration, playful and anarchic emissaries from the subconscious depths of selfhood. Maxwell's effervescent figure implies that the Hamiltonian actualisation of knowledge, as it comes to consciousness, is not only mysterious but creative.

Maxwell's earlier reading of Paris's *Philosophy in Sport* would have encouraged the idea he makes explicit in his 1873 essay on Plateau, of bubbles as a definitive principle of scientific play. It is unsurprising that he identifies himself and his peers with children blowing bubbles for, as Simon Schaffer observes, from the 1850s bubbles were seen to hold 'clues to the basic structures of matter', and consequently became a keen object of research.[76] Those who are most clearly 'of the same family as those Etruscan children' include Paris's friend Faraday and William Thomson, who urges us to 'Blow a soap-bubble and look at it – you may study all your life, perhaps, and still learn lessons in physical science from it.'[77]

Maxwell was reading Plateau in 1856, the year he wrote 'Recollections'. He refers to Plateau's experiments on compound colours in a paper from August of that year, while in his winning essay for the 1857 Adams prize, submitted in December 1856 on the prescribed topic of the stability of Saturn's rings, he considers the analogy that his continental contemporary makes of them with the centrifugal dispersal of oil as it spins in a mixture of alcohol and water, 'the actual breaking up of fluid rings in the beautiful experiments of M. Plateau'.[78] 'Recollections of Dreamland' follows the bold pattern of Plateau's speculations by finding a model for complex phenomena in the viscous behaviours of liquids.

Contemporary with 'Recollections of Dreamland', Maxwell's jocular exposition of puns and analogies in the 'Analogies' essay offers a paradigm case for the poem's epistemology as, predicated upon the distinction between the unwitting 'author' of the pun and 'the publisher' who 'find[s] it out,' it locates the principles of unconscious and conscious knowledge in the mediate principle of language. Truth is seen as relational and latent in language, intrinsic to the riddling phenomena of puns and analogies, which accordingly require an act of mind to bring them to consciousness. While this recalls Aristotle's methodological principle of *ta legomena*, his idea that all our knowledge is embodied in language usages and can accordingly be retrieved from them, it functions through forms of condensation. A principle that Freud will recognise later as integral to the dream-work, jokes and the treacherous utterances that become known as Freudian slips,[79] such condensation is akin to the 'powers and thoughts within us' that in 'Recollections of Dreamland' dreams are seen to convey to consciousness. We can, as the 'Analogies' essay observes, utter puns unconsciously, so that they play with us much as dreams do. This parallel is demonstrated by *Alice's Adventures in Wonderland* and *Through the Looking-Glass*, each of which takes the form of a dream that proceeds through various puns that are either native to language, such as the example of the flower-beds cited in Chapter 2, or synthesised from it, as occurs with such portmanteau words as '*slithy*' ('lithe and slimy').[80]

The pun's subversive surfeit of meaning, the irreducible ambiguity that gives it a kinship to the condensed imagery of dreams, exercises and enlarges Maxwell's sense of hermeneutic possibility, the imaginative field in which he models and theorises phenomena by, for instance, the reciprocal means of analogy. Nonsense elements to puns, dreams and indeed other expressions are recognised as departures from, or affronts to, our normative realms of conscious thought and the meanings it intends. Maxwell recognises the unconscious mind as a deep dimension to selfhood that precludes us from coinciding with ourselves completely. It means that we are always by degrees strangers to ourselves, just as the object world is found to be a stranger to the persona of 'Reflex Musings', the assertive self that tries to be self-identical with it, to find '*Self* in every form I greet'. Scientific knowledge becomes possible with the persona's recognition of this radical estrangement, his acknowledgement that the inquiries he makes of nature must necessarily strike either it or himself as 'Questions strange'. These disjunctures open out the self and its relation to nature to chance and contingency, an intrinsic otherness that is a precondition for radical new discovery, and diametrically opposed to the cosy self-identical reiterations of pedantry, tautology and solipsism.

TRUTH AND BEAUTY

Figured playfully not only as bubbles but also as sparkling light, 'a guiding radiance [that] gleams', dreams are attributed with powers of enchantment in 'Recollections of Dreamland'. Their most unfathomed and unfettered forms are 'the spells of aimless fancies'. Such *a priori* 'wandering fancies' offer an intriguing parallel to the *a posteriori* 'wandering sense entrancing' phenomena described in the final stanza of 'Reflex Musings':

> Yes, the faces in the crowd,
>   And the wakened echoes, glancing
> From the mountain, rocky browed,
>   And the lights in water dancing –
>   Each, my wandering sense entrancing,
> Tells me back my thoughts aloud,
>   All the joys of Truth enhancing
> Crushing all that makes me proud.

The turning-point in this stanza is marked dynamically with a dash. On one side of it, the first quatrain catalogues the phenomena of reflection described in the earlier stanzas, while on the other, the poet draws upon his experience of them to account for knowledge as a process of reflexion. Its pivotal line, 'Each, my wandering sense entrancing', describes an enchantment of the senses, as the poet, no longer focused upon his reflexions, loses himself in apprehensions of the object world, much as in the later poem the persona does in the 'wandering fancies' of the dream state. The *OED* notes that the verb 'entrance' can be traced through 'trance' to the Latin *transire*, 'to pass over, cross', an etymology that links it with the noun 'entrance'. Indeed such a pun appears to be at work (or play) here. In contrast to the sterile trances of solipsism and narcissism, in which 'all that makes me proud' simply answers to 'my thoughts aloud', such 'entrancing' of the senses is presented as 'Truth enhancing'. It is seen to confirm, and indeed facilitate access to, the truth of the object world. Rather than mental reflexion folding phenomena back solipsistically to its own terms, the mind is bent reflexively by phenomena to the autonomous being of the material world. This surrendering of self to the experience of the physical world is the lesson of the poem's third stanza, after which the phenomenon that the poet describes self-interestedly in the second verse as the reflection of his 'stooping form' in 'the darksome well' is read conversely and brightly as objective physical phenomena, 'the lights in water dancing'.

The delightful 'entrancing' of the senses indicates a principle of aesthetic pleasure, reminiscent of that which Maxwell associates with his earliest scientific endeavours. Indeed it formalises the principle of his original and abiding appreciation of curves as coincident instances of truth and beauty. Questioned in 1873 for the English eugenicist Francis Galton's *English Men of Science*, Maxwell describes as one of his 'Mental Peculiarities' being 'delighted with the forms of regular figures and curves of all sorts'.[81] His first scientific paper, the 1846 'Observations on Circumscribed Figures' introduced in Chapter 2, sprang from his youthful interest in ovals and other such geometrical and ornamental forms, which was stimulated by papers given by the painter and decorative artist David Ramsay Hay that he attended with his father at the Edinburgh Society of the Arts in 1845–6. Campbell records that Hay's 'attempt to reduce beauty in form and colour to mathematical principles had attracted considerable attention amongst scientific men'. He adds that at this time 'Such ideas had a natural fascination for Clerk Maxwell, and he often discoursed on "egg-and-dart", "Greek Pattern", "ogive", and ... on the forms of Etruscan urns',[82] an example of which, embellished with representations of the curious curvilinear phenomenon of soap bubbles, he returns to in his 1873 essay on Plateau. Like Hay,[83] Maxwell sought to draw and describe mathematically the curves from which such abstractly beautiful figures and patterns are formed. His 1846 paper furnishes an elegant and indeed definitive mechanical method for drawing a perfect oval, which was, as J. J. Thomson notes, subsequently found to have 'been anticipated by no less famous a mathematician than Descartes'.[84]

Maxwell reiterates his aesthetic appreciation of curves in his 1854 essay 'Has Everything Beautiful in Art its Original in Nature?', where he finds both the mathematician and John Ruskin united in their 'admiration of the Ellipse', a convergence of science and aesthetics that is encapsulated in a subsequent reference he makes to them here as 'bicentral sources of lasting joy, as the wondrous Oken might have said'. The essay theorises this pleasurable apprehension of beauty and truth in the following terms: 'It is a universal condition of the enjoyable that the mind must believe in the existence of a law, and yet have a mystery to move about in.'[85] The terms of law and spacious mystery suggest the conscious and subconscious principles, and their epistemologically promising interactions, outlined in 'Reflex Musings' and 'Recollections of Dreamland'. Maxwell's formula corresponds to the Kantian idea of 'free' beauty, in which the unifying principle of law or regularity demanded by the understanding coincides with the freedom of play ('a mystery to move about in') required by the imagination. This results in, as Kant puts it, a 'proportionate accord' or 'harmony of the cognitive faculties' that produces an aesthetic 'feeling of pleasure'.[86] The principle was available to Maxwell through Hamilton's lectures: 'In the case of Beauty, – Free Beauty, – both the Imagination and the Understanding find occupation; and the pleasure we experience from such an object, is in proportion as it affords to these faculties the opportunity of exerting fully and freely their respective energies.'[87] While for Kant 'free' beauty offers an assurance of the existence of noumena but no knowledge of it, Maxwell makes no such distinction. The 'entrancing' of the mind in 'Reflex Musings' not only suggests the psychological state induced by an instance of Kantian 'free beauty' but, as noted earlier, it is also seen to confirm and enrich the apprehension of truth, 'All the joys of Truth enhancing'.

'Has Everything Beautiful in Art Its Original in Nature?' not only finds in geometry instances of coalescent truth and beauty, but also in mechanics. Rivers and mountains, it may be recalled, are seen to yield pleasure through the 'apprehension of their fitness as the forms of flowing and withstanding matter'. The wildness of the river's flowing water nonetheless keeps within the bounds of the riverbed, while the mountain steadfastly maintains a resolute and sublime unity against all the forces that may accost it. As such they can be seen to hold in tension principles of unity and free play, 'a law, and ... a mystery to move about in'. They humour the Kantian principle of 'free' beauty accordingly as they instance Newton's Third Law of Motion, which states that for every action there is a corresponding equal and opposite reaction. With the accession of aesthetic pleasure that proceeds from such instances the mind is seen to warmly congratulate rivers and mountains for being superlative instances of a fundamental truth of physics.

The 'enjoyment' that Maxwell sees to supervene upon the mind's recognition of an instance of truth and beauty springs from the purely formalist criterion of relation, which Maxwell describes as fundamental to scientific method:

Whenever [men] see a relation between two things they know well, and think they see there must be a similar relation between things less known, they reason from the one to the other. This supposes that although pairs of things may differ widely from each other, the *relation* in the one pair may be the same as that in the other. Now, as in a scientific point of view the *relation* is the most important thing to know, a knowledge of the one thing leads us a long way towards a knowledge of the other.[88]

This corresponds to the emphasis that Hamilton's scientific method places upon 'formal similarities between relations rather than of material analogies between the things related',[89] and indeed encompasses mathematical and physical principles of formal continuity and invariance.[90]

Rhyme, which describes a set of relations between sounds incarnated as different words, offers an emblem for this principle of analogy, the formal principle of relation between sets of terms that Maxwell specifies in the passage. Similarly, each case of physical reflection, whether of light or sound, offers an image that differs from its original whilst preserving the relations between its parts. Parallel to the pattern of rhyme, and in contrast to the simple chiming of solipsism, analogy makes difference or novelty apparent and accessible to the mind, recognisable and knowable, by drawing together disparate terms through formal and semantic affinities. The poem's allegory of analogy is enacted with aesthetic formality in the final stanza of the poem through the artificial acoustic reflections of rhyme, as the objective charms of phenomenal echoes 'glancing' and the reflected lights 'dancing' translate into the subjective 'entrancing' of the senses and then, 'All the joys of Truth enhancing', facilitate the recognition and enjoyment of knowledge, the discrimination of 'Words of Truth from words of Pride'. The rhymes themselves elicit the aesthetic 'sense entrancing' quality they include as one of their terms. They also complete the trochaic tetrameter lines, as they furnish a supplement to the catalectic lines, much as the mountain's echoes do for the persona's utterances. Indeed, like the composite sound of the echo, their rhyme consists of two terms, a stressed and an unstressed syllable, while the rhyme in the clipped alternate lines consists of a stressed syllable only. The more stolid rhymes belonging to the catalectic lines – 'crowd', 'browed', 'aloud' and 'proud' – suggest the 'words of Pride' that several of them allude to, the residue of the self-reference with which the poet approaches the object world in the early stanzas, along with the mountain's stern judgement of them. With their long vowel sounds terminating wilfully with a thud on the final uniform consonant, they fall like stones amongst the resonant and lilting trochaic rhymes, 'Words of Truth' that suggest the gentle play of light on water or the reverberant cadences of echo. The curtailed sounds represent a fixed point, a principle of law, denuded of the resonant open sounds, 'a mystery to move about in', that complete the trochee in the other lines.

The differentiated unity of analogies, 'Truth enhancing' correspondences described by the final stanza of 'Reflex Musings', has its formal counterpart in the relations of rhyme, while these instances of truth and beauty coincide in the acoustic reflection of the echo, as thought returns to the mind 'aloud', in a form that is altered and vivified by its encounter with material nature. The words that the persona shouts find their corresponding term, of both analogy and rhyme, in 'the wakened echoes'. 'Reflex Musings' depicts nature, 'the hollow mountain-side', as resolute in its independent being but nonetheless responsive to apt physical and, by analogy, conceptual projections that we can throw against it, to as it were sound its depths and discover its laws.

Such projections suggest Whewell's hypothetico-deductivism, in which inductive truths are seen to disclose themselves as we apply to phenomena an apt 'theory' or 'Conception'. Exploring such 'inductive philosophy', Maxwell writes to Litchfield in June 1855 that he is 'grinding out "appropriate ideas", as Whewell calls them'. He explains the process with imagery that correlates with the dynamic Hamiltonian dialectic of the poet's echoing questions in 'Reflex Musings'; 'and by dint of knocking them against all the facts and ½-digested theories afloat, I hope to bring them to shape.'[91] A year later, the object world of nature is imaged in Maxwell's 'Analogies' essay, much as it is in the third and fourth stanzas of 'Reflex Musings', as a great stone mass, the depths of which are penetrated and distinguished, in accord with Whewell's philosophy of science, by theory, which is described by analogy with a carefully calibrated optical instrument: 'The dimmed outlines of phenomenal things all merge into another unless we put on the focussing glass of theory and screw it up sometimes to one pitch of definition, and sometimes to another, so as to see down into different depths through the great millstone of the world.'[92] Space furnishes the common medium for mind and matter in this mechanistic analogy, as in that of 'Recollections of Dreamland'. While the poem dramatises the mysterious depths and imaginative freedom that can be drawn upon in creating theories, the essay describes how such theories can discriminate the corresponding depths of the object world.

CRYSTALS

The formal economy by which Maxwell appreciates the pun and the analogy as reciprocal forms in both 'Reflex Musings' and the 'Analogies' essay, a playful dynamic of semantic condensation and explication, is focused at the conclusion of the essay in an intense instance of optical reflection:

And last of all we have the secondary forms of crystals bursting in upon us, and sparkling in the rigidity of mathematical necessity and telling us, neither of harmony of design, usefulness or moral significance, – nothing but spherical trigonometry and Napier's analogies. It is because we have blindly excluded the lessons of these angular bodies from the domain of human knowledge that we are still in doubt about the great doctrine that the only laws of matter are those which our minds must fabricate, and the only laws of mind are fabricated for it by matter.[93]

Maxwell's account builds upon the work of Whewell in establishing mathematical crystallography during the 1820s and early 1830s, and, more particularly, invokes the use of spherical trigonometry that his successor to the Cambridge chair of mineralogy, W. H. Miller, brought to this study during his long tenure from 1832 to 1870. A form of 'mixed' or 'applied' mathematics, modern crystallography parallels Maxwell's science of physics, the inspiration for Whewell's model of induction in which, as he observes, *a posteriori* and *a priori* science coincide: 'so that we obtain, combined with all the physical reality of the external world, all the permanence and symmetry of mathematical relations'.[94]

In accord with Hamiltonian dynamism, the image of 'forms of crystals bursting in upon us' recalls Maxwell's 1853 poem 'To F.W.F.', in which the light of scientific 'Truth' is described as having 'burst and entered'.[95] The raw immanence of crystalline forms has no need of the mind's projections, such as the natural theological considerations that preoccupy the 'Hymn', nor utilitarian and other moral motives. By articulating a space that is understood as both physical and metaphysical, the 'sparkling' of the crystals illustrates Hamilton's reciprocal Secundo-primary qualities, in which the mind both feels as it knows and knows as it feels, balancing principles that are developed correspondingly in Maxwell's formula for 'the great doctrine', a vindication of the analogy presented by the title 'Reflex Musings: Reflection from Various Surfaces', 'that the only laws of matter are those which our minds must fabricate, and the only laws of mind are fabricated for it by matter'.[96] The poem's echoes come to consciousness as a potential that is actualised both as thought and as matter in motion, the resonant vibrations of which are exemplified analogously, synaesthetically, in the epiphanic luminosity of the sparkling crystals, a concentrated play of light. The crystals provide a rich analogy for the poem's model of knowledge as a dialectic of endless reflection, in which the complementary terms of 'Reflex Musings: Reflection from Various Surfaces' face each other like two mirrors, each constantly offering the other instances of correspondence and resistance. Scientific discovery is experienced through this teasing *mise-en-abyme* effect, 'All the joys of Truth enhancing', as poetic reverie.

As Maxwell indicates in his description of the crystals 'sparkling in the rigidity of mathematical necessity', the abstract language of mathematics furnishes the medium through which the forms of thought and of nature can be translated and equated. Exploring and testing the epistemological principle of analogy during the 1850s in his poems and his Apostles essays, Maxwell makes it the ground for his scientific methodology and practice in the adventurous and imaginative models he develops in his subsequent series of papers 'On Physical Lines of Force' (1861–2). 'Physical research is continually revealing to us new features of natural processes', Maxwell declares in his 1870 Liverpool address, 'and we are thus compelled to search for new forms of thought appropriate to these features.' In contrast to the heuristic use of analogy in 'On Faraday's Lines', where he uses 'mechanical illustrations to assist the imagination, but not to account for the phenomena', 'On Physical Lines' attempts a mechanical *explanatory* correlate of such phenomena and their equations with an ether model of molecular vortices and 'idle wheel' particles; 'According to our hypothesis, the magnetic medium is divided into cells, separated by partitions formed of a stratum of particles which play the part of electricity.' Maxwell writes that he has devised his model 'to bring out the actual mechanical connexions between the known electro-magnetic phenomena'.[97] 'The nature of this mechanism is', as he explains to Tait in 1867, 'to the true mechanism what an orrery is to the Solar System.'[98] He explicates the underlying principle in the Liverpool address:

The characteristic of a truly scientific system of metaphors is that each term in its metaphorical use retains all the formal relations to the other terms of the system which it had in its original use. The method is then truly scientific – that is, not only a legitimate product of science, but capable of generating science in its turn.[99]

Such a system of formal relations is predicated upon punning. The symbolic expressions of mathematics, such as Napier's equations, are cases in which 'two [or more] truths lie hid under one expression'. They can yield other applications, facilitate further analogies, and so reciprocally enrich themselves as puns. This further capacity was famously demonstrated by Maxwell's discovery of the Electromagnetic Theory of Light as an 'unexpected consequence' of his molecular vortices model in the third part of 'On Physical Lines'.[100] The unwitting 'author' of a pun, he then 'find[s] it out', becomes its 'publisher'. A particularly audacious attempt to draw together the laws of the mind and the laws of nature, Maxwell's mechanical and mathematical models, the sets of analogies he develops between 'the luminiferous medium' and 'the magneto-electric medium' yield the equations named after him, to establish that *'light consists in the transverse undulations of the same medium which is the cause of electric and magnetic phenomena'*.[101]

In his poetry and his science alike Maxwell endeavours to recognise registers of reality, 'different depths', by applying to phenomena a spectrum of analogies: his various forms of Physical Analogy and geometrical modelling, explanatory mechanical analogies, mathematical quantities, 'an allegory about incompressible elastic solids' he finds in a paper by Thomson, along with the 'parables, fables, similes, metaphors, tropes, and figures of speech' he outlines at the start of his essay on 'Analogies'.[102] Truth is accordingly addressed by such means in 'A Student's Evening Hymn' as personal and spiritual, and a week earlier, in 'Reflex Musings', as quotidian and scientific. Similarly, Maxwell's advocacy of puns, indeed his playful use of them in the 'Analogies' essay as the original premise for his epistemological inquiries, testifies to the suppleness and wit with which he allows his thought to find relations amongst its objects. It places him, poised between poetry and science, with such early thinkers as Heraclitus and Lucretius, who similarly explore nature through word-play.[103] Puns allow Maxwell to move between different worlds of ideas and experience, not only along what he refers to later in his essay as the 'focusing glass of theory', but also in the comic associations he makes in his later light verse for his colleagues in the Red Lion Club. The Lions met after the day's work at the annual BAAS meetings for informal dinners, where, as the following chapters observe, comic verses and songs would flow anarchically like 'empty bubbles' through the Association's earnest currents of thought and discussion.

CHAPTER 4

Red Lions: Edward Forbes and James Clerk Maxwell

Formally relinquishing his youthful ambition to be a poet in 1829, Rowan Hamilton addresses the muse in 'To Poetry' with a request that she not abandon him, 'though my life be now / Bound to thy sister Truth by solemn vow'. The character of Poetry that he wishes to retain is specified in the opening stanza as 'Thou who keepest festival / In the mind's ideal hall'.[1] He describes her sibling Science four years earlier in a letter to Arabella Lawrence, a friend of Coleridge's he knew from Maria Edgeworth, as the supremely aloof mistress of such a mental space, for it 'sits enthroned in its sphere of isolated intellect, undisturbed by passion, unclouded by doubt'. The sociability he attributes to Poetry, albeit of an oddly *a priori* kind, implies an argument that he makes in his letter to Lawrence, in which this 'rival principle' is seen to compensate for the austere isolation of Science; 'man is not a creature of intellect alone', indeed, he maintains, 'His heart is even more important than his mind; he was made to be a social creature, and his second duty is to love man. Now I think that poetry is eminently qualified to strengthen and refine the links which bind man to his kind.' While their professional commitment may, as Wordsworth argues, preclude scientists from mastering the art of poetry, Hamilton makes a plea for its special importance as a means of facilitating the social connections and sympathies which he sees science as denying them. He also considers the BAAS to have its *raison d'être* in encouraging such sociability. Indeed, in his Secretary's address to the 1835 meeting in Dublin, he argues that this is the great contribution the Association makes to furthering the ultimately solitary pursuit of scientific discovery: 'if it be inquired how ... we as a Body hope to forward Science, the answer briefly is: that this great thing is to be done by us through the agency of the social spirit – we meet, we speak, we feel *together now*, that we may afterwards the better think, and act, and feel *alone*'.[2] Hamilton's prescription for the advancement of science would also be realised, somewhat erratically, by members of the BAAS later in the decade, with the formation of the Red Lion Club.

Amateur forms of versifying proved to be especially convivial for the Victorian scientists who gathered during the meetings of the BAAS for the Red Lion Club dinners, where various forms of pastiche, lampoon and doggerel would be read out or sung. The geologist and palaeontologist Edward Forbes is attributed with founding the original Club during the 1839 meeting at Birmingham. He and other young scientists, mainly fellow naturalists from Section D, 'finding the dull conventionality of the "ordinaries" insupportable, started simple dinners of their own', cheap meals of beef and beer 'enlivened by joke and song' at a local tavern, The Red Lion.[3] In 1842 Forbes took the chair of botany at King's College, London, where two years later, as he explains to his friend and fellow geologist Andrew Crombie Ramsay, he and the other London-based Lions 'succeeded in establishing a monthly meeting and feed of Red Lions at the "Chesire Cheese" [mutton-chop house] in Fleet Street'.[4] For the following decade the Metropolitan Club provided opportunities for young London scientists, along with Lions visiting from the provinces and international guests, to meet socially on the third Thursday of each month (Figure 2).[5]

Writing from the Oxford meeting of the BAAS in 1894, Huxley recalls the Red Lions dinner at the 1851 Ipswich meeting, which he attended with his friend Tyndall:

Being young with any amount of energy, no particular prospects, and no disposition to set about the ordinary methods of acquiring them, we could conduct ourselves with perfect freedom; and we joined very cordially in the proceedings of the 'Red Lion Club', of which I had become a member in London, and which had been instituted by that most genial of anti-Philistines, Edward Forbes, as a protest against Dons and Donnishness in science. With this object, the 'Red Lions' made a point of holding a feast of Spartan simplicity and anarchic constitution,

Figure 2 Insignia of the Metropolitan Red Lion Association, early 1850s. Imperial College, London.

with rites of a Pantagruelistic aspect, intermingled with extremely unconventional orations and queer songs, such as only Forbes could indite, by way of counterblast to the official banquets of the Association, with their high tables and what we irreverently termed 'butter-boat' speeches.[6]

Corresponding to the hierarchies of the BAAS, the Red Lion Club developed its own mock-castes of a presidential Lion King, ordinary members, 'cubs' and catering 'jackals'. As his old friend the chemist George Wilson, a founding member of the Lions from the 1839 meeting, and the young geologist Archibald Geikie record in their biography of Forbes, the members developed various playful protocols of 'signifying their approval or dissent by growls and roars more or less audible, and, where greater energy was needed, by a vigorous flourishing of their coat-tails'.[7] In a diary entry on the 'grand Red Lion meeting' at Ipswich, Huxley reports that, in replying to a toast made to his health, the ornithologist Bonaparte, the discoverer of Lear's Macau and 'singularly like the portraits of his uncle', 'told us that as he was rather out of practice in speaking English, he would return thanks in our fashion, and therewith he gave three mighty roars and wags'.[8] Forbes was in the habit of writing comic verse, and established the tradition of singing and reciting such work at the Red Lions' gatherings, 'the subjects being usually taken from some branch of science', Wilson and Geikie observe, 'and treated with that humour and grotesqueness in which he so much delighted'. Indeed he 'never failed to chant at their annual dinner at least one new song'.[9] Rather than the decorous Cartesian theatre in which Hamilton had placed it, poetry's festival found its home amongst the scientists at the raucous Red Lion feasts.

Forbes offers an account of the alternative association early in his rambling poem from 1851, 'A Yawn from a Red Lion', which was 'suggested by a "Dinner" at [the celebrated French chef Alexis] Soyer's Restaurant':

> Some Lions once, in friendly, lordly mood
> Agreed to have a dinner, and a spree:
> To sing 'Together let us range –',
> And leave their wives behind them; – for a change,
> And as for food,
> The Jackal to 'all that' – of course would see.
>
> These Lions were of British Breed; the roars
> Of some had echoed through the World; and all
> Had roared to purpose, more or less,
> Or they had not been Lions, as I guess;
> For many scores
> In Lions' skins, have only brayed, withal.[10]

Most of the sixty-one verses of Forbes's poem are dedicated to describing the meal, an extravagant exception to the Lions' usual simple dinners. On this occasion Soyer included the Red Lions in 'A Grand Banquet for the *literati* of all nations' he held on Thursday, 15 May 1851 for over three hundred British and international journalists who had gathered to cover the Great Exhibition, amongst them the unknown (and unemployed) Karl Marx. A dinner that, as the chef remarked, was 'the reverse of an *omelette soufflée*', which is 'puffed to be eaten: now the dinner is eaten to be puffed',[11] it was held to publicise his grand Baronial Hall restaurant. Regarded by Soyer as a group of famous scientists, the Red Lions were seen to add to the tenor of the event. Ramsay records the evening in his diary:

> At six went down with Forbes to the Red Lions at Soyer's. It appeared that he had a great dinner to the Press, etc., of all nations, and having made no provision for us, he dodged us into dining with them in the great hall. His first request was that we should dine at the same hour to save his cooks. There were all the Reds of note, including Owen, [R. G.] Latham, Dr Smith, etc. etc. He appointed the best places at table for us, and made his people ply us with all sorts of good dishes and wine. It was a splendid joke. In the garden was a huge oven, in which half an ox was roasted. At a signal the covers were removed, and it was wheeled on to great dishes on a hand-barrow. Twelve cooks carried it, and a brass band marched before playing 'The Roast Beef of Old England', while all the guests came up behind laughing ... The whole thing was so cleverly done that, save Latham, perhaps, all of us took it as a joke and laughed prodigiously.[12]

The evening concluded with a firework display and the release of a fire balloon in the garden.

The reference that Forbes's poem makes to a 'lordly mood' recalls the aristocratic patronage that W. Vernon Harcourt and other founders of the BAAS saw as essential to the success of the new society, which from the first meeting at York provided participants with lavish gourmandising, including venison and turtles from country estates. Brewster had early seen the advantages of aristocratic patronage for the new Association as it contested the authority of the Royal Society, although he was also keen to exclude from it those he regarded as dilettantes and charlatans. Whewell was similarly wary of what he calls 'the crowd of *lay* members' that diluted the Royal Society.[13] Both men accordingly suggested that membership of the BAAS be restricted to practising research scientists who had published at least one scientific memoir. However, Harcourt and his allies prevailed in their argument that the fledgling Association needed to open itself to a wider, larger and more influential, membership in order to be effective. The records for the Association show that its early membership was dominated by Anglican clergy, with the next largest group being medical doctors, then aristocrats, followed by academics, with smaller numbers of parliamentarians and government officers.[14]

In a diary entry for 19 April 1849, Ramsay records that Owen abstained from attending a Royal Society meeting in favour of an evening with the Lions, where he 'made a most humorous speech, contrasting the pleasure of sitting in this snowy night, so cosy and merry round the table, with the horrors of the Royal Society then sitting'.[15] Owen's preference is not entirely facetious. With all their tail-wagging and roaring, Forbes's Red Lions nonetheless represented the future of the BAAS as a professional society that would value research merit over Anglican and aristocratic privilege. As Forbes puts it in his poem, the Lions 'roared to purpose'. They make real contributions to science. Others who 'have only brayed', wear the lion's pelt like the donkey in Aesop's fable, or impotent impersonators of Herakles. They are merely British Asses, members of the British Association who produce little or no scientific research. Indeed Forbes may be making a joke at his own expense, for the president's chair from which he was reading his poem was probably draped with such a skin, as it was in January 1854, when Tyndall's friend Thomas Archer Hirst records in his journal that 'The skin of a veritable lion forms the back and canopy of the chair; his hind legs serve as arms for resting the elbows, and they are so arranged that the chairman can use the lion's paw to beat the table and command attention, or applaud at will.'[16]

Looking back in 1894, Huxley notes 'the evolution of "Red Lionism" into respectability'. He and his peers, with their ideas of professional merit-based research science, had prevailed: 'The last time I feasted with the "Red Lions" I was a Don myself', Huxley writes, and 'the dinner was such as even daintier Dons than I might rejoice in.'[17] Nonetheless, throughout most of the second half of the nineteenth century, the permissive carnivalesque environment of the Red Lions dinners continued to facilitate the writing and reading of humorous verses, often fuelling new factional and ideological differences within British science, whilst promoting an overarching culture of professional fraternity and, 'leav[ing] their wives behind them', masculine solidarity.

'THE FATE OF THE DO-DO'

The Red Lions share with the BAAS, of which they are, of course, all members, a professional exclusivity. Having defined science by Newtonian paradigms of physics and mathematics, no provision was made in the early formation of the Association's sections for such current claimants as phrenology and animal magnetism, while there was only 'scant encouragement of anthropology and medicine' and a 'reluctant encouragement of agriculture and geography'.[18] Using mockery and occasionally personal attacks, the Red Lions directed their verses to the repudiation of elements they felt had no place within professional science.

The polemical deployment of a principle of nonsense in Red Lions poems to affirm certain theories and practices as legitimate science, and especially to dismiss others, is nicely demonstrated by Forbes's 1847 poem 'The Fate of the Do-do'. Forbes was well placed to play with disciplinary boundaries and diverse theories, being, as the young Huxley writes of his friend and mentor in 1851, both 'a man of letters and an artist' and a scientist with broad interests and accomplishments: 'More especially a Zoologist and a Geologist than a Comparative Anatomist, he has more claims to the title of a Philosophic Naturalist than any man I know of in England.'[19] The 1847 BAAS meeting at Oxford included a long lecture on the dodo by Forbes's friend Hugh Strickland. This prompted a lively discussion about the animal the following morning, in which Bonaparte argued that it was only a chicken, while most of the company agreed it was a form of giant pigeon.[20] Geikie and Wilson describe 'The Fate of the Do-do' as 'a long ornithological romance containing a humorous account of the debate', which, in deference to the Lions dinner for which it was written, the poem supplements with accounts of Dutch sailors eating the animal. First published in *The Literary Gazette*, its eleven stanzas were subsequently edited by Forbes's friend John Hughes Bennett to the five that focus upon the scientific debate, 'the best in the song', to which he gives the title 'The Song of the Dodo'.

Following upon the delegates' informal discussion, the 'Song of the Dodo' presents various theories about the bird, a native of Mauritius that became extinct in the seventeenth century. As the Caucus-race in Carroll's *Alice in Wonderland* serves to highlight, the dodo lends itself to nonsense treatments. A flightless bird is almost an oxymoron, an absurdity that is articulated in the preposterous traditional representations of the dodo as fat, stupid and clumsy, as for instance in the early painting by Jan Savery that inspired Carroll and his illustrator Tenniel. 'The Fate of the Do-do' attributes the bird's discovery to 'Vasco de Gama', and indeed the etymology of its name has long been associated with the Portuguese word for silly, crazy, simple or foolish, as well as the English *dolt*, while its Latin title was *Didus ineptus*.

The dodo is notorious for being both grossly corporeal and extinct. Forbes's 'Song' begins by observing that the bird is rendered all the more absurd by being spectral:

> Do-do! Although we can't see him,
> His picture is hung in the British Museum:
> For the creature itself, we may judge what a loss it is,
> When its claws and its bill are such great curiosities.
> Do-do! Do-do!
> Ornithologists all have been puzzled by you.[21]

The claws and bill referred to here, and again in the longer version as having been 'saved from the poultry stew', describe the only remnants of a taxidermed specimen of the animal, an early bequest to the Ashmolean museum in Oxford that, having grown dusty and mouldy, the Board of Visitors ordered to be burned in 1755. With so little physical evidence extant, many during the first half of the nineteenth century doubted that the creature had ever existed. The remainder of Forbes's poem is dedicated to the dodo's spectral afterlife in the competing theories of eminent scientists.

The second stanza of 'The Song' is dedicated to the anatomist and zoologist Henri de Blainville (1777–1850), the professor of natural history at the French Academy of Sciences, whose lectures at the Jardin des Plantes Forbes had attended in 1836 and 1837. His 1829 *Mémoire sur le Dodo* places the creature in the vulture family. The third stanza introduces John Edward Gray (1800–75), the Keeper of Zoology at the British Museum from 1840 until 1874 who, amongst other books, wrote the text for *Gleanings from the Menagerie and Aviary at Knowsley Hall*, for which Lear had provided the plates. Gray not only 'Doubted what Mr de Blainville did say', but like many of his peers, some of whom thought the remnant foot and bill of the Ashmolean specimen to be fake, doubted the existence of the bird itself and hence the original reports of it by Dutch sailors, holding 'that', as the poem puts it, 'the bird was a vile imposition, / And that the old Dutchman had seen but a vision'. Like Carroll, Grey presents the dodo as a fiction. The final theory that the poem discusses is Strickland's, 'that our *rara avis* was only a pigeon', a thesis that he and Alexander Gordon Melville would defend in their book *The Dodo and its Kindred* (1848). Owen had argued in 1846 that the dodo was 'an extremely modified raptor', but having co-written the report on the 1847 BAAS meeting, which was published in *The Literary Gazette* alongside 'The Fate of the Do-do', he contributed some further Red Lion verses in a footnote to this poem, which begin by acknowledging Strickland's thesis: 'Of all the queer birds that ever you'd see, / The Dodo's the queerest of *Columbidae*.'[22]

The final stanza of Forbes's poem, which is headed '*Moral*' in the original version, reflects upon the competing theories of the bird and their proponents:

> Do-do! alas! there are left us
> No more remains of the *Didus ineptus*,
> And so, in the progress of science, all prodigies
> Must die, as the palm-trees will some day at Loddige's,
>   And, like our wonderful do-do,
> Turn out not worth the hullabaloo.[23]

While the *OED* traces the first metaphoric use of the word dodo to describe a redundant person or institution to 1886, Forbes's poem applies this analogy to both the scientists and their theories, grouping them all together as freaks of nature through its pivotal reference to 'all prodigies'. Much play is made on the word Dodo, which is here pronounced '*doo-doo*', so that each verse of the poem opens with what sounds like an urgent invocation to action, to solve the puzzle of the bird ('Do-do!'), while the final stanza concludes that 'the progress of science' ultimately constitutes much ado about nothing. The forced rhyme with 'Loddige's', the plant nursery famous throughout the eighteenth and nineteenth centuries for its rare plants, makes light of the ultimate fate of scientific prodigies, which are characterised here as exotics, like the nursery's palm-trees, primitive forms that in their rareness threaten to become extinct. The poem's *memento mori* extends to scientific ideas, for its juxtaposition of contradictory theories about the Dodo, including one that regards it as a fiction, demonstrate that at least two of them must be doomed. Furthermore, their exposition in chronological sequence enacts the pattern by which theories are contested and superseded 'in the progress of science'. New theories arise, rendering the old as nonsense. The equivocal word 'prodigies' embraces this relation, as in the field of scientific theory yesterday's marvels, 'like our wonderful do-do', can turn out to be quaint monsters with no future, destined for the most resolute of deaths, species extinction. The vanity of science is assured here by the pessimistic meta-induction from past falsity, which draws from the evidence of past scientific theories that have failed and been superseded the conclusion that all new theories will in turn also prove to be false. Much as Maxwell reiterates later with his entropic signature, the absurdity that inheres within life as its necessary condition is replicated in the precarious lives of scientific theories, which constantly threaten to collapse into nonsense and nonentity.

'TO THE CHIEF MUSICIAN UPON NABLA'

Maxwell attended the Edinburgh meeting of the BAAS in 1850,[24] and other meetings later in the decade, although he did not join any of the Red Lions dinners. This may have been because of the factional nature of the Club, which was still dominated by Section D, and the decline it suffered after Forbes's death in 1854. The Lions had, however, revived by the time the BAAS returned to Edinburgh in 1871. The Lion King that year was Maxwell's friend the Scots engineer and physicist W. J. Macquorn Rankine, whom their young colleague Oliver Lodge identifies with the second great period of the Club's flourishing; 'what were considered the palmy days of the Red Lion Club, when Professor Rankine was in great form and sang a number of his own jovial songs on the occasion, his appearance being rather "Jove-ial" too'.[25] For most of the following decade, until the year before his death in 1879, Maxwell would exercise his playful principles of nonsense in a series of poems on British scientific culture, which he addressed to his peers and often recited at the annual dinners.

Lodge recalls that from the 1870s, when he first began attending the Red Lions dinners:

It was the custom to parody not only the President's address, but one or other of the most popular of the Evening Lectures. They were the only things that were attended by the whole Association: the sectional proceedings, though they might occasionally lend themselves to parody, would not have been understood by the majority of the audience; but the others were public property. The fun was good-natured enough, and comic experiments were performed, recalling and, of course, guying the actual experiments which had been sometimes brilliantly shown at the real lecture.[26]

Maxwell wrote several such parodies in verse, including versions of Tyndall's Belfast address, 'British Association, 1874: Notes of the President's Address' and an evening address that Tait gave at Glasgow two years later, 'Report on Tait's Lecture on Force:—B.A., 1876', while he follows his own evening 'Discourse on Molecules' at the 1873 Bradford meeting with 'Molecular Evolution', the subject of Chapter 7. The remainder of the present chapter and the next focus upon 'To the Chief Musician upon Nabla: *A Tyndallic Ode*', the poem he read to the Red Lions at the 1871 Edinburgh meeting of the BAAS. Dedicated to Tait, this poem parodies a genre of popular public lectures that John Tyndall had been delivering very successfully at the Royal Institution in London since 1853. Including the entertaining experimental demonstrations around which these lectures were organised, the '*Tyndallic Ode*' also provides a unique record of the Lions' 'comic experiments'.

Tait concludes an essay on 'Clerk-Maxwell's Scientific Work', which he wrote for *Nature* after his friend's death in 1879, with a tribute to his poetry, where he remarks that; 'No living man has shown a greater power of condensing the whole marrow of a question into a few clear and compact sentences than Maxwell shows in these verses. Always having a definite object, they often veiled the keenest satire under an air of charming innocence and *naive* admiration.' Clearly demonstrating these qualities of richness, precision and disingenuously poised satire, '*A Tyndallic Ode*' targets what Tait describes at the close of his essay as the 'vain-babbling, pseudo-science, and materialism' of the age.[27] Indeed, the '*Ode*' is written as a parody of the lyric in Tennyson's poem 'The Brook', an aptly running joke that identifies the materialist Tyndall's prodigious lecturing with the anthropomorphised 'babbling brook' of the original verses, which declares; 'I chatter, chatter, as I flow', and at other points discloses similar capacities to 'murmur' and 'bicker'.[28]

Maxwell gave Tait the manuscripts of his two earliest versions of the '*Tyndallic Ode*'; a short one in pencil written on notepaper for the Edinburgh meeting that, Knott writes, 'was evidently dashed off by Maxwell in the B. A. Reception Room', and a more definitive seven-stanza version in ink, that 'seems to have been written the same evening'.[29] The focus upon Tyndall's popular science demonstrations in the four-stanza version offered a genial object of satire for its broad audience of his peers, while the more vicious additional verses, which take aim at his materialism and system-building, were apparently intended only for Tait. Read out at the Red Lions dinner, the shorter four-stanza version was published by *Nature* in its reports from Edinburgh as one of the 'papers . . . contributed to this section':

> I come from fields of fractured ice,
> Whose wounds are cured by squeezing,
> They melt and cool, but in a trice,
> Grow warm again by freezing;
> Here in the frosty air the sprays,
> With fern-like hoar frost bristle,
> Their liquid stars, their watery rays,
> Shoot through the solid crystal.
>
> I come from empyrean fires,
> From microscopic spaces,
> Where molecules with fierce desires
> Shiver in hot embraces;
> The atoms clash, the spectra flash,
> Projected on the screen,
> The double D, Magnesian b
> And Thallium's living green.
>
> This crystal tube the electric ray
> Shows optically clean,

> No dust or cloud appear – but stay:
>   All has not yet been seen:
> What gleams are these of heavenly blue,
>   What wondrous forms appearing?
> What fish of cloud can this be, through
>   The vacuous spaces steering?
>
> I light this sympathetic flame,
>   My slightest wish to answer,
> I sing, it sweetly sings the same,
>   It dances with the dancer;
> I whistle, shout, and clap my hands,
>   I hammer on the platform,
> The flame bows down to my commands,
>   In this form and in that form.[30]

Edinburgh was, as *Nature* reminded its readers in reporting the 1871 meeting, 'the den of the great "Red Lion" Forbes', who as a young man had studied medicine at the university and at the end of his life was made its professor of natural history. Maxwell was on his home turf with his Scots scientific friends, principally Tait, the professor of natural philosophy at Edinburgh. '*A Tyndallic Ode*' was accompanied on the night (and later on the same page of *Nature*) by another composition, a song entitled 'The British Ass'. A renewal of Forbesian ritual, it made the night memorable for Geikie, who in his autobiography recalls 'that Alexander Nicolson ... produced and sang to the tune of "The British Grenadiers" his verses on the British Association, which have often enlivened subsequent dinners of the "Lions".'[31] It is unsurprising that Geikie, who was president of the Geology Section at Edinburgh, makes no mention of Maxwell's recital. In contrast to Nicolson's toast to the BAAS, the humour of which is, like its simple tune, inclusive and accessible ('We find the trace of the monkey's face / In the gaze of the British Ass!'), Maxwell's '*Tyndallic Ode*' is a subtle and often esoteric complex of in-jokes directed to Tait and his other peers in Section A.

While the title '*A Tyndallic Ode*' announces the object of the poem's parody, its more oblique dedication 'To the Chief Musician upon Nabla' refers to Tait through a quaternion operator that Rowan Hamilton discovered. Tait's assistant William Robertson Smith dubbed the quaternion 'Nabla', as the shape of the symbol used to represent it, an inverted delta, resembles the Assyrian harp of that name mentioned in the Bible.[32] Building upon this conceit, the title of the poem in the longer of the original manuscript versions is accordingly given not in English but in Hebrew.[33]

Four days before the poem was read to the Lions, Tait had used his presidential address to Section A to promote quaternions, focusing in particular upon the Nabla operator: 'But, as it seems to me that mathematical methods should be specially valued in this Section as regards their fitness for physical applications, what can possibly from that point of view be more important than Hamilton's ∇?'[34] Listening to the address Maxwell would have found his friend once again, as he puns in his letters to him, 'harping on that Nabla'.[35]

One group of Maxwell's poems from the 1870s focuses on Tait and his lectures, another on lectures by Tyndall. 'To the Chief Musician upon Nabla: *A Tyndallic Ode*', which is probably the earliest of them, serves to introduce not only both groups of poems, but also the respective tribes of scientists that Tait and Tyndall often led in battle against one another. Its title does not so much bring together two figures that Maxwell's poetry elsewhere treats separately, as pit them against one another. Juxtaposing its factional terms on either side of the colon, the title recalls the oppositional structure of 'Reflex Musings: Reflection from Various Surfaces', without, however, its impartial poise and balance, the possibility of arriving at a resolution between them. Tait represents a fraternity of Scots Presbyterian physicists and engineers, dubbed the 'North British' natural philosophers, Tyndall a close group of London-based scientific naturalists known as the 'Metropolitans'.

METROPOLITANS AND NORTH BRITONS

Huxley, who with Tyndall would become a *de facto* leader of the Metropolitans, first met Forbes at the BAAS meeting in Southampton in September 1846, just before he sailed to Australia and New Guinea as the scientific assistant surgeon with the *Rattlesnake* expedition. Although they had met only once, Forbes corresponded with Huxley while he was away, furnished him with advice and introductions, and saw his research papers through the press.[36] Huxley began attending the Red Lions' gatherings in London soon after the expedition returned in October 1850.[37] The extent of his involvement with the Club can be gauged by the elaborate and amusing illustrations he made for a long poem Forbes had written about his teaching at the Royal School of Mines in Jermyn Street, 'Soliloquy Composed by a R[ed] L[ion] who came down to Anderton's July 15 1852', copies of which were printed, presumably for distribution among the members of the Metropolitan Club.[38] It is likely that, as well as going with him to the Red Lion dinner in Ipswich in 1851, Tyndall accompanied Huxley to some of the subsequent monthly gatherings in London, and that the title of 'Metropolitans' used for their group follows from such early associations with the London Red Lions. Like Forbes's original Red Lions, the Metropolitans drew together young ambitious scientists, mainly from Section D of the BAAS and for the most part without financial means, who asserted scientific merit and autonomy as the governing criteria for professional science against establishment privilege and prejudice.

Notwithstanding the disparaging connotations of the phrase 'North Britons', which opponents of the Scots physicists may have savoured and mobilised on occasion, such terminology is retained in recent work on the group, most notably in Crosbie Smith's definitive study, *The Science of Energy: A Cultural History of Energy Physics in Victorian Britain*. That this epithet was not considered to be as offensive during the Victorian period as it is in our own age of Scottish devolution and independence movements is clear from the choice that the new Free Church of Scotland made in 1844 to name its journal the *North British Review*. It is also notable that, as Smith observes, letters from members of the group to one another were often addressed from 'N[orth] B[ritain]'.[39] Tait and his allies became identified with the *North British Review*, which had been launched as an alternative to the secular Scots *Quarterly* and *Edinburgh* reviews, through a series of polemical essays they contributed to it, many of them attacking Tyndall and his materialism.

Maxwell demonstrates an early propensity for factionalism in the satirical poems he wrote at Cambridge, with such verses as 'A Vision' and '*Lines written under the conviction*' displaying a partisan loyalty to what he sees as Scots sense and integrity against Cambridge pretence and pedantry. Tait concurs with this assessment of the two systems in an address he gave to the Edinburgh University Graduates in 1866, while Maxwell reiterates it on behalf of them both in his late poem 'To the Additional Examiner for 1875'.[40] The times that Maxwell and Tait spent at Cambridge evidently brought into relief the distinctive legacies of the Scots educational system, their shared experiences and intellectual values that, with their Presbyterian religious beliefs, set them apart from their English contemporaries.

Drawn together by their shared histories and cultural identity, and pitting themselves against Tyndall and his Metropolitans, Maxwell and Tait, along with their old friends Fleeming Jenkin and Balfour Stewart (who was at Edinburgh University in 1845–6), formed a band of close friends during the 1860s and 1870s within the North British network. The origins and identity of this powerful though informal group can be traced to the work of William Thomson, professor of natural philosophy at the University of Glasgow, and Rankine, Regius Professor of Engineering at the same institution, who from the early 1850s pioneered the concepts and terms for the new science of energy. Other members of the North British group include William's brother James, who was professor of engineering at Queen's College, Belfast, and the Manchester natural philosopher and independent pioneer of the energy concept, James Prescott Joule.

Like the North Britons, the Metropolitans developed strong and abiding loyalties to one another through shared histories and values. Tyndall, the son of a Belfast policeman, had an ordinary school education to the age of seventeen. Similarly, Huxley, whose father was a mathematics teacher who fell on hard times, had only two years of poor quality formal schooling, although he evidently benefited from his father's influence and tuition. Tyndall and Huxley first met on the way to the 1851 BAAS meeting at Ipswich and quickly became good friends and professional allies. Both would rise precipitously in the world of science, largely through their own efforts.

The great catalyst for Tyndall's scientific career was a Friday Evening Discourse he delivered at the Royal Institution on 11 February 1853, which, as a letter from Huxley dated precisely two weeks later documents, received high praise from, amongst others, the electrical researcher and businessman John Peter Gassiot 'at the Philosophical Club (the most influential scientific body in London)'. But 'So much for glory, – now for Economics.' Huxley urged Tyndall to 'hammer your reputation while it is hot', and most of his letter is devoted to finding a 'fine opening' for his friend, who was working as a teacher at Queenwood College, Hampshire at the time. Indeed he suggests to Tyndall that 'you ought to be looking to Faraday's place', that is, to effectively run the Royal Institution. The Institution depended financially upon giving public lectures, and Huxley recognised Tyndall's advantages here: 'What they want, and what you have, are *clear powers of exposition* – so clear that people may think they understand even if they don't.'[41] Tyndall was appointed professor of natural philosophy at the Institution, taking the position in September 1853.

Huxley and Tyndall each brought to their friendship a dowry of two other close scientific friends. Tyndall had met the fifteen-year-old Hirst in 1845, when he was twenty-five and both were working for a Halifax firm of engineering surveyors. Two years later, at Queenwood College, where he had become the mathematics master, Tyndall met the new chemistry teacher Edward Frankland. In October 1848 the two men travelled to Germany to pursue studies in the physical sciences at the relatively inexpensive Philipp University at Marburg. Along with chemistry (which he took with Robert Bunsen) and physics, Tyndall also studied mathematics, although his subsequent scientific researches would abjure this discipline in favour of experimental approaches. He graduated as doctor of Philosophy early in 1850, the year that Hirst went to study at Marburg. Huxley, Joseph Hooker and George Busk, who were all surgeons with experience working in the navy, sought to establish themselves as naturalists, with Huxley achieving spectacular success and recognition for his research during the early 1850s. By the middle of the decade the members of both groups had good positions as academic and research scientists, and were mixing together socially. Tyndall was a great mountaineer, and during the summer of 1856 Huxley, Hirst and Frankland joined him in the Swiss Alps. Herbert Spencer, William Spottiswoode and John Lubbock were drawn into the group in the late 1850s and early 1860s, an accession that completed the company that would band together in 1864 as the X Club. 'Besides personal friendship', Hirst writes in his journal entry describing the first meeting of the X Club, 'the bond that united us was devotion to science, pure and free, untrammelled by religious dogmas.'[42] While Huxley insists that the X Club was formed as a dinner club for friends to meet, its members supported one another as they made their way in British science, so that it became over the following decades, in McLeod's phrase, one of the 'informal *élites*' of science.[43]

The North Britons and the Metropolitans were both outsiders from the Anglican scientific establishment and progressive forces within Victorian science, while their respective religious convictions and their class origins set them apart from one another. Maxwell was the son of a landowner, the younger brother of a baronet, while Tait's father was secretary to the fifth duke of Buccleuch. As noted earlier, both men received excellent formal educations. James and William Thomson's father was a professor of mathematics, first at Belfast and then at Glasgow, who took full responsibility for the education of his sons through to matriculation, which in William's case occurred when he was ten. The brothers both studied at Glasgow, with James graduating to pursue further studies in engineering, while William left before completing his Scots degree to go to Cambridge, where he became second wrangler in 1845. Jenkin offers an exception here, having had an impoverished childhood and youth, while conversely the X Club included men from privileged Anglican backgrounds, such as the baronet and banker Lubbock, the businessman Spottiswoode, printer to the queen, and Hooker, the son of the Regius Professor of Botany at Glasgow.

As their members' vigorous participation in the BAAS meetings and committees demonstrates, both groups were dedicated to the professionalisation of science, although they often differed significantly in their understandings of what this goal meant. For many mid- and late-Victorian scientists, including the Metropolitans, professionalisation required, as Frank Turner has argued, naturalistic science to usurp the culturally dominant role of the Christian Church,[44] whereas the more socially conservative and Presbyterian North British physicists focused upon eliminating what they regarded as amateur practices, by giving energy physics a mathematical basis, working to standardise its scientific measurements and reforming university science curricula. Both, however, as Smith has observed, competed with one another for authority over the British scientific jurisdiction, 'a monopoly on "truth", once the preserve of the old Oxbridge Anglican lions'.[45]

In an essay he wrote about Tyndall after his friend's death in December 1893, Huxley recalls that he was 'almost the earliest' of those 'to whom, indeed, I have found the old shikaree's definition of a friend, as "a man with whom you can go tiger-hunting", strictly applicable'.[46] Close bonds of trust and shared adventures, akin to those of mountaineering, inform this analogy, which suggests Huxley and Tyndall's campaigns to outwit various cunning and dangerous adversaries, principally the Anglican establishment but also the North Britons, and presumably claim their pelts as trophies. Matching Huxley and Tyndall, on the North British side Tait was something of a street-fighter, engaging directly in public polemics with the Metropolitans over particular scientific issues, while Maxwell took a more aloof and often ironical stance to such factional politics.

The North Britons seem to have had it all their own way at the 1871 Edinburgh meeting, with William Thomson as the President of the meeting, Tait the president of Section A (with their friend the mathematician James Joseph Sylvester vice president), Jenkin the president of Section G (Mechanical Science), and Tyndall away in the Alps. Delivered to an audience of Red Lions that included Rankine as 'Lion King' and a full complement of ebullient North Britons, the '*Ode*' effectively offers the absent Tyndall as a sacrifice to its addressee.

GLACIAL RELATIONS

Maxwell's '*Ode*' presupposes, and participates in, a history of factional battles that Tyndall and Tait led, most vociferously during the 1860s and early 1870s. While at the start of its lyric Tennyson's brook declares, 'I come from haunts of coot and hern',[47] Maxwell's Tyndallic lecturer traces his origins to icy expanses: 'I come from fields of fractured ice.' The Metropolitans and the North Britons originally took sides in the late 1850s in a controversy over glacier motion between James Forbes and Tyndall. Forbes argued that glaciers move in a liquid viscous flow, while Tyndall maintained that they slide over their terrain, with parts of them breaking and reforming to accommodate themselves to the changing contours of the rock beneath them, a theory that Maxwell's opening line invokes in its characterisation of Tyndall's 'fields of . . . ice' as 'fractured'. Indeed, as Chapter 5 documents, the controversy was still alive at the time of the 1871 Edinburgh meeting, with Tyndall's absence serving as a reminder of it, as he was in the Alps seeking further evidence to defend his theory against a recent attack.

Tait and Maxwell were taught by Forbes in the 1840s, the decade of his great Alpine expeditions and researches on glaciers. Although the young men were not public participants in his early debate with Tyndall, they retained a strong loyalty to their old teacher and defended his position in subsequent conflicts. Succeeding Forbes in his Edinburgh chair in 1860, Tait later became one of the authors and editors of his teacher's *Life and Letters* (1873). Having first met Forbes at the age of fifteen, Maxwell was, Tait recalls, 'a particular favourite' of the professor's, who also took an active interest in guiding his student's later choices of Cambridge University and Trinity College, and in following his progress there.[48] Campbell remembers Maxwell saying to him in 1869, the year after their teacher's death, that he 'loved James Forbes'. He also comments on a series of experiments on jellies and gutta-percha that Maxwell embarked upon at Edinburgh University, that 'it seems probable that they were immediately suggested by Forbes's Theory of Glaciers, which had recently called attention to the whole question of the difference between solid, liquid and "viscous" bodies'. Maxwell invokes the controversy in a joke he makes at Cambridge about the contents of a meal-ark, a large chest used to contain cereal meal, saying that it 'may be called a fluid as much as a glacier'.[49]

The principle of regelation, the bonding that occurs when pieces of ice come into contact with one another, was crucial to Tyndall's theory, as it described the way that broken pieces of glacier could be re-set in new forms. His colleague Faraday argued in an 1859 paper that regelation was facilitated by a liquid layer that forms on ice at pre-melting temperatures, while James Thomson, with some help from his brother William, maintained that the key to the phenomenon was his discovery that pressure between bodies of ice will result in a rise in temperature at their point of contact. Believing that his brother's pioneering work had not received adequate recognition in Tyndall's theory of glaciers, William Thomson became a vocal public ally of Forbes in his debate with the Metropolitans.

Writing in his *Memoir* of Ramsay, with whom Tyndall made a scientific expedition to the Alps in 1858, Geikie describes the late 1850s and early 1860s as the period of 'the ice-fever in geology'.[50] Tyndall had developed his theory of glaciers when he was in Switzerland with Huxley, Hirst and Frankland during the summer of 1856. Indeed the 1857 paper in which he introduces the theory is co-authored with Huxley. Hopkins, the famous Cambridge private coach who tutored Maxwell and Thomson, also publicly defended Tyndall's theory. When it came to protecting one's reputation and attacking that of others, those who lived away from London were, as Forbes was aware, seriously disadvantaged.[51] Tyndall privately accused Forbes of plagiarism and publically denigrated his researches,[52] while Huxley and Frankland worked to defeat his nomination in 1859 for the Royal Society's Copley Medal, which by legitimating the liquid theory would have correspondingly damaged their friend's reputation. Huxley wrote to Frankland, a Fellow of the Society who had recently received its Royal Medal, suggesting that Forbes had adopted as his own Canon Rendu's theory of glaciers. J. S. Rowlinson notes that Tyndall repeated this charge a year later in *The Glaciers of the Alps*, where he uses four of the quotations from Rendu's 1841 memoir he had made in the earlier letter to Frankland, and heads the chapter on the flow theory of glacier movement 'Rendu's Theory'.[53] Forbes replied with a pamphlet, to which Tyndall did not respond.

The unfinished controversy over glaciers was soon accompanied by a further, more radical and consequential, conflict between the Metropolitans and the North Britons over thermodynamics. In his 1862 lecture for the Royal Institution 'On Force', Tyndall attributed the German Robert Mayer with the discovery and full appreciation of the energy concept, claiming that he had arrived 'at the most important results some time in advance of those whose lives were entirely devoted to Natural Philosophy', a reference to Joule's work in the 1840s. The Thomson brothers' work on thermodynamics was also slighted, with William being described as having simply applied his 'admirable mathematical powers to the development of the theory [of energy]'.[54] Furthermore, as Knott puts it delicately, Tyndall's paper 'in several particulars corresponded curiously with Tait's own inaugural lecture of 1860 [as professor at Edinburgh]'.[55]

Tyndall strikes at the heart of the North British *raison d'être* by attacking Joule's claim to priority in discovering and experimentally establishing the energy concept.[56] 'On Force' marks the decisive first volley in his long campaign to claim energy physics for Metropolitan naturalism and materialism, much as Huxley was at the time working to assimilate Darwinian biology to the cause. Such efforts galvanised and mobilised the North British faction. Tait led the charge with a series of public essays and letters targeting Tyndall. Energy would either be placed in the service of

Metropolitan agnostic naturalism or retained within North British natural philosophy, as Smith puts it, 'in harmony with, though not subservient to, Christian belief'.[57] Having staked his claim over 'the fields of . . . ice' in the first stanza of Maxwell's *'Ode'*, the Tyndallic persona opens the second by declaring a similarly original and proprietorial interest in the phenomena of thermodynamics: 'I come from empyrean fires'.

With so much at stake, Tait and Thomson wished to reach a wide general readership for an essay on 'Energy' they wrote to oppose Tyndall's account of its discovery. They accordingly chose to publish in the magazine *Good Words*, which was edited and produced by Church of Scotland evangelicals. Appearing in the October 1862 issue, it stimulated a correspondence between Tyndall and Tait in the *Philosophical Magazine* from 1863–4. In the first of these letters Tyndall suggests that the authority of his opponents is compromised by their use of a popular magazine to discuss serious scientific problems, to which Tait responds provocatively that if *Good Words* was not appropriate for airing such matters than neither were certain popular lecture series at the Royal Institution.[58] In 1864, Tait developed his polemic further in two articles for the *North British Review*, on the 'Dynamical Theory of Heat' and 'Energy', each of which he left unsigned.[59] The first of these includes a review of Tyndall's *Heat considered as a Mode of Motion* (1863), 'the criticisms being', as Knott notes, 'mainly in regard to the history of the subject'.[60] The articles for the *North British Review* were revised and expanded, and their authorship disclosed, in his *Sketch of Thermodynamics* (1867).

Maxwell comments in an 1865 letter to his friend H. R. Droop that 'The only man I know who can make everything the subject of discussion is Dr. Tyndall. Secure his attendance and that of somebody to differ from him, and you are all right for a meeting.'[61] Tyndall was a notorious controversialist, and his public debate with Tait over the energy concept was overshadowed only by the more general furore over his 1874 Belfast address. It did not, however, prevent the two men from working together, along with Balfour Stewart, on a BAAS Committee charged with repeating and extending J. D. Forbes's experiments on the thermal conductivity of iron, which produced its single 'Provisional Report' in 1869. They met again over Forbes's work in 1873, this time, however, in 'a controversy of some bitterness'[62] over his contribution to glaciology, which was provoked by Tyndall's *Forms of Water* (1872). Tyndall repeated and added to his earlier criticisms, suggesting that Forbes derived his ideas not only from Rendu, but also from the Swiss naturalist Louis Agassiz. Tait addressed these charges in *The Life and Letters of J. D. Forbes*, which was co-authored and edited by J. C. Shairp and A. Adams-Reilly in 1873. Tyndall's response, 'Principal Forbes and his biographers' for the *Contemporary Review*, prompted an enthusiastic defence of Forbes by Ruskin, which he presented as a review of *The Forms of Water*.[63] A strong ally of Tait's in this battle, and a minor contributor to *The Life of Forbes*, Maxwell observed acidly of Tyndall's essay that 'The person most injured in the *Contemporary Review*, (next to J[ohn] T[yndall]) is Principal Shairp, who has been subjected to textual criticism and also to comparison with the author of the article.'[64] Tyndall and Tait embarked upon a further correspondence in *Nature*, which was stopped by its editor Norman Lockyer, apparently at Hooker's urging, because of its 'personal tone'.[65]

Tait charges Tyndall with failing to acknowledge fully and justly the contributions of other researchers in the fields of glaciology and thermodynamics he discusses. His faithful disciple Knott traces the motive behind Tait's public polemics to a tribal loyalty that entails a strong antipathy to Tyndall: 'Strong in his likes he was also strong in his dislikes. With true chivalry he fought for the claims of his friends if these were challenged by others. It was this indeed which led him into controversy. Thus arose the controversies with Tyndall concerning the history of the modern theory of heat and Forbes' glacier work.'[66] Indeed Tait appears to have been obsessed with Tyndall. The scrapbook that members of Tait's family assembled shows that he collected newspaper and magazine articles on Tyndall and his controversies with him, to which he occasionally added exclamation marks and sarcastic remarks.[67] Tait discusses his standing amongst his scientific contemporaries in a notebook entry, where he concludes that 'In a list which does not include Rowan Hamilton, I have no right to appear. In a list which includes Tyndall and Agassiz, I should be ashamed to appear.'[68] His identification with Hamilton accords with the title Maxwell gives him of 'Chief Musician upon Nabla'. In their postcards and letters to one another, Thomson, Tait and Maxwell routinely use the notation T'' to refer to Tyndall, so ranking him after Thomson (T) and Tait (T' or OT') as 'a second order quantity'.[69]

Tait's various collaborative publications with W. J. Steele, Philip Kelland, A. Crum Brown, Thomson, Stewart and others indicate that he enjoyed being part of a community of research scientists. Although Tyndall had his own allies in Huxley and the other members of the X Club, nearly all of his legion publications are single-authored. Especially in comparison to Tait, Tyndall's approach to scientific work is individualistic. This is clear also from his championing of Mayer in the energy controversy as the solitary Carlylean hero and romantic man of genius.[70] While the rank of 'Chief

Musician' implies collaborative work with other players, who may also be 'harping on that Nabla', or on other instruments or operators, the '*Tyndallic Ode*' entails a single declarative speaker, an Alpine adventurer and lone lecturer, the keynote of whose song is the personal pronoun: 'I come from fields of fractured ice'. The second stanza also begins with the persona asserting his solitary selfhood, which this time hails from antithetical but analogously inhospitable and unsociable extremes of heat and claustrophobic space: 'I come from empyrean fires, / From microscopic spaces'. The next chapter looks more closely at this poem's subtle fossil record of British scientific culture at the beginning of the 1870s.

CHAPTER 5

Popular science lectures: 'A Tyndallic Ode'

An unsigned manuscript poem, 'Profes[sor] Tait's Christmas Dream', makes fun of Tyndall's 1863 Royal Institution Christmas lecture, 'Electricity at Rest and Electricity in Motion', as 'a hideous dream' in which 'all was Tyndallized':

> There on a platform stood the fiend
> His wide mouth grimaced in smile
> Upon his right, the electric light
> And on his left a [galvanic] Pile
>
> And lo! The well dressed multitudes
> Pressed forward in profusions
> While scientific beggars sat
> On the door of the Institution.[1]

The poem is not in Tait's or Maxwell's hand, but is clearly marked as Tait's property, with the request 'Return O.T'' above the title and his address at the end of the document. It includes references to Joule and Thomson and evidently circulated amongst the incipient North British group, as it is annotated with light-hearted comments in various hands, including one signed by Tait ('T'), 'Can you guess the author? I *think* I can.' Its droll style suggests that the poem was written by Maxwell, who was working at King's College London at the time. The use of the distinctive verb, to *Tyndallize*, also points to Maxwell, who appears to have coined it.[2] 'Profes[sor] Tait's Christmas dream' is probably the earliest of the extant poems that Maxwell wrote for Tait mocking Tyndall's lectures.

'To the Chief Musician upon Nabla: *A Tyndallic Ode*' resumes the concerns with science education that Maxwell airs precociously in some of his early Edinburgh and Cambridge verses. Tyndall's public lectures were famous for their demonstrations, the practical element of experiment that Maxwell valued so much in his Edinburgh degree and found lacking at Cambridge. His 1870 Royal Institution lecture series on 'Electrical

Phenomena and Theories' was, writes *Nature*, 'made as interesting as all his lectures are, by the ingenuity and completeness of the experimental illustrations'.[3] The use of such demonstrations became especially engaging for Maxwell in March 1871, when he was appointed the first Cavendish Professor of Experimental Physics, charged with introducing practical courses into the Mathematical Tripos and overseeing the design, construction and equipping of the new university physics laboratory. His inaugural lecture, which he gave on 25 October, has a special interest in 'experiments of illustration'.[4] Maxwell would have had occasion to think about such experiments in relation to the genre of the '*Tyndallic Ode*' not only in writing his poem in August, but earlier in the year as he prepared his own lecture for the Royal Institution, 'On Colour Vision',[5] which he delivered on 24 March.

In a letter about the Royal Institution lecture to his friend C. J. Monro, dated 15 March, Maxwell identifies the popular lecture form with Tyndall, much as he will in his poem, and writes that he is adapting himself to meet its requirements: 'I have been so busy writing a sermon on Colour, and Tyndalizing my imagination up to the lecture point.'[6] His Royal Institution 'sermon' suggests a preacher and a passive congregation, an audience ill-equipped to independently evaluate the material presented to them. The reference to 'Tyndalizing my imagination' recalls Tyndall's paper 'On the Scientific Use of the Imagination', which Maxwell had attended at the previous year's BAAS meeting at Liverpool and read a couple of months later.[7] The exercise itself appears to have required him to translate the main points of his lecture 'On Colour Vision' into forms of what he refers to elsewhere as 'striking illustrations'.[8] These included a picture borrowed from Huxley of the rods and cones structure of the back of the eye, and a series of experimental demonstrations, in which he projected onto a screen the prismatic spectrum and light passed through various chemical solutions, including chromium chloride, which gave audience members the opportunity to discover whether or not they had a 'yellow spot' on their retina.[9]

Popular lectures were a topical subject at the Edinburgh BAAS meeting, which appointed a 'Committee on Science Lectures and Organisation'. Among the thirty members of the committee were Maxwell, Tait, William Thomson, Huxley, Lockyer, Jenkin, Stewart, Joule and William Roscoe.[10] The professor of chemistry at Owens College Manchester, Roscoe had recently organised a series of public lectures in the northern city, which had been given by such eminent scientists as Huxley, Lockyer and the physicist and astronomer William Huggins. Their publication earlier in 1871 as *Science Lectures for the People* prompted an editorial by Lockyer in the

1 June issue of *Nature* that urged 'a common action' to disseminate such lectures throughout Britain, so as to facilitate 'a powerful union capable of forcing the claims of science before the Government of the country'.[11] Lockyer had also been conducting a campaign in *Nature* in 1870–1 for scientific lectures for women,[12] to which Maxwell contributed a verse postscript with the first of his 'Lectures to Women on Physical Science'.[13] Tait presented a 'Course of Experimental Physics' to the Edinburgh Ladies' Educational Association in 1868–9, which Maxwell's favourite cousin Elizabeth Cay and both of Tait's sisters attended.[14] With so much interest in public lectures on science, the BAAS committee was formed with a view to immediate action: 'To consider and report on the best means of advancing Science by Lectures, with authority to act, subject to the approval of the Council, in the course of the present year if judged desirable.'[15]

Presented in *Nature* alongside reports of the Edinburgh meeting, '*A Tyndallic Ode*' offers a burlesque contribution to the meeting's discussion of 'the best means of advancing Science by Lectures'. The poem is organised as a composite form of the popular lecture series, with the stanzas it draws together furnishing renditions of various lecture demonstrations that, as Gillian Beer observes, correspond to particular passages in Tyndall.[16] Indeed the sixth stanza in the later version of the poem published in Campbell and Garnett, a verse on the siren that Maxwell added in August 1874, is given its own title, 'On the atmosphere as a vehicle of sound', which, as the reference that accompanies it acknowledges, was taken directly from a paper that Tyndall had read to the Royal Society earlier in the year.[17]

A SCIENCE EISTEDDFOD

'To the Chief Musician upon Nabla: *A Tyndallic Ode*' juxtaposes Tait and Tyndall on either side of the dividing colon, like boxers in their corners of the ring, although here the competition is rather an eisteddfod, as they are each identified with a lyric form. Tyndall had sent one of his poems to Maxwell as recently as June 1871, while in the same month Maxwell sent to Tait, at his request, a copy of Sylvester's poem, 'Tasso to Eleonora'.[18] Maxwell was evidently accustomed to thinking of poetry as a correlative for the distinctive styles of science that he and his peers practised.

Aligning himself in his 1871 Edinburgh address with recent discussions of mathematics and imagination, which Sylvester established in his sectional address at Exeter in 1869 and Maxwell replied to at the 1870 meeting, Tait snubs the unmathematical Tyndall's recent discourse 'On the Scientific Use of the Imagination'; 'The flights of the imagination which occur to the pure mathematician are in general so much better described in his formulae than in words.'[19] For Maxwell such concentrated forms of expression make Tait's mathematics akin to lyric poetry, an analogy that is registered in the term he coins for a set of quaternion equations that use the ∇ operator, nablody, which adds to its stem a suffix derived from *ōidē*, the Greek word for a poem intended to be sung: 'This letter is called "Nabla",' he explains to Campbell in October 1872, 'and the investigation a Nablody.'[20] The usage is warranted by the pedigree Hamilton gives to his quaternions, which were, he writes in 1855, 'born, as a curious offspring of a quaternion of parents, say of geometry, algebra, metaphysics, and poetry'. Indeed, glossing them as an algebra of four dimensions, consisting of the three of space and the fourth of time, he goes on to declare that he formulated his best account of them in his 1846 poem 'The Tetractys':

I have never been able to give a clearer statement of their nature and their aim than I have done in two lines of a sonnet addressed to Sir John Herschel:

'And how the one of Time, of Space the Three,
Might in the Chain of Symbols girdled be.'[21]

Maxwell's 'Nablody' casts such quaternion equations as a distinct genre, a usage that is demonstrated in the following remark, from the close of a letter he wrote to Tait in May 1872 on a problem in orthogonal surfaces: 'It is neater and perhaps wiser to compose a nablody on this theme which is well suited for this species of composition.'[22] Described ambiguously in taxonomical terms as a 'species of composition', the genre encompasses the synthetic efforts of both the mathematical sciences and the arts. In this punning Pythagorean conceit, the nablody, literally the song of the Nabla, is seen to furnish Tait's 'theme' with an apt and elegant form. In a postcard to Tait from October 1872 Maxwell describes quaternions as a language system in need of a formal grammar: 'The great want of the day is a Grammar of 4nions in the form of dry rules as to notation & interpretation... Contents, Notation, Syntax, Prosody, Nablody.'[23] The nablody represents its quaternion equations as a hybrid form that recalls Maxwell's 1854 poem, 'A Problem in Dynamics',[24] where mathematical formulae and poetry are drawn together as analogously concentrated semantic forms and economies.

Maxwell's conception of quaternions as having their own poetic forms and prosody corresponds to the cross-pollinating theories and idioms of mathematics and poetry that his friend Sylvester, another devotee of Hamilton's quaternions, had recently elaborated in his book *The Laws of Verse* (1870). Maxwell had evidently read this book at the time he delivered his Liverpool address in September 1870, as he cites a long footnote that Sylvester added to the text of his corresponding 1869 address at Exeter for its publication in the volume. The footnote casts recent sectional addresses, beginning with Spottiswoode at Dundee in 1867, as a progressive discussion about the nature of mathematics and physics, which leads Maxwell at Liverpool to take up its invitation to complete 'the Tetralogy' by providing 'a discourse on the Relation of the two branches (Mathematics and Physics) to, their action and reaction upon, one another'.[25]

By identifying Tait's mathematical practices with musical composition and performance, the title of 'Chief Musician upon Nabla' associates him with two ancient cultural traditions, that of the Pythagoreanism that Maxwell was attracted to as a young man, and that of the Bard, the heroic romantic figure who composes and sings verses recording history, laws and facts, which he would usually accompany on the harp. Maxwell was familiar with this tradition from his youthful reading of Walter Scott's poetry and the Ossian poems, and his efforts at writing verse in this vein, such as a prize poem on the death of Douglas he composed when he was fourteen, a copy of which Tait retained throughout his life.[26] In his last letter to his friend, 'purporting', Knott writes, 'to be a soliloquy or self-communion by Tait himself', Maxwell offers a mock history that places 'Nabla' in this tradition: 'I had heard that this harp had been called by a name like this. But not in all Wales could such a harp be found, nor yet the lordly music which has not been able to come down through the illimitable years.' The vision described here occurs 'under the invocation of the holy ALBAN', that is Francis Bacon, Viscount Saint Alban, and being written down, and its page turned, the author finds the word's mirror image, each letter and their order reversed, to yield the strange revelation, 'impressed by the saint himself', of NABLA.[27] Tait is revealed by this ingenious portent as a successor to Bacon.

The etymology of the suffix that Maxwell adds to the name of Hamilton's quaternion operator for his coinage 'nablody' is shared by the English word ode, the form of poetry identified with Tyndall in the poem's title. The ode, an extended and exalted address in rhyme that was written originally to be sung, gives way here to a modern genre of public performance, the popular scientific lecture that in the 1860s Tyndall made his own. Hermann von Helmholtz famously dubbed him 'poet of science' in 1864, a soubriquet that the spirit world immediately adopted, as Tyndall discovered in scrutinising a séance later that year, in which the letters of the phrase were spelled out in a series of table-rappings.[28] Maxwell appears to have considered the public and occasional nature of the Pindaric ode to furnish the most apt analogy for Tyndall's lectures. Originating as a choric performance for the victors of athletic contests, the Pindaric ode suggests that the parallel genre of the '*Tyndallic Ode*' honours scientific achievement after the event, rather than itself making a contribution to it.

While the conceit of 'the Chief Musician upon Nabla' is Pythagorean, its musical reference being a metaphor for complex mathematics, the '*Tyndallic Ode*' consists of only words. Considering the current popular interest in physics and the problems of meeting and satisfying it, Tait observes in his Edinburgh address that the formulae by which 'the pure mathematician' expresses himself make his science inaccessible to 'outsiders', while 'several of the highest problems of physics are connected with those simple observations which are possible to the many', so that 'even the most abstruse branches of physics, as yet totally incapable of being popularized, attract the attention of the uninitiated'.[29] In so doing he reiterates a distinction that Maxwell makes a year earlier in his sectional address at Liverpool, between mathematics, which he sees to be the keystone of professional science, and the popular public interest in the subject: 'But the great majority of mankind are utterly unable, without long training, to retain in their minds the unembodied symbols of the pure mathematician, so that, if science is ever to become popular, and yet remain scientific, it must be by a profound study and a copious application of those principles of the mathematical classification of quantities which, as we have seen, lie at the root of every truly scientific illustration.'[30] This is the crux of the North British opposition to Tyndall: he is not a wrangler. While the North Britons placed their science of energy on a rigorously mathematical footing, as Thomson and Tait's 1867 *Treatise on Natural Philosophy* and Maxwell's 1873 *A Treatise on Electricity and Magnetism* demonstrate most fully, Tyndall's physics is experimental and avoids mathematical modelling. Indeed he states that his lectures spring from 'a desire to interest intelligent persons who may not possess any special scientific culture'.[31]

A SONG FROM THE CHIEF MUSICIAN: TAIT'S 1876 LECTURE ON FORCE

Tait presented a series of public lectures in Edinburgh in spring 1874, which were published two years later as *Lectures on Recent Advances in Physical Science*. He explains in the preface to the book that he offered them 'at the desire of my friends, – mainly professional men, – who wished to obtain in this way a notion of the chief advances made in Natural Philosophy since their student days'.[32] The idea, and much of the audience, came from members of the Evening Club, which had been founded in 1869 by Tait and thirteen other 'well known citizens' of Edinburgh, including his friend and colleague David Masson, the professor of rhetoric and English literature, other academics and several prominent lawyers. By 1874 there were 150 members, including 'physicians, clergymen, teachers, ... bankers, commercial men, publishers, engineers, etc'.[33] An audience of such men could be addressed as *de facto* peers, as they came with an understanding of rudimentary scientific principles and methods, the disciplinary foundations that Maxwell emphasises in his Liverpool address, as Tait does also in the lecture on Force he delivered to the BAAS meeting at Glasgow on 8 September 1876, which was reprinted later that year in the second edition of *Recent Advances*. In his review for *Nature* of the first edition, Maxwell is careful to distinguish between his friend's lectures to such educated men and various forms of introductory and merely entertaining scientific lectures:

In lectures of this kind, therefore, we are not to look for the elaborate exposition and reiterated inculcation by which the facts and methods of science are impressed upon the minds of beginners. Still less are we to expect the forcible language and striking illustrations by which those who are past hope of being even beginners may be prevented from becoming conscious of intellectual exhaustion before the hour has elapsed. We are rather to listen to one who has climbed high on the hard and slippery peaks of science as he points out the grand features of the prospect to those who stand on a lower level but yet on the same solid foundation with himself.[34]

These remarks indicate the class-based difference between the conception of lectures that Maxwell and Tait shared, as a gentlemanly genre they knew from their universities, and Tyndall's practice, which derived from the fervent forms he attended at Mechanics' Institutes in Preston, Halifax and Huddersfield during the 1840s.

It is unsurprising given the self-motivated and autodidactic origins of Tyndall's and Huxley's educations that they should wish to broaden scientific education inclusively. To this end they worked to introduce science curricula at all levels of formal education and wrote journal articles and reviews for such periodicals as the *Westminster Review* and the *Contemporary Review*. They also offered a range of public lectures, including lectures to working men at the Royal School of Mines and the Royal Institution. Maxwell also taught classes for artisans, first while he was an undergraduate and a Christian Socialist in the 1850s, helping to establish the Cambridge Working Men's College, and subsequently with regular evening classes in Aberdeen and London until 1865, when he returned to live in the family home at Glenlair.[35] His 1871 textbook *Theory of Heat* and his 1876 *Matter and Motion*, which was published by the Society for Promoting Christian

Knowledge, also testify to his abiding commitment to educating the wider public in the foundations and findings of contemporary physics.

Maxwell offers a candid opinion of popular lectures in general, and those of Tyndall in particular, in the annotations he made in September 1873 to the proof sheets for Tait's 'Tyndall and Forbes', the letter he wrote to *Nature* answering Tyndall's *Contemporary Review* essay: 'Principal Forbes and his biographers'. At one point, Tait comments that 'Dr. Tyndall has, in fact, martyred his scientific authority by deservedly winning distinction in the popular field.' While Maxwell granted Tyndall some scientific credibility, having, for example, reviewed and recommended two of his papers for publication in the *Philosophical Transactions* in 1866,[36] he found Tait's gentlemanly tact provoking:

> Can a man do *good* service in popularising certain parts of science and *thereby* lose his claim to scientific authority? If a man has a claim to scientific authority the only way he can lose it is by writing bosch. If he writes it in a dry manner it is bad enough, but the harm is confined to students. But if he seasons it for the public, and the public swallow it (like the Saturday Reviewer) then it is a sad misuse of words to say that this is a useful work.
>
> Unless indeed it was a good work when the D[evi]l invented popular tunes, because the pious were thereby enabled to set hymns thereto. If so, are you prepared to write an orthodox Libretto of the Tyndallic lectures containing the spirit which gives liveliness, and avoiding the letter which would pluck any man?[37]

In disapproving of the concession that Tait's comment yields to Tyndall (which he in turn glosses as 'the product of mere ignoble spite'[38]), Maxwell invokes his familiar analogy with music, implicitly contrasting the sublime Pythagorean music that professional science strives to recognise mathematically, with the demotic (and indeed demonic) 'popular tunes' of the Tyndallic odes. Tait makes a similar but much more crass musical comparison in the opening lecture to his 1877 natural philosophy class, as he explains to an evidently receptive audience that the 'ignorance' and 'rhetorical display' exhibited by 'popular science' 'destroy a taste for pure science in the same way that one who has cultivated a passion for nigger melodies is incapable of appreciating the higher music. (Laughter.)'[39]

With his address on 'Force', Tait takes the concern over popular lectures he shares with Maxwell to the BAAS, where, in clarifying the meaning of the concept, he lists various abuses of it by his peers within the Association. This made Tait something of a hero for Maxwell, who writes an apotheosis of his friend in his verse version of the paper, 'Report on Tait's Lecture on Force:— B.A., 1876'. It begins by warning 'Ye British Asses' that Tait is expert at recognising 'bosh', the writing of which the poet describes in his earlier letter as 'the only way' to lose 'scientific authority':

> For Tait comes with his plummet and his line,
> Quick to detect your
> Old bosh new dressed in what you call a fine
> Popular lecture.[40]

Tait's lecture was published only in a bowdlerised form. 'The raciest and most critical passages', Knott recalls, 'were omitted' from the version printed in *Recent Advances* and Tait's *Scientific Papers*.[41] Lodge, who attended the lecture, writes of the version published in *Nature* that 'the sting was taken out of some of the paragraphs, or they may have been omitted altogether, but his language at the time was strong'.[42] Unpublished during his lifetime, Maxwell's verse paraphrase provides a unique record of the unexpurgated lecture: 'The real fun of the lecture', Knott testifies, 'is well shown forth in the humorous verses which Maxwell sent to Tait a few days later, with the heading "For P. G. Tait but *not* for Ebony" – meaning *Blackwood's Magazine*.'[43]

As if following the precedent he set at Edinburgh in 1871, Tyndall was absent from the Glasgow meeting, as Huxley was also, while Tait and Maxwell were once again on home ground with their North British peers. The mood of the evening lecture was reminiscent of a Red Lions gathering: 'Professor Tait commenced his lecture', reports *The Edinburgh Courant*, 'amid loud and hearty cheering. It was vain, he said, to expect that more than the elements of science could be made popular.'[44] Tait begins his lecture by alluding to the value that the BAAS meetings put upon 'novelty or originality', before arguing that 'definiteness and accuracy' are more important qualities: 'In fact, without them there could not be any science except the very peculiar smattering which is usually (but I hope erroneously) called "popular".'[45] The scare quotes here quarantine what is for Tait and Maxwell the definitive Tyndallic epithet.

While in writing his paper Tait must have considered Spencer's ambiguous use of the word 'force' in the public controversy they had over Newton's Laws of Motion in 1874,[46] contemporary newspaper articles he preserved report that the particular target of his lecture was Tyndall's provocative 1862 lecture 'On Force'. Both *The Edinburgh Courant* and *The Northern Whig* quote an extract in which Tait chastises Tyndall for failing to acknowledge the role played by gravitation in an account he gives of lifting a weight: 'Perhaps nothing so rich, and yet so sad, as this had been published as science, since ... Lord Brougham [wrote in a book that was recalled and pulped] ... that a porter carries a load on his shoulders rather than in his hands, because it is thus lifted at least twice as high about the earth, and gravity diminishes as the square of the distance from the attracting body increases. (Laughter.)'[47] This is evidently an example of what Knott refers to as 'The raciest and most critical passages' that were excised from the original lecture:

In these Tait let himself go to the intense amusement of many of his audience and to the horror of some who did not quite appreciate the form Tait's humour occasionally assumed. Lord Brougham and Professor Tyndall, though not explicitly named, were singled out as having been guilty of carelessness of diction in the expression of scientific truth; and the audience were startled when Tait capped his exposure of the recent President of the British Association by the question, 'Are these thy gods, Oh Israel?'[48]

The Northern Whig records that Tait 'pitched headlong into Professor Tyndall and his agnosticism'.[49] The account that Tyndall gives of the ancient 'mob of gods' at the opening of his Belfast address, as anthropomorphic representations of nature that science has demystified,[50] is sharply turned back on him here. Tait's quotation from Exodus 32.8 could express what he considers to be the unwarranted esteem in which Tyndall is held, or else identify him with Bacon's doctrine of the Idols. In particular, it accuses him of making an idol of the term 'force', just as Aaron cast the golden calf for the Israelites to worship and make oblations to. Like the ancients he discusses in his address at Belfast, Tyndall is charged with construing such terms in his own image, reifying them as a personal pagan pantheon and sacrificing scientific truth to a popular lecture cult.

Tait's muted argument continues in the published text of his lecture by noting that the indefinite and undecidable belong to popular expression and culture: 'In popular language there is no particular objection to multiple meanings for the same word', indeed, he continues, 'their existence is one of the most fertile sources of really good puns, such as those of Hood, Hooke, or Barham'.[51] Maxwell is of course not averse to puns, and as Barri J. Gold notes, his poem makes a series of such plays on current uses of the word 'force'.[52] Such puns readily disclose and contrast Tyndall's uses of the term as merely ambiguous nonsense:

> Whence comes that most peculiar smattering,
>   Heard in our section?
> Pure nonsense, to a scientific swing
>   Drilled to perfection?

> That small word "Force" is made a barber's block,
>   Ready to put on
> Meanings most strange and various, fit to shock
>   Pupils of Newton.[53]

Popular science, which Tait describes as a 'very peculiar smattering' of science, is glossed here as a polished and suggestive simulacrum, a nonsense that Maxwell's punning military analogy defines as a counterfeit discipline. The superficiality and mendacity of the 'Pure nonsense ... / Drilled to perfection' is elaborated upon with the grooming metaphor of artificial hair, as novel meanings are invented and placed upon the term 'force' like wigs on the barber's wooden model. Substantive in itself, the term is nonetheless drained of intelligent meaning, made into a servile blockhead to support the demonstrator's displays of falsity and vanity. Cast ambiguously at the end of its line, the poised phrase 'fit to shock' suggestively recapitulates the forced fit of the 'shock' of false hair, the Tyndallic 'Meanings most strange' that obscure 'That small word "Force"'. They are then disclosed to be 'fit to shock / Pupils of Newton', to affront directly the canonical tradition of British science, grounded in Newtonian mathematics and physics, upon which the BAAS was originally founded by Whewell and his peers.

Against the broad sweep of opportunistic and confusing abuses of the term 'force' he catalogues, those meanings that 'like a nose of wax, / Suit each occasion', Tait is seen to heroically assert the definite and definitive North British value of mathematical expression, 'the use of conventional symbols'.[54] While the language Tyndall uses to define the term 'force' is vague and allusive, ostensibly 'poetic', the mathematical composition that 'the Chief Musician upon Nabla' employs to the same end yields a consequence of genuine lyric poetry, an epiphany:

> But see! Tait writes in lucid symbols clear
>   One small equation;
> And Force becomes of Energy a mere
>   Space-Variation.
>
> Force, then, is Force, but mark you! not a thing,
>   Only a Vector;
> Thy barbèd arrows now have lost their sting
>   Impotent spectre!
> Thy reign, O Force! is over. Now no more
>   Heed we thine action;
> Repulsion leaves us where we were before,
>   So does attraction.

> Both Action and Reaction now are gone.
>   Just ere they vanished,
> Stress joined their hands in peace, and made them one;
>   Then they were banished.
> The Universe is free from pole to pole,
>   Free from all forces.
> Rejoice! ye stars – like blessed gods ye roll
>   On in your courses.[55]

The chemist Ilya Prigogine and the philosopher of science Isabelle Stengers cite these lines to illustrate the way that, 'especially since Hamilton, mathematical physics has abandoned Newtonian representation'.[56] The Hamiltonian, a value for the total energy of a system, appreciates changes in the position and velocity of bodies through the dynamics of potential and actual energy. Tait regards the Newtonian idea of force as a reification of the local properties of a dynamic system of energy, seeing it as 'not a thing', a causal agent acting upon bodies at particular points, but a consequence derived from the whole, 'a mere / Space-Variation'. The phrase recognises Tait's Hamiltonian conception of force as mathematically precise, an application of the quaternion operator with which he was identified by his closest colleagues: 'What do you think of "Space-variation" as the name of Nabla?' Maxwell asks his friend in a letter from December 1873.[57]

'ICE REDUCED TO LIQUID FLOWERS'

'*A Tyndallic Ode*' introduces the genre of the popular scientific lecture as a type of parlour game. The first and second stanzas proceed in the manner of a riddle, as the persona invites the reader to discover the places he comes from and guess who he is. While the title of the '*Ode*' means that the persona's identity is only a mock-mystery, the riddle that the poem nonetheless persists with prepares the way for its satirical representation of Tyndall and his lectures by de-familiarising him. Coming from places characterised by extremes of temperature and spatial magnitude, vast arctic 'fields of fractured ice', 'empyrean fires' and 'microscopic spaces', Tyndall is introduced as an alien but intriguingly paradoxical creature.

The identity of the persona is established efficiently by the icy setting of the first stanza, which points to Tyndall's current whereabouts in the Swiss Alps[58] and furnishes the phenomena for some of his best-known experimental demonstrations:

> I come from fields of fractured ice,
>   Whose wounds are cured by squeezing,
> Melting they cool, but in a trice,
>   Get warm again by freezing.
> Here, in the frosty air, the sprays
>   With fern-like hoar-frost bristle,
> There, liquid stars their watery rays
>   Shoot through the solid crystal.

'[O]ne hunter-like climbing the giddy Alpine heights of Science', as his hero Thomas Carlyle describes the figure of the scientist in *Sartor Resartus*,[59] Tyndall was well known for his annual mountaineering trips to the Swiss Alps, where he wrote several lectures that, as the first line of the poem suggests, he delivered upon his return.

'His muscle makes him so that he delighteth in his own legs', observes *Vanity Fair* in an 1872 'Men of the Day' article on Tyndall, 'and he scales virgin Alps one after another, for the pleasure of the exercise as well as for the study of natural phenomena.'[60] The conflation that the first stanza of the '*Tyndallic Ode*' makes of the Alps with Tyndall's experiments is true to his experience and practice in such works as *The Glaciers of the Alps* (1860) and *Hours of Exercise in the Alps* (1871), which are divided into two parts, a narrative of his mountaineering adventures in the Alps, of the kind he published separately in *Mountaineering in 1861: A Vacation Tour* (1863), followed by scientific observations and experiments suggested by the phenomena he encounters. The poem recognises the strategy by which, as Bruce Hevly observes, 'Alpinist-scientists such as Forbes and Tyndall presented themselves as arguing from firsthand experience on the subject of glacier mechanics and appealed to deference due to them as men who had undergone a rigorous experience on behalf of science.'[61] Maxwell, however, introduces Tyndall as forsaking the Edinburgh meeting not for precipitous mountains but for mere fields of broken ice. He is not like Tait, 'one who has climbed high on the hard and slippery peaks of science'.

The place that the Tyndallic persona comes from demonstrates the curious behaviour and beautiful forms that water molecules exhibit in freezing. The ideas and experiments alluded to in this stanza, most notably those in the first quatrain that recapitulate his theory of glacier motion, can be found in *The Glaciers of the Alps*, the lecture series *Heat considered as a mode of Motion* and, most decisively and provocatively, in the early essays on glaciers that Tyndall reprinted in 1871 in *Hours of Exercise in the Alps*, which, published a few months before Maxwell wrote his poem, would have revived memories of the founding factional dispute over Forbes's work.[62]

The opening stanza of the poem pointedly describes James and William Thomson's theory of regelation. Tyndall recognises and explains the principle in his 1865 essay 'Helmholtz on Ice and Glaciers', which was reprinted in *Hours of Exercise*. He writes that 'the freezing-point of water must be lowered by pressure', so that, under the force of the 'squeezing' that the poem specifies, the ice draws upon its own latent heat to melt into liquid water, the temperature of which is below 0°C. Then, escaping and momentarily relieving the pressure, the water solidifies again and so attains the higher temperature of the freezing point, in conformity with the surrounding ice:[63] 'Melting they cool, but in a trice, / Get warm again by freezing.' Tyndall also acknowledges Faraday's discovery of regelation, but claims credit for 'the application of the fact to the formation and motion of glaciers'.[64] The process by which he believes glaciers are created and move, as mechanical forces of tension break what he characterises as brittle ice, and pressure draws the pieces together into new forms,[65] is described as '*fracture and regelation*' in *Hours of Exercise*[66] and correspondingly in Maxwell's poem as 'fractured ice / ... cured by squeezing'.

A note dated April 1871 that Tyndall added to one of the essays in *Hours of Exercise* concedes that his theory of glacier formation received a blow the previous year, when William Mathews, one of the founders of the London Alpine Club, showed that a plank of lake ice could be permanently bent without breaking, a demonstration of plasticity that appeared to vindicate Forbes's viscous theory of glacier motion.[67] Tyndall reasserts his theory in the wake of Mathews's findings by reprinting his early essays in *Hours of Exercise*. The review of the book in *Nature*, which appeared in the 13 July issue, a few weeks before the BAAS meeting, notes that its latter part 'is devoted to a *résumé* of the "viscous" and "regelation" theories of glacier motion; a controversy which can hardly yet be regarded as concluded' due to the experiments of Mathews, 'to which Prof. Tyndall briefly alludes'. The reviewer deprecates both Tyndall's theory of glacier formation and his 1870 Liverpool discourse on Imagination by 'classing the ice ploughs of past ages as among the efforts of the scientific imagination'.[68]

Tyndall was evidently unsettled by Mathews's findings, conspicuously forsaking the BAAS meeting to travel to Switzerland to repeat his experiments on '*fracture and regelation*' using glacier ice. The reason for his absence would have been widely circulated to the other members of Section A through his old friend Hirst, who was to report the results of Tyndall's latest experiments to the meeting. In the event these results arrived 'a few hours too late'[69] to be presented. Calculated to remind the audience why he was not with them, the opening lines of the '*Ode*' suggest a dramatic late return from the Alps, as having missed the day's BAAS meeting Tyndall reports his findings in an impromptu address to the Red Lions dinner.

The remaining couplets of the first stanza of the '*Tyndallic Ode*' each sketch an attractive aspect of the Alpine environment. The second of these, in which 'liquid stars their watery rays / Shoot through the solid crystal' of the ice, describes the pattern that Tyndall displayed in one of his most celebrated demonstrations, for the 1863 lecture series on *Heat considered as a Mode of Motion*, which Tait had reviewed. The effect is produced by projecting lamp-light through a block of ice that has been cut parallel to its planes of freezing, so that the heat of the beam absorbed by the ice discloses its crystalline forms as it contracts in melting and its 'architecture is taken down'. A lens casts these 'liquid stars' magnified on a screen for the audience's appreciation: 'Here we have a star, and there a star; and as the action continues, *the ice appears to resolve itself into stars, each one possessing six rays, each one resembling a beautiful flower of six petals* ... as the action continues, the edges of the petals become serrated, and spread themselves out like fern leaves upon the screen' (emphases added).[70]

In his 1864 draft review of Tyndall's book, Tait cites 'Ice reduced to liquid flowers' as an example of rhetoric that 'is not indeed desirable, for the credit of British Science'.[71] The central part of the extract, which I have placed in italics, not only appears to be the direct source for Maxwell's lines, but is also quoted as an explanatory note to a set of verses by a member of Tyndall's audience, his friend the barrister Frederick Pollock. Tyndall and Huxley had appointed Pollock literature editor of *The Reader: A Review of Literature, Science, and Art*, a weekly journal that Masson had started in 1863 and the X Club took over in December 1864.[72] Pollock, whose translation of Dante's *Divine Comedy* into blank verse was published by Chapman and Hall in 1854, had been writing mock Valentine verses to Tyndall since 1863. The poem prompted by the extract on the 'liquid stars' was written for Valentine's Day 1865 (Figure 3), and published in *The Reader* as 'The Ice Flower':

> Within the ice,
> In strange device,
> A sleeping beauty, I
> Thy coming wait,
> At happy date,
> To bring my destiny.
>
> When through my frame
> The electric flame
> Its radiant pulses sends,
> I rise from death;
> Thy fervent breath
> My glacial fetters rends.[73]

Figure 3 Frederick Pollock's 1865 Valentine for Tyndall, 'The Ice Flower', watercolour, ink and metallic paper. Royal Institution of Great Britain, London.

'Folded in frost' it weeps 'With frozen tear', whilst 'Under thy glow / My petals blow, / Ecstatic with delight'. It accordingly hopes not to be returned to its 'frigid fate: / Dissolved by you, / In raptures new / May I ne'er regelate!'[74]

Pollock and Maxwell, friend and foe alike, both recognise and respond to Tyndall's appeals to aesthetics in his description of the melting ice. Such appeals occur often in his lectures, serving to encourage and justify Maxwell's lyric analogy of the Tyndallic Ode. Both he and Pollock would, for instance, have known the conclusion Tyndall draws from his example of the ice flower: 'Nature "lays her beams in music", and it is the function of science to purify our organs, so as to enable us to hear the strain.'[75]

Following the drama of paradox and the beauty of pattern that the first stanza of Maxwell's poem finds in freezing water, the subsequent stanzas similarly forsake *physis* for *aesthesis*, enacting science as spectacle. The Friday evening discourses at the Royal Institution that Faraday organised and often presented himself were formal affairs frequented by elite scientists and London's aristocracy. Writing in January 1851, George Eliot declares Royal Institution lectures to be 'as fashionable an amusement as the Opera'.[76] Iwan Rhys Morus places these prestigious lecture series amongst a wide range of scientific presentations that were offered to the Victorian public, which also included demonstrations of industrial applications and sensational optical and electrical spectacles.[77] As the Royal Institution depended upon attracting large audiences to survive financially, Faraday laboured for reports of the Friday evening discourses to be given not only in such specialist journals as the *Quarterly Journal of Science* but also in periodicals that reached a broader audience, such as the *Literary Gazette* and the *Athenaeum*.[78] Similarly dedicated to the success of the Institution, Tyndall became a celebrity lecturer, who was, for instance, asked to provide signed photographs of himself for the emperor of Brazil.[79]

Maxwell's poem charges Tyndall with simply offering popular entertainments for a new mass audience, edifying amusements akin to contemporary exhibition halls, galleries, dioramas and panoramas, museums, theatre and magic lantern shows. Indeed N. D. McMillan and J. Meehan give several examples 'of theatrical performance' that Tyndall embedded in his lectures, including a well-practised knack of saving a deliberately dropped glass flask, and lighting the chairman's cigar with a beam of infra-red radiation.[80] Bernard Lightman also cites this gimmick, along with a demonstration of meat cooked by the 'invisible rays' of reflected heat, both of which, along with several of Tyndall's lecture topics, Pepper copied for the middle-class mass audience of his Royal Polytechnic Institution.[81] Pepper's homage would not have helped Tyndall's reputation with Maxwell and his peers. Hirst similarly notes his friend's theatrical excesses, recording in his journal an occasion when he thought Tyndall's 'experiments were repeated too much in showy ways. Strings whose simple vibrations were perfectly visible and instructively so, were made to vibrate in blue light, red light, green light, purple light &c &c.'[82]

While in his letter to Monro Maxwell figures his Royal Institution lecture 'On Colour Vision' as a type of preaching, he insinuates humorously that its venue is simply another place of entertainment, indeed an immoral place of temptation and intoxication: 'I have not asked you if you wish to go to [my] sermon on Colour for I do not think the R. I. a good place to go to of nights even for strong men. I have however some tickets to spare.'[83] Indeed he appears to have fled the place after his lecture. A plaintive note from Tyndall asks Maxwell 'Why . . . did you run away so rapidly. I wished to shake your hand before parting.'[84]

SENSATION AND SCIENCE

Electrical phenomena had since the eighteenth century provided the most sensational and popular material for scientific entertainments. The final quatrain of the second stanza and the stanza that follows it in the version of the '*Tyndallic Ode*' published in *Nature* focus upon peculiar optical and electrical effects:

> The atoms clash, the spectra flash,
>   Projected on the screen,
> The double D, Magnesian b
>   And Thallium's living green
>
> This crystal tube the electric ray
>   Shows optically clean,
> No dust or cloud appear – but stay:
>   All has not yet been seen:
> What gleams are these of heavenly blue,
>   What wondrous forms appearing?
> What fish of cloud can this be, through
>   The vacuous spaces steering?

Having established his identity earlier in the poem as an emissary from a curious scientific world of fire and ice, the lecturer draws from this realm a series of mystificatory and proprietorial demonstrations. These are enhanced by his questions, which are rhetorical rather than heuristic, evidently meant to heighten the drama and spectacle of the demonstrations rather than precipitate scientific understanding of their phenomena.

The Tyndallic lectures exemplify the formula for the popular lecture, of 'forcible language and striking illustrations', that Maxwell gives in his review of Tait's *Recent Advances*. In his 1870 Liverpool address to Section A, he identifies the first of these, a coercive rhetorical power, with his rival's corresponding address to the 1868 Norwich meeting, declaring that in listening to it he had been 'carried by the penetrating insight and forcible expression of Dr Tyndall'. Similarly, in the preface to the *Theory of Heat*, written in the same year as the '*Tyndallic Ode*', Maxwell offers an equivocal endorsement of 'Professor Tyndall's work on "Heat as a Mode of Motion", in which the doctrines of the science are forcibly impressed on the mind by well-chosen illustrative experiments.'[85] Tait offers a less circumspect account of such commanding qualities in his 1864 review of *Heat considered as a Mode of Motion*: 'Grandiloquence, especially when rising almost to the style of the modern sensational school of fiction-writers, is not adapted even to popular science; true scientific language is ever calm and dignified, and we fear the worst when we hear of magnetic needles moving as if "inspired by a sudden affection" for the audience, medals "struck dead by the excitement of the magnet", and other catastrophes too numerous to mention.'[86]

Tait renewed these criticisms of Royal Institution lectures in *Nature* a month before Maxwell wrote and recited his particular 'Libretto of the Tyndallic lectures'. His article 'Sensation and Science', which led the 6 July issue, opens by observing that:

The morbid craving for excitement, which is characteristic of mental indolence, as well as of effete civilisation, has led to the introduction of Sensation (as it is commonly called), not merely into our newspapers and novels, but even into our pulpits. It could not be expected that our popular scientific lectures would long escape the contamination.

This article, the first of two Tait wrote on the subject, discusses the Reverend Professor Haughton's recent lectures at the Royal Institution on 'The Principle of Least Action in Nature', which we are told 'were racy (*i.e.* sensational) in the extreme'. Tait corrects a series of scientific errors, each of which he regards as 'merely another proof that we are dealing with Sensation where we looked for Science'.[87] Illustrating the charge, he quotes a passage from one of the lectures, in which Reverend Haughton speculates about renting a farm, refusing to pay the rent, 'and in due time shooting my landlord, and ... dissecting him at my leisure'.[88] The second of the essays discusses *On the Nature of Comets* by

J. C. F. Zöllner, 'a work of exceptionally high merit as a mere literary composition' that contains '[h]undreds of racy passages'.[89]

The charge of showmanship that Maxwell makes against Tyndall's lectures is more easily sustained by the form in which they were originally delivered, which involved a formal bifurcation of content and performance, than the published versions by which we know them. The review in *Nature* of his 1870 Royal Institution lecture series *On Electrical Phenomena and Theories* notes approvingly that for the sake of 'expediency' Tyndall would provide his audience with printed notes, a 'systematic outline' of each lecture, which allowed him to focus upon the presentation of his elaborate experimental demonstrations: 'By this procedure the Professor is able to give full attention and time to each step of his illustrative demonstration, without being hampered with the need of telling everything that he has marked out beforehand.'[90]

The closing lines of the second stanza of the '*Ode*' sketch a demonstration in which the chemical elements Sodium, Magnesium and the recently discovered Thallium are placed in a clean flame to yield a light strong enough to allow a spectroscope to throw each of their spectra onto a screen for the lecture audience to see. Displayed here as bright lines, the 'double D' and 'Magnesian b' lines were first observed as distinctive dark lines in the solar spectrum by Joseph von Fraunhofer in 1814. The 'double D' lines register two out of the eight sodium lines of the spectrum and the 'Magnesian b' three of the seven Magnesium lines. While Fraunhofer had noted the coincident position on the spectrum of the double D dark lines and the yellow bright lines produced by a candle flame, it was left to such subsequent workers in the field as George Stokes, Gustav Kirchhoff, Thomson and Stewart to make sense of this by establishing that each element has a distinctive frequency of light waves. Lockyer explains that Kirchoff 'discovered this very remarkable fact, that gases and vapours have the power of absorbing those very rays which they give out when in a state of incandescence',[91] hence the phenomena of correspondent dark and bright lines along the spectrum that define their chemical composition.

The history of these discoveries would have been fresh in the minds of Maxwell and his peers at the Edinburgh meeting of the BAAS from Thomson's Presidential address, which includes a long discussion describing the crucial role that the double D lines played in the development of spectroscopy, from Fraunhofer to the most recent researches of Lockyer and Huggins.[92] Such optical effects are not, however, contextualised in the demonstration described by the poem, in which each of the 'spectra flash', their phenomena following one another in rapid succession. Not only is it, as Maxwell observes in February 1873, 'difficult, especially in these days of the separation of technical from popular knowledge, to expound physical optics to persons not professedly mathematicians', but 'it is extremely easy to show such persons the phenomena, which are very beautiful in themselves, and this is often accepted as instruction in physical optics'.[93]

The third stanza of Maxwell's poem focuses upon the interactions of light with gases and vapours that the second stanza alludes to with its references to the Fraunhofer lines. The rather accommodating terms in which it describes its experiment suggests a popular demonstration using a vacuum tube, such as the Geissler tube, that contains rarefied gas or vapour through which high-voltage electricity can be discharged to produce strange and entertaining effects of coloured light. Tyndall had demonstrated such experiments recently in his lecture 'Electric Discharge through rarefied Gases and Vapours', one of the series *On Electrical Phenomena and Theories*, the subject of the review mentioned earlier. Details of the demonstration described in the '*Ode*', however, indicate that Maxwell's principal source was an important scientific paper Tyndall had delivered to the Royal Society in October 1868, 'On a New Series of Chemical Reactions produced by Light'. This introduces an experiment Tyndall became famous for, his demonstration of the 'artificial sky', which he had presented at the Royal Institution on 15 January 1869 in his lecture 'On Chemical Rays of the Light of the Sky'. Pollock attended the lecture, and sent Tyndall the following couplet a 'day or two afterwards':

Dialogue between Urania, the Mother of the Heavens, and Celestine, the Nymph of the Blue Sky.

URANIA. Tell me, sweet Celestine, what makes you pout?
CELESTINE. O! mother dear! that Tyndall's found me out.[94]

Tyndall specifies a long glass tube for his experiment, '2.8 feet long and of 2.5 inches internal diameter'.[95] Through one end, which is sealed with glass, a strong electric beam is directed along the axis of the tube, while at the other, small tubes connect it to an air pump and allow mixtures of purified air and chemical vapours to be introduced. The effect of the electric light passing through the tube is to produce what Tyndall describes as 'Curious clouds' of various distinctive colours, as specific chemical vapours absorb parts of the light spectrum. Amongst these he produces '*Sky-blue by the decomposition of Nitrate of Amyl*',[96] although this particular effect occurs also with vapours of other chemical particles of comparable size, which offer the eye similar wavelengths of light. Maxwell acknowledges Tyndall's 'sky in a bottle' experiment, as Ruskin calls it,[97] with his reference in the

'*Tyndallic Ode*' to the 'gleams ... of heavenly blue'. The final lines of the third stanza focus upon Tyndall's preoccupations with the shapes of the luminous 'clouds', which he sees to assume 'perfect flower-like forms' and various animal shapes.[98] Maxwell refers to another of these experiments in his 1870 address to the BAAS at Liverpool: 'we may see Dr Tyndall produce from a mere suspicion of nitrate of butyle an immense cloud'.[99] Amongst other such experiments, an example furnished by 'Hydriotic Acid' suggests the source for Maxwell's 'fish of cloud'. 'The development of the cloud was like that of an organism', Tyndall records in 'Chemical Reactions produced by Light', 'Once it presented the shape of a fish, with eyes, gills, and feelers.'[100]

DARK SCIENCE AND LIGHT

The third stanza of the '*Tyndallic Ode*' describes an apparition. It begins with the sealed tube being scrutinised scientifically by the merciless glare of 'the electric ray' and declared 'optically clean', as if to demonstrate definitively that there is nothing up the lecturer's crystal-clear sleeve. Nevertheless, 'gleams' and then 'wondrous forms' of luminescent cloud emerge in its 'vacuous spaces'. This sequence echoes Tyndall's account in 'Chemical Reactions produced by Light': 'For a moment the tube was *optically empty*, nothing whatever was seen within it; but before a second had elapsed a shower of liquid spherules was precipitated on the beam, thus generating a cloud within the tube. This cloud became denser as the light continued to act, showing at some places a vivid iridescence.'[101] The conventional rhetoric of showmanship in the poem – 'but stay: All has not yet been seen' – almost announces the lecturer's sleight of hand, for the vigilant and trusty 'electric ray' that initially demonstrates the emptiness of the tube is precisely the agent that produces the curious phenomena it subsequently appears to disclose. This is explained in Tyndall's account of the experiment, familiar to both Maxwell and his original audience, from 'On the Scientific Use of the Imagination':

Therefore, I say, sharply and definitely, that the components of the molecules of sulphurous acid are shaken asunder by the ether waves. Enclosing the substance in a suitable vessel, placing it in a dark room, and sending through it a powerful beam of light, we at first see nothing: the vessel containing the gas is as empty as a vacuum. Soon, however, along the track of the beam a beautiful sky-blue colour is observed, which is due to the liberated particles of sulphur. For a time the blue grows more intense; it then becomes whitish; and from a whitish blue it passes to a more or less perfect white.[102]

For many of the demonstrations recounted in 'On a New Series of Chemical Reactions produced by Light', Tyndall rather theatrically drains and fills the tube in darkness, making the effect of introducing the light all the more dramatic and mysterious. This is especially the case when, as occurs with the 'vacuous spaces' described in the poem, the greater rarefaction of the gas further obscures the causal role played by the light, for as Tyndall observes 'The rapidity of the action diminished with the attenuation of the vapour.'[103]

With their transcendent 'gleams . . . of heavenly blue' and organic shapes of luminescent cloud, the Tyndallic demonstrations in Maxwell's poem suggest spiritualist manifestations, such as the coloured auras of 'the Odic or Psychic force' that the chemist Carl von Reichenbach proposed in the 1840s and 1850s.[104] Apart from some minor variations in punctuation, the first quatrain of the stanza describing these events remains unchanged in the extant versions of the poem. The second, however, was subjected to successive revisions, which forsake Tyndall's references to the phenomenon produced in the tube as a cloud in favour of more mysterious and supernatural imagery. The first of the following extracts is from the original version of the poem given to Tait, the second from the 1874 revision that Campbell and Garnett chose for their edition:

> What gleam is this of heavenly blue
>   What wondrous form appearing,
> What mystic fish, what whale, that through
>   The etherial void is steering!

> What gleams are these of heavenly blue?
>   What air-drawn form appearing,
> What mystic fish, that, ghostlike, through
>   The empty space is steering?[105]

'The etherial void', which supersedes 'The vacuous spaces' of the original short version, turns upon the equivocal contemporary understanding of the ether concept that Maxwell notes in his 1853 Apostles essay 'Idiotic Imps', in which its use in optics is coopted by the spiritualists as 'the invisible medium through which the communion of the sensitive takes place'.[106] It accordingly provides an apt medium to propagate a 'mystic fish', the apparition that is described in the subsequent version of the poem as 'ghostlike'.

Campbell records being at a séance with Maxwell in Edinburgh during their Christmas vacation in 1850–1.[107] 'I see daily more & more reason to believe that the study of the "dark sciences"', Maxwell writes to Litchfield in August 1853, 'is one wh[ich] will repay investigation. I think that what is called the proneness to superstition in the present day is much more significant than some make it.' In 1851 another of his peers at Trinity, the future archbishop of Canterbury Edward White Benson, made Cambridge the focus for British interest in spiritualism by establishing the Ghostlie Guild, which later became the Society for Psychical Research. Spiritualism is often glossed in relation to evolutionary biology, and Victorian scientism generally, as a displaced form of religious belief, a new assurance of an afterlife that is apparently demonstrated empirically by the phenomena of séances. For Maxwell it is a 'dark science', not so much a rival for Christianity as a threat to the enlightened Victorian project of professionalising science: 'they profess to treat of laws which have never been investigated ... imitating the phraseology of science ... combining its facts with those which must naturally suggest themselves to a mind unnaturally disposed'. More specifically, he observes, they 'were or pretended to be physical sciences', and, as he illustrates with the example of the ether, 'Their language was imitated from popular physics.' Maxwell accordingly reads the 'proneness to superstition in the present day' as an index of current attitudes to science generally and physics in particular. The success of the 'dark sciences' shows that the public combine great credence in science with ignorance of its foundational principles and methods, just as he sees popular lectures on science to demonstrate in the 1860s and 1870s. Such audiences are, Maxwell observes directly after his discussion of *Villette* in the 1853 letter to Campbell, unable to independently assess claims made in the name of science, dark or otherwise:

Faraday's experiments on Table-turning, and the answers of provoked believers and the state of opinion generally, show what the state of the public mind is with respect to the *principles* of natural science. The law of gravitation and the wonderful effects of the electric fluid are things which you can ascertain by asking any man or woman not deprived by penury or exclusiveness of ordinary information. But they believe them just as they believe history, because it is in books and is not doubted. So that facts in natural science are believed on account of the number of witnesses, as they ought! I believe that tables are turned; yea! and by an unknown force called, if you please, the vital force, acting, as believers say, thro' the fingers.[108]

The credence given to spiritualism reciprocally demonstrates a failure to understand physics, the science whose ideas and idioms it so unashamedly parasitises: 'The prevalence of a misdirected tendency proves the misdirection of a prevalent tendency.'[109] The '*Tyndallic Ode*' enacts the converse case of this relation by presenting popular demonstrations of physics as impersonations of spiritualism, 'the popular occult sciences', as he refers to them in his 1853 letter to Reverend Tayler.[110] Indeed, as both Tyndall and Maxwell would have been aware, this was precisely what Pepper was doing at the Royal

Polytechnic Institution, where 'Pepper's Ghost' and other supernaturally themed optical illusions accompanied his hugely popular lectures on light.[111] Jill Howard observes that Tyndall was also apt to present science as occult mystery, especially when lecturing to juvenile audiences. She cites the following example from his 1862 lectures on light: 'I think, in order to enable you to distinguish one beam from the other, I will produce a little magic, a little necromancy: I will colour these two discs for you by that wonderful property of interference.'[112] In describing the 'ghostlike' light effects of the Tyndallic demonstration, Maxwell may also have been prompted by 'On the Scientific Use of the Imagination', the subject of his poem's final stanzas. Invoking an earlier scientific sense of the word 'occult', used by Newton and his peers for phenomena not amenable to direct observation, Tyndall writes that such optical experiments illustrate 'some of the more occult features and operations of this agent [i.e., light]'.[113] While Maxwell sees such popular demonstrations as furnishing their audience with only the illusion of scientific knowledge, when they are heightened by mystificatory and pseudo-scientific theatricality, as occurs in the Tyndallic lectures, they are seen to function as dark science, to retard both the advance of the public's scientific understanding and the professionalisation of science itself.

The style of Tyndall's lectures is accordingly seen by Maxwell to be at odds with his well-known public stance against spiritualism, which he reaffirms in 1871 by including his earlier essay 'Science and the "Spirits"' in his popular collection *Fragments of Science for Unscientific People*.[114] While Tyndall was a vigorous campaigner against spiritualism, which he saw as an affront to his scientific values, the lecturer in Maxwell's poem presents himself as a medium who is able to harness ostensibly occult powers and commune with a mysterious other world, of microscopic spaces and empyrean flame. As a purveyor of signs and wonders to those who, as Maxwell and Tait believe, are not sufficiently equipped with the fundamental scientific principles to understand and assess his demonstrations, the lecturer in the '*Tyndallic Ode*' is equated with the ambiguous and similarly popular figure of the spiritualist. The rhetoric and imagery that the poem applies to the lecture demonstrations means that, as with other apparently occult phenomena of the day, a question hangs over them as to whether they are genuine spiritualist phenomena, or else can be explained naturalistically, either by science or as fraudulent conjuring tricks.

The confusion between physics and spiritualism that Maxwell observes in 'Idiotic Imps' and charges popular lecture demonstrations with facilitating in the '*Tyndallic Ode*', reached deep into professional science itself in the year the poem was written. Indeed 1871 was a very good year for spiritualism. 'One scarcely can sit down without being involved in a noisy "knock-conversation" with a dining-room table, an accordian, or a towel-horse', observes the Hungarian-born revolutionary, spy and historian G. G. Zerffi in his 1871 book *Spiritualism and Animal Magnetism*.[115] A large committee that the London Dialectical Society had appointed to investigate spiritualist phenomena produced its long-awaited report in this year. The result of eighteen months' work, it vindicated the spiritualists, concluding that 'motion may be produced in solid bodies without material contact, by some hitherto unrecognised force'.[116] The year also saw the eminent chemist and physicist William Crookes overcome his scepticism to recognise the existence of what he called a 'psychic force', a finding that, as will be seen later, commanded a great deal of attention from his colleagues at the BAAS meeting at Edinburgh.[117]

Crookes focused his scientific efforts, as the Dialectical Society had done also, although not exclusively, upon the séances held by the most celebrated medium of the age, Daniel Dunglas Home. The Scottish-American Home had been in London since 1855, conducting séances for British aristocracy and European royalty, novelists, poets, artists, scientists, politicians and legion professionals, who were confronted with tables moving, ectoplasmic hands, floating bodies and musical instruments apparently playing without human contact. Home had been investigated before by numerous stage conjurors and a series of scientists, including Brewster in 1855 and the pioneer of telegraphy Cromwell Varley in the 1860s. Tyndall had battled with Home and lost, after he accused him in an 1868 letter to the *Pall Mall Gazette* of having not allowed Faraday to investigate his séances. Home produced correspondence showing that on the contrary Faraday had no inclination to attend.[118] None of the earlier inquiries into Home's séances had proven him a trickster, while most recently the work of the Dialectical Society and of Crookes had vindicated his spiritualism with investigations of unprecedented scientific thoroughness and prestige. According to the *Spiritual Magazine* '1871 had', as Peter Lamont puts it summarily, 'seen more progress, perhaps, than any other year since the advent of Modern Spiritualism'.[119]

WILLIAM CROOKES

Maxwell's specialist audience of Red Lions would have recognised a subtle but topical allusion to spiritualism in the second stanza of his poem:

> The atoms clash, the spectra flash,
> Projected on the screen,

> The double D, magnesian b
> And Thallium's living green.

Their colleague Crookes was well known and celebrated as the discoverer of Thallium, having disclosed the element spectroscopically in 1861, while from 1870 he became notorious for his experiments in psychic phenomena.[120] Maxwell's reference to the spectroscopic projection of Thallium's intense bright line invites associations with its discoverer's notorious forays into the spirit world, so that the 'living green' of the element accordingly suggests an impersonation of verdant life by dead matter, a punning convergence of the two senses of spectrality. The claims of 'dark science' to scientific legitimacy, which Maxwell had dedicated his 1853 Apostles essay to refuting, had resurfaced at Edinburgh, this time through the efforts of one of his most respected colleagues in Section A. '[O]ne of the severest tests of a scientific mind', he observes in his 1878 review of Stewart and Tait's *Paradoxical Philosophy*, 'is to discern the limits of the legitimate application of scientific methods.'[121]

Crookes's article, 'Experimental Investigation of a New Force', which was published in the July 1871 *Quarterly Journal of Science*, documents a séance held by Home on 31 May, to which he had brought along Huggins as a witness.[122] A couple of letters on 'The New Psychic Force' in the 3 August issue of *Nature*, which was current at the time that Maxwell wrote his poem and read it to the Red Lions, reminds its readers of the pair's attendance at Home's séance.[123] While the mention of Thallium in the '*Tyndallic Ode*' invokes a clear association with Crookes, who attended the Edinburgh meeting, the specific references to the double D lines that accompany it suggest the work on the spectroscopy of gases and vapours in the sun's atmosphere that Huggins developed in the 1860s. Thomson discoursed on the nature and importance of this work in his Presidential address, reminding the delegates that the Sodium lines furnished a standard point of reference for such researches.[124] Extending his reach to the stars and nebulae, Huggins established that the latter were clouds of luminous gases, the phenomena of Tyndall's experiment writ large. Crookes's 'Experimental Investigation' documents corresponding effects, including 'a luminous cloud', which he claims to have seen 'hover over a heliotrope on a side table, break a sprig off, and carry a sprig to a lady'.[125]

Stewart responded to Crookes's article with a respectful but sceptical rejoinder, 'Mr Crookes on the "Psychic" Force', which led *Nature* for 27 July, the current issue for the week that the Edinburgh meeting began. The next issue, of 3 August, included the two letters to the editor mentioned earlier, which were written in response to Stewart's article and published directly after the reports of the BAAS gathering. Indeed one of the correspondents closes his letter with the challenge that 'Mr Crookes but repeat any one of the experiments at one of the evening *soirées*' at the meeting.[126]

The timing and sensational nature of Crookes's 'Experimental Investigation' would have ensured that it was well discussed by the delegates at Edinburgh. Its title refers to a notorious experiment that Crookes made at the Home séance in May, which Zerffi refers to in the extract cited earlier. Crookes took to the gathering a new accordion he had purchased, which he ostentatiously secured from physical influences by isolating it in a cage of wood that he wound with string and insulated copper wire, through which an electric current was passed. Despite these encumbrances the accordion apparently managed to float within its cage and then play chords and melodies. Counter-naturalistic phenomena were deemed by Crookes to have occurred in circumstances that allowed of no interfering by the demonstrator. While Crookes awards scientific legitimacy to spiritualism with this experiment, Tyndall is presented conversely in Maxwell's poem as delegitimising science with a demonstration in which mysterious 'air-drawn' apparitions appear within the ostensibly tamper-proof 'crystal tube', a hermetic parallel to the elaborate basket Home places around the accordion.

The indulgence extended to spiritualists by such eminent peers as Crookes, Huggins and A. R. Wallace, who also attended several of Home's séances, prompted comments in various BAAS addresses and papers at Edinburgh. Tait's address to Section A includes a scathing mention of spiritualism, and a further reference that, like Maxwell's '*Ode*', parallels 'the comparatively harmless folly of the Spiritualist' with 'the pernicious nonsense of the Materialist', each being seen to demonstrate that 'great ignorance almost necessarily presumes incapacity'.[127] Tait's counterpart in the biology section, Professor Allen Thomson, concludes his president's review of recent developments in his discipline with a long paragraph headed '*Spiritualism*', in which he regrets 'that a few men of acknowledged reputation in some departments of science have lent their names, and surrendered their judgment, to the countenance and attempted authentication of the delusive dreams of the practitioners of spiritualism, and similar chimerical hypotheses'.[128]

Crookes addresses Stewart's essay in passing, along with other less public responses to his 'Experimental Investigation of a New Force', in 'Some Further Experiments on Psychic Force', which was published in the *Quarterly Journal of Science* on 1 October 1871. Over half of Crookes's essay tells of his peers turning against him and his spiritualist experiments after the appearance of his 'Experimental Investigation', and his consequent difficulties in presenting further spiritualist researches to both the Royal Society and the BAAS. He cites Thomson's dismissive comment that no such inquiries 'can deserve the name of study or investigation'.[129]

In his reply to W. B. Carpenter's 'Spiritualism and its Recent Converts', Crookes recalls that this energetic opponent accosted him for over an hour at Edinburgh.[130] He does, however, appear to have received some support at the meeting, for, as he records in his October 1871 article, 'At the urgent request of gentlemen on the committee of section A I communicated a paper ... to the British Association, in which I recounted some of the experiments described in the present paper.' The section referred it to another committee, on behalf of which Stokes wrote a report that did not 'see much use discussing the thing in the sections', but which in a conciliatory gesture suggested that a small committee could be formed to investigate the subject. The high reputation that Crookes had as a scientist meant that some of his peers at the meeting not only excused his spiritualist investigations but defended them, as for instance Professor Challis does in commenting on his colleague's scrutiny of the Home séance: 'In short, the testimony has been so abundant and consentaneous that either the facts must be admitted to be such as are reported, or the possibility of certifying facts by human testimony must be given up.'[131] Such a sympathetic ethos, no doubt inflected as badinage, is likely to have prevailed amongst Crookes's sectional peers when, removed from the more serious contexts of the day's work at the BAAS, they met as Red Lions.

Maxwell's correspondence with his friends during the early 1870s includes some gentle jokes at Crookes's expense, which can be taken as symptomatic of further verbal mirth-making over his spiritualist activities at this time. So, for example, in a reply to a lost letter in which Maxwell asks for information about the solar spectrum, Huggins, referring to a particular spectral line, assures his friend in October 1874, 'I will inquire about the ghost's name when I see Crookes.'[132] Maxwell makes a similarly ironic remark about his colleague's scientific work in a letter to Tait from late April 1874. Indeed its subject is one of Crookes's most important discoveries. Having noticed odd effects in the vacuum balance he used to ascertain the molecular weight and properties of Thallium, Crookes conducted further experiments that led him in 1875 to invent the radiometer that is named after him. This consists of an evacuated glass bulb in which four light metal vanes, the sides of which are alternately polished and blackened, were discovered to spin freely when subjected to radiant heat. This sensational effect was theorised by Maxwell to occur as molecules within the low-pressure environment leave the dark surfaces, which absorb more of the radiant heat, with greater momentum than those leaving the cooler polished surfaces.[133] Maxwell was greatly interested in the device, and the earlier experiments leading to its invention, as they corroborated the kinetic theory of gases he had been developing since the late 1850s. In 1874 Maxwell refereed a paper by Crookes on some earlier demonstrations of this phenomenon for the Royal Society, and subsequently witnessed it himself at the Society's premises on 22 April 1874. These experiments with the effect of radiant heat upon a pendulum poised within a low-pressure environment furnished the occasion for a comment he made in his letter to Tait that 'They whip spirits all to pieces.'[134] Still sustained by Huggins, Maxwell and Tait as a joke three years later, such conceits would have been fresh and irresistible at the time of the Red Lions dinner in 1871.

'SOUNDING FLAMES, &C.'

The role of the lecturer as magus that emerges from the account of the experiment with the 'crystal tube' is further developed in the stanza that follows it, which closes the original version of the '*Ode*' published in *Nature*:

> I light this sympathetic flame,
>   My slightest wish to answer,
> I sing, it sweetly sings the same,
>   It dances with the dancer;
> I whistle, shout, and clap my hands,
>   I hammer on the platform,
> The flame bows down to my commands,
>   In this form and in that form.

This stanza is a rendition of the sixth lecture in Tyndall's series on *Sound*, which he delivered at the Royal Institution in 1867, 'Sounding Flames, &c.' As its title indicates, Tyndall's lecture documents several types of fire that allow sound waves to be visualised. It explains that the effect results from perturbations that cause the flame to be continuously extinguished and relit, 'an incipient flutter' that makes it more sensitive to the vibrations around it: 'When I now sound a whistle, the flame jumps visibly.' Among his more subtle demonstrations of such flames, Tyndall shows the peculiar sensitivity of what he calls the 'vowel flame' by reading to it a stanza from Spenser's *Faerie Queen*.[135]

Insistently reiterating the first-person pronoun with which the poem began, the lecturer presents not the physical phenomena but himself as the focal point and decisive agent in his experiments. 'An unpleasant tone of egotism pervades the whole', Tait writes in an excised passage from his draft review 'Professor Tyndall on Heat and Motion', 'and one's patience is sorely tried when, on reading the description of some well-known experiment, one finds, instead of the name of its author recorded once for all, the continual "I must devise".'[136] Complementing Tyndall's insistent first person, another of Pollock's valentines has the sensitive flame describe the experiment from her own perspective:

> Where, but to serve a wanton game,
> You played your antics with my flame;
> And called me out to please a crowd,
> And shouted my deep secret loud;
> Showing me how I could languish
> Or could utter song in anguish;
> Or could at your accents tremble –
> Do all things, except dissemble!
> Most base your triumph was, though short
> You extinguished me in sport;
> In a dark house I am lying
> Feeble, pale, as if a-dying:
> But sir, I am not gone out quite;
> I may leap again to light
> And burst my iron prison bar
> And rush upon you from afar
> And clasp you in a hot embrace
> And of the traitor leave no trace.[137]

While Pollock's seventy-nine line poem documents Tyndall's original demonstration, Maxwell's stanza is evidently drawn from the published account. Tyndall concludes his demonstration by playing a mechanical music box, to which, he writes, once again substantiating Tait's charge of 'Grandiloquence', that 'The flame behaves like a sentient creature; bowing slightly to some tones, but curtseying deeply to others.' The performance is echoed in the poem, as 'The flame bows down to my commands' and, less specifically, 'dances with the dancer'. Similarly, the activities catalogued in the fifth and sixth lines, 'I whistle, shout, and clap my hands, / I hammer on the platform', correspond suggestively to those that Tyndall brings to bear upon the candle flame; 'I may shout, clap my hands, sound this whistle, strike this anvil with a hammer.'[138]

The fourth stanza of Maxwell's poem closes the first published version of the poem with the lecturer's apparently conclusive demonstration of his power. The luminous imagery of the third stanza and the mysterious occult powers it alludes to are quickened and intensified in the description of the persona's mastery of fire in the fourth. The choice that Tyndall makes to be a popular lecturer suggests a Faustian pact, as hailing 'from empyrean fires' and having earlier summoned mysterious spectacles from another world, he now exercises complete control over elemental phenomena, having apparently subdued and mastered the diabolical element of fire, and with it his audience.

CHAPTER 6

John Tyndall and 'the Scientific Use of the Imagination'

Physics and mathematics became increasingly dependent upon the imagination during the early to mid-Victorian period. Such hypothetical entities as the luminiferous ether, the energy principle, the electromagnetic field and the irreducible particles of atoms and molecules became staples of physics at this time, each marking a shift from positivist experiment to *a priori* analysis and speculation that is registered definitively in the discipline's mathematisation during the 1860s. In mathematics itself, non-Euclidean geometry was becoming known in Britain during this decade through the belated publication in 1867 of the German mathematician Bernhard Riemann's Habilitation lecture 'On the Hypotheses which Lie at the Bases of Geometry' (1854). Non-Euclidean geometry offered an especially formidable challenge to the imagination, as it directly affronts our intuitive experience of space by positing the existence of four dimensions and other, variant and local, forms of space. Indeed Riemann theorises space as extending to n-dimensions, that is, any number of dimensions. Having established that an axiomatic system was not necessary for finding new types of space, Reimann proposed analytical procedures instead. 'The imagination', as Hans Reichenbach observes, 'is thus given conceptual support that carries it to new discoveries.'[1]

Sylvester discusses Riemann at length in his 1869 presidential address to Section A at Exeter, and declares that the combined authority of such non-Euclidean peers demands that he develop his powers of imagination to their level: 'If Gauss, Cayley, Riemann, Schalfli, Salmon, Clifford, Krönecker, have an inner assurance of the reality of transcendental space, I strive to bring my faculties of mental vision into accordance with theirs.' Sylvester cites the analogy of some two-dimensional bookworms that C. F. Gauss devised to argue for the existence of non-Euclidean spaces: 'as we can conceive beings (like infinitely attenuated bookworms in an infinitely thin sheet of paper) which possess only the notion of space of two dimensions, so we may imagine beings capable of realising space of four or a greater number of dimensions'.[2]

Akin to the supremely superficial home of the bookworm, a two-dimensional world literally furnishes the platform for Edwin Abbott's novella *Flatland: A Romance of Many Dimensions* (1884). Flatland is occasionally and sensationally impinged upon by a preposterously three-dimensional sphere, which leads its narrator, 'A. Square', to consider the possibility of yet more dimensions. Sylvester read *Flatland* soon after its publication, and writes to his closest friend and colleague Arthur Cayley, the first Sadleirian Professor of Pure Mathematics at Cambridge, in November 1884 that he had been recommending it to his students at Oxford 'in order to obtain a general notion of the doctrine of space of *n* dimensions'.[3] The book is dedicated by its two-dimensional author to 'SOLID HUMANITY', 'The Inhabitants of SPACE IN GENERAL', 'So [that] the Citizens of that Celestial Region / May aspire yet higher and higher / To the Secrets of FOUR FIVE OR EVEN SIX Dimensions / Thereby contributing / To the Enlargement of THE IMAGINATION.'[4] Also published in 1884, Charles Hinton's essay 'What is the Fourth Dimension?' similarly assists its readers to make the imaginative leap from the third to the fourth dimension by analogy with the familiar transition from the second to the third.[5]

Carroll's *Alice in Wonderland* and *Through the Looking-Glass* describe situations that furnish suggestive correlates for the ostensibly preposterous new cosmologies that the mathematical and physical sciences were generating at this time. Indeed, as Beer demonstrates in her lecture 'Alice in Space', and no doubt elaborates in her forthcoming book on Carroll, they offer witty representations of such discombobulating ideas.[6] The physicist George Gamow takes his cue from Carroll for his *Mr Tompkins in Wonderland* (1940) and its sequel *Mr Tompkins Explores the Atom* (1944), companions to a modern mathematics and physics that is seen to defy the commonsense understandings of an educated general readership, and so require a series of imaginative allegories and parables for its exposition. Gamow's account traces this audaciously counter-intuitive science, which culminates in Einstein's theory of relativity and quantum theory, back to Riemann's non-Euclidean principle of curved space and 'Maxwell's Demon' in the late 1860s and early 1870s. While Gamow is able, with the benefit of hindsight, to place these ideas as starting points for an outlandish but coherent episode in the history of science, initial reactions to such bold developments were often uncertain and confused, the first experiences of the vast counter-intuitive cosmologies that still characterise mathematical physics. Increasingly uncoupled from experience, the scientific imagination appeared to be embarking upon a period of irresponsible and dangerous adventurism.

In the spirit of Hamilton's dictum that 'imagined possibility affects us otherwise than believed reality',[7] Sylvester, striving to imagine non-Euclidean space through his 'faculties of mental vision', insists in his Exeter address that 'actuality is not cancelled or balanced by privation [of empirical fact]'.[8] Similarly, the concluding paragraph to Hinton's 'What is the Fourth Dimension?' opens unapologetically with the statement: 'It *is*, of course, evident that these speculations present no point of direct contact with fact.'[9] The sober professional science that the BAAS had been instrumental in establishing in the preceding decades looked like it was being undermined by a spate of rash apriorism.

One of the mathematicians Sylvester lists in his 1869 address as having 'an inner assurance of the reality of transcendental space', George Salmon, responded to his colleague's confidence in him by promptly denying any such capacity: 'I do not profess to be able to conceive of *affairs* of four dimensions,' he is quoted in *Nature* as saying, adding sarcastically, 'I advise you to believe whatever Sylvester tells you, for he has the power of seeing things invisible to ordinary mortals.'[10] Salmon implies that Sylvester is not so much a scientist as a spiritualist medium. His remark is cited in a letter entitled 'Transcendent Space', by Sylvester's friend the Shakespeare scholar Clement Ingleby, who, in a further note published under the same heading a month later, speaks similarly of 'The ghost of a fourth dimension, which had haunted [Rowan] Hamilton's Triplets.'[11]

The idea of the fourth dimension was adopted opportunistically by spiritualism, just as it had earlier appropriated that of the luminiferous ether. Oscar Wilde's 1887 short story 'The Canterville Ghost: A Hylo-idealistic Romance' testifies to the currency that the idea had achieved, as its eponymous spirit finds refuge in non-Euclidean space: 'There was evidently no time to be lost, so, hastily adopting the Fourth dimension of Space as a means of escape, he vanished through the wainscoting.'[12] As was noted in Chapter 5, scientists of the calibre of Crookes, Huggins and Wallace were defending the scientific credibility of spiritualism in the 1870s, while Stewart and Tait were using Continuity, energy physics and non-Euclidean geometry as a springboard for their cosmological speculations in *The Unseen Universe* (1875) and its sequel, *Paradoxical Philosophy* (1878), which similarly posit a transcendental world able to fulfill the functions of the Christian afterlife. The 1870s presented something of a crisis of legitimacy for British professional science, which centred on the place of imagination in science.

MOLECULAR MACHINES AND LASCIVIOUS BODIES

The well-known caricature of Tyndall that *Vanity Fair* published in April 1872 depicts him as the great public lecturer, looking earnestly and forthrightly at his audience. 'No. 43' in the fashionable magazine's 'Men of the Day' series, it is a portrait of the prelapsarian Tyndall, the respected scientist and Alpine adventurer, before his incendiary 1874 Belfast address led to him being cast before the public as a godless materialist. His *Heat considered as a Mode of Motion* had reached its fourth edition in 1870, the year in which he also published *Nine Lectures on Light* and *Seven Lectures on Electrical Phenomena*, followed by *Fragments of Science for Unscientific People* and *Hours of Exercise in the Alps* in 1871. He is, however, identified in the *Vanity Fair* article with only one of his writings, 'that famous lecture, which most people have read, on "The Scientific Use of the Imagination"'. Originally delivered as the evening discourse for the 1870 Liverpool meeting of the BAAS, the paper was published by Longmans in book form in November as *Essays on the Use and Limit of the Imagination in Science* and also reprinted in *Fragments of Science*. *Vanity Fair* does not, however, discuss the content of Tyndall's discourse, the third edition of which appeared in 1872 under the title *Scientific Use of the Imagination, and other Essays*. Rather, it adopts the title of the essay as the caption for the caricature, a cipher for Tyndall's popular success, his reputation for distinctive powers of poetic expression and dramatic spectacle.

Tait kept a copy of the *Vanity Fair* caricature, under which he wrote, evidently from memory, a line from Edward Young's poem *Night Thoughts*: 'Pigmies are pigmies still – though perched on Alps' (Figure 4). He also took the opportunity to fill in its shadowy lower region, the void space below the surface that its subject's hands lean upon, with his reply to a letter from Tyndall, a concise sample of the viciousness that was precipitated by their public controversies:

Dr Tyndall says, among other lively things, some of which have since been retracted, 'Mr Tait ... when publically hoisted by his own petard, retired to void his venom against me in the anonymous pages of the *North British Review*.' I need only remark on this point that the charge, if deserved, is really in the first place directed not against me but against the editor who permitted such an abuse of his pages. These tremendous words are, after all, merely Dr Tyndall's grandiloquent mode of admitting that I had detected and pointed out several ridiculous blunders in his work on *Heat [considered as a Mode of Motion]*.[13]

Tait evidently saw this as a *coup de grâce* that countered the *Vanity Fair* image, the emblem of the public esteem and indeed celebrity that his

Figure 4 'Pigmies are pigmies still – though perched on Alps.' A page from P. G. Tait's Scrapbook, James Clerk Maxwell Foundation, Edinburgh.

opponent enjoyed at this time through his discourse on the scientific imagination.

A knowing letter that Maxwell wrote to Tait on 14 November 1870, in which he refers mockingly to details of Tyndall's discourse and requests that his friend send him a copy, presumably the volume that Longmans published in that month, shows that they had both attended the reading at the BAAS and were exercised by it.[14] Tait wrote a review of this edition for *Nature*, which was published anonymously on 16 March 1871 under the title 'Imagination in Science'. It opens uncompromisingly by implying that the popular lecturer, having long appealed to the imagination of his audience with his showy demonstrations, now offers a rationale for such amateurs to practise science themselves: 'Professor Tyndall will eventually have much to answer for. He has lent his authority to the admission of imagination in the pursuit of science, and there is every prospect that people whose imaginative faculty is stronger than their habit of observation will give us all plenty to do.' Tyndall is charged with undermining modern scientific method and sanctioning arbitrary *a priori* approaches to natural phenomena: 'Are we to live, scientifically, in the same way as alchemists and astrologers did in the Middle Ages? and are we to ignore all that Bacon and Newton have done for us?'[15]

Essays on the Use and Limit of the Imagination in Science also reprinted Tyndall's 1868 presidential address to Section A at the BAAS meeting in Norwich, under the title 'Scope and Limit of Scientific Materialism'. It appears again a few months later with the discourse in his *Fragments of Science*, this time as the 'Scientific Limit of the Imagination'. Maxwell replied to Tyndall's Norwich address in his corresponding address as sectional president at Liverpool on 15 September 1870, while Tyndall read his discourse on the scientific imagination on the following evening. As was noted earlier, Tait in turn makes some pointed references to Tyndall and the imagination a year later in his presidential address to Section A at Edinburgh.

The 1874 President's address to the BAAS at Belfast was the most notorious, but not the first, occasion that Tyndall used to present his materialist account of the natural world. He argues in the Norwich address that atoms are held together by innate forces, a principle that drives his materialist system yet derives from romantic metaphysics; 'In fact, throughout inorganic nature, we have this formative power, as Fichte would call it – this structural energy ready to come into play, and build the ultimate particles of matter into definite shapes.'[16] Rather than have the eminent scientist formally address the BAAS as its President, 'the poet-philosopher sings' in Maxwell's verse paraphrase, 'British Association, 1874: *Notes of the President's Address*'. The poem presents Tyndall as the Germanising bard and enthusiast for Carlyle, whom he eulogises in closing the Address, as it records in mock wonder the lecturer's account of atoms: 'How he clothes them with force as a garment, those small incompressible spheres!'[17] Presented as a folly of the imagination, akin to that of the Emperor's New Clothes, Tyndall's principle of molecular force is conflated with the philosophy of clothes expounded by Teufelsdröckh, the hero of Carlyle's *Sartor Resartus*.

'This tendency on the part of matter to organize itself' is described in the Norwich address as 'all-pervading' and, more provocatively still, as 'Incipient life'. Atoms cohere as molecules and then as more complex self-organising aggregates, including plants and animals capable of reproducing themselves and of evolving into various further forms through the Darwinian principle of natural selection. Early in the address, Tyndall notes a formal correspondence between the Egyptian pyramids and the crystalline aggregates that salt molecules assume. He draws from this the observation that while 'the final form of the pyramid expressed the thought of its human builder' there is no need to invoke a creator in the case of the 'salt-pyramids', for 'these molecular blocks of salt are self-posited, being fixed in their places by the forces with which they act upon each other'. Tyndall moves by analogy from the crystals of salt, 'a dead mineral[,] to a living grain of corn',[18] and from there storms all of organic nature, barely stopping short of explaining human thought, to sketch a scrupulously deterministic cosmology that has no need of a divine Creator and Designer.

In his 1870 President's address to the BAAS at Liverpool, Huxley lends qualified support to Tyndall's materialist hypothesis at its most contentious point, where it insists upon a transition from inorganic matter to organic life-forms. In 1828 the German chemist Friedrich Wöhler had become the first person to synthesise an organic compound, urea, and so demonstrate the radical continuity between inorganic and organic substances, which Tyndall makes axiomatic and indeed animate in his conception of matter as 'Incipient life'. Huxley coins the term 'Abiogenesis' in his address to describe the spontaneous generation of organic life from inorganic matter, declaring that 'if it were given me to look beyond the abyss of geologically recorded time to the still more remote period when the earth was passing through physical and chemical conditions, ... I should expect to be a witness of the evolution of living protoplasm from not living matter'. He also argues that the 'prodigious strides' being made by 'organic chemistry, molecular physics, and physiology' make it 'the height of presumption for any man to say that the conditions under which matter assumes the properties we call "vital" may not, some day, be artificially brought together'.[19] Nonetheless, aware of the controversy caused by Tyndall's Norwich address, Huxley is, as he puts it later, 'wise and prudent',[20] and careful not to be provocative. He accordingly holds back from making any final assertion of his doctrine, declaring tactfully that 'Belief, in the scientific sense of the word, is a serious matter, and needs strong foundations.'[21]

Thomson used his President's address at the 1871 Edinburgh meeting of the BAAS to draw attention to his predecessor's measured remarks on abiogenesis, congratulating Huxley for a rigorous argument *against* the principle. In so doing he effectively pits Huxley's address against the comprehensive cosmology that Tyndall had elaborated a few days after it in his discourse on the scientific imagination, where he defends and extends the thesis of the 1868 Norwich address. Indeed the paper audaciously traces the most sophisticated forms of life to the original hot gases from which, according to the nebular hypothesis, the solar system was precipitated: 'not alone the more ignoble forms of animalcular or animal life, not alone the nobler forms of the horse and lion, not alone the exquisite and wonderful mechanism of the human body, but the mind itself – emotion, intellect, will, and all their phenomena – were once latent in a fiery cloud'. More extravagantly still, Tyndall declares that many advocates of the nebular hypothesis, which was originally advanced in the late eighteenth century by Kant and independently by Pierre Simon Laplace, 'would probably assent to the position that at the present moment all our poetry, all our science, and all our art – Plato, Shakespeare, Newton, and Raphael – are potential in the fires of the sun'.[22] A year later, the eponymous scientist-poet of the '*Tyndallic Ode*' claims this most ancient pedigree for himself:

> I come from empyrean fires,
>   From microscopic spaces,
> Where molecules with fierce desires,
>   Shiver in hot embraces.

These lines offer a contracted but unexpurgated version of a summary account that Maxwell gave a year before, early in his 1870 Liverpool address, of his predecessor's Norwich address: 'I have been carried by the penetrating insight and forcible expression of Dr Tyndall into that sanctuary of minuteness and of power where the molecules obey the laws of their existence, clash together in fierce collision, or grapple in yet more fierce embrace, building up in secret the forms of visible things.'[23]

Acknowledging the hypothetical nature of atoms and molecules, Maxwell discerned the existence and nature of these submicroscopic particles statistically, through their collective activity, an approach that he had recently extended with his thought experiment. From the vantage point of this cautious approach, Tyndall's speculative attribution of an innate force to his atoms and molecules is unwarranted and extravagant, indeed an anthropomorphic imposition that Maxwell brings forward by identifying it with conflict and passion, predicates that accordingly cast the particles as fictitious characters. Drawing an analogy between the activity of bodies in physics and human sexual behaviour, akin to that of 'In Memory of Edward Wilson', the Tyndallic molecules are seen to embrace in a semblance of sexual intercourse, which, as the means by which complex life-forms are engendered, accordingly assures the reader of their capacity for abiogenesis, for breaching the great ontological divide between inorganic and organic matter. Most basically and basely, especially in the more focused form offered to the Red Lions, the image is, as Gowan Dawson observes, calculated to evoke in its original audiences the dubious sexual associations and interests that coloured Tyndall's reputation from the 1860s.[24] Indeed the lascivious molecules are used to casually characterise Tyndall's entire cosmology, summarily identifying his metaphysical materialism with licentious hedonism.

Maxwell observes in his 1871 inaugural lecture as the Cavendish Professor, that 'No one has as yet seen or handled an individual molecule ... but the idea of the existence of unnumbered individual things, all alike and all unchangeable, is one which cannot enter the human mind and remain without fruit.'[25] Similarly, Tyndall reminds the audience for his 1870 discourse on the imagination that atoms and molecules belong 'behind the dropscene of the senses', in what he later clarifies as 'a region where things are intellectually discerned':[26]

It cannot be too distinctly borne in mind that between the microscope limit and the true molecular limit there is room for infinite permutations and combinations. It is in this region that the poles of the atoms are arranged, that tendency is given to their powers, so that when these poles and powers have free action and proper stimulus in a suitable environment, they determine first the germ and afterwards the complete organism.[27]

Dramatised by Maxwell in the 'microscopic spaces' of his poem's riddling second stanza, Tyndall's 'region' was for the Metropolitans and the North Britons alike a space of speculation that each sought to colonise. As well as generating new sciences, the Victorian scientific renaissance also critiqued and extended long-established disciplines, giving them a new fluidity, which in the case of physics the factions fought to channel definitively along their own lines. The debate over irreducible particles during the late 1860s and early 1870s furnished the battle between the Metropolitans and the North Britons with its most radical and consequential focus.

Maxwell endows such particles with a Pythagorean quality of commensurability in his 1873 'Discourse on Molecules', remarking that 'Each molecule, ... throughout the universe, bears impressed upon it the stamp of a metric system as distinctly as does the metre of the Archives at Paris', a uniformity and immutability that gives to it 'as Sir John Herschel has well said, the essential character of a manufactured article, and precludes the idea of its being eternal and self-existent'. Unobtrusively renewing Paley's watchmaker argument for the industrial age, the 'Discourse' closes by acknowledging 'Him who in the beginning created, not only the heaven and the earth, but the materials of which heaven and earth consist'.[28] Tyndall, on the other hand, fills the 'region' he discerns with self-organising atoms and their deterministic propensity to form molecular and organic coalitions. Indeed he not only subverts the Argument from Design with the comparison he draws in the Norwich address between the man-made Egyptian pyramids and the spontaneous formation of salt crystals, but begins his discussion by considering the watch as an analogy for objects in nature, a pointed reference to Paley's foundational argument.[29] Tyndall acknowledges that the 'motion of the [watch] hands may be called a phenomenon of art, but', he writes, 'the case is similar with the phenomena of nature' not because they too are instances of artifice that imply a maker, but because a 'force' impels them both, for like the watch, natural phenomena 'also have their inner mechanism, and their store of force to set that mechanism going'.[30]

Tyndall's atomic forces are made to do for all matter what Darwin's principle of natural selection achieves only for organic nature: provide a naturalistic mechanism that effectively inverts the natural theological argument, so that evidence of pattern and design is accordingly seen to demonstrate not the necessity of a Creator but, on the contrary, nature's radical independence from such a principle. Maxwell, however, having read Max Müller's 1856 essay 'Comparative Mythology',[31] presents Tyndall's cosmology in 'A Tyndallic Ode' as nothing more than a creation myth that, grounded in the 'empyrean fires' of solar mythology and corroborating rumours of his sexual preoccupations, the bachelor has developed by analogy with human procreation.

TYNDALL'S 'SCIENTIFIC IMAGINATION'

Tyndall works to secure his foundational principle of molecules theoretically in his discourse on the imagination, an effort that Maxwell burlesques in the final stanzas of the '*Tyndallic Ode*'. He begins his discourse, much as the

Tyndallic persona of the '*Ode*' opens his verse lecture a year later, by announcing that he has come from an icy Alpine world to deliver it: 'I carried with me to the Alps this year the heavy burden of this evening's work.'[32] Several pages of personal recollections follow, which, as the letter that Maxwell sent to Tait in November 1870 shows, the two friends regarded as an object of fun: 'As for T″ who took his Bain to gnaw in the Alps along with the Farbenlehre he cured his distress by applying to the Sortes Bainales. Send his lecture.'[33] The first line of the '*Tyndallic Ode*' evidently renews this private joke between the poem's author and its dedicatee.

'I took with me two volumes of poetry,' Tyndall recalls early in his discourse, 'Goethe's "Farbenlehre", and the work on "Logic" recently published by Mr Alexander Bain.' In November 1870 he added a footnote to this passage, in which he replies to a critic who 'does not see the wit' of describing Goethe and Bain as poetry: 'Nor do I', he declares archly. Stimulated by these books, and applying a familiar arachnoid trope for artistic production, he conceives of his discourse similarly as a work of poetry and imagination; 'all that remained to me was to fall back upon such residues as I could find in the depths of my consciousness, and out of them to spin the fibre and weave the web of this discourse'. Paralleling Tyndall's oppositional relation to the North British mathematical physicists, Goethe is appreciated as a Promethean romantic 'genius' whose work 'broke itself in vain against the philosophy of Newton'. Tyndall was, however, most engaged by the first volume of Bain's *Logic*, which had been published earlier in 1870. Indeed, he finds in it something of a 'self-help' book:

'The uncertainty where to look for the next opening of discovery brings the pain of conflict and the debility of indecision.' Such was my precise condition in the Alps this year; in a score of words Mr. Bain has here sketched my mental diagnosis.

Defining logic as 'the Science of the Laws of Thought', Bain's book opens with a long disquisition on psychology, the field in which he made his name as the author of *The Senses and the Intellect* (1855) and, from 1876, the founding editor of the journal *Mind*. Tyndall likens himself to 'a sick doctor', and a further extract from Bain to the 'sorely needed ... prescription of a friend':

He said, 'Your present knowledge must forge the links of connection between what has been already achieved and what is now required.' In these words he admonished me to review the past and recover from it the broken ends of former investigations. I tried to do so. Previous to going to Switzerland I had been thinking much of light and heat, of magnetism and electricity, of organic germs, atoms, molecules, spontaneous generation, comets, and skies. With one or another of these I now sought to re-form an alliance, and finally succeeded in establishing a kind of cohesion between thought and Light.[34]

Despite being like a newspaper horoscope, accommodatingly vague yet nonetheless prescriptive, the statement Tyndall extracts from the *Logic* is representative for Bain's characterisation of the mind as actively associationist, a psychology that he develops from J. S. Mill and endeavours to ground scientifically in physiology and empirical method. Although Bain explicitly repudiates materialism, by effectively explaining mental activities as material processes his work was easily and widely identified with this position. This ideological affiliation informs Maxwell's punning reference to Bain as a bone for Tyndall 'to gnaw', and makes him an object of derision here.

At the time Maxwell wrote his letter to Tait, Bain's *Logic* was notorious among their circle for its conflations of energy with force and momentum,[35] which Robertson Smith had recently noticed. On 2 November Tait devoted a section of his opening address to the University of Edinburgh, which was published a month later in *Nature* as 'Energy, and Prof. Bain's Logic', to correcting Bain's confusion of these fundamental concepts.[36] Hence Maxwell's quibbles on the 'Sortes Bainales', in which the sorts or categories Bain presents in his book are dismissed as *banal*, ideas that offer Tyndall releases or exits (*sorties*) from his 'distress', a cure that he discovers in the Alps, like a course of immersions at the thermal baths (*bains*).

In reading Bain, Tyndall comes to understand his malaise as a crisis of cohesion amongst his ideas, a failure to establish necessary relations between the elements he catalogues, principles that constellate loosely in his doctrine of Scientific Materialism. Indeed, the task he sets himself of connecting 'thought and Light' is, as the following analogy he gives for it indicates, a synecdoche for justifying his grand system: 'How, for example, are we to lay hold of the physical basis of light, since, like that of life itself, it lies entirely without the domain of the senses?' His answer is 'the power of Imagination'.[37]

Tyndall's evening discourse describes a familiar romantic pattern. It opens with its hero anguished and existentially isolated amidst sublime nature, akin to the Alps of Percy Bysshe Shelley's 'Mont Blanc' (and Mary Shelley's *Frankenstein*), only to find help, like the persona of Coleridge's 'Dejection; an Ode', through 'what nature gave me at my birth, / My shaping spirit of Imagination':[38]

The scientific imagination, which is here authoritative, demands as the origin and cause of a series of ether waves a particle of vibrating matter quite as definite, though it may be excessively minute, as that which gives origin to a musical sound. Such a particle we name an atom or a molecule.

Just as musical sound resonating through the medium of the air can be traced to visibly vibrating materials, such as violin strings in a pizzicato, particular wavelengths of light, which register in eyesight as specific colours and were believed to be propagated by a luminiferous ether, are, Tyndall argues, understood by the scientific imagination to have their source in the vibrations of imperceptible atoms of matter. In making such connections the imagination is, according to Tyndall, guided by the principle of the continuity or unity of nature, which he outlines in defending his materialist cosmology at the conclusion of the discourse:

Those who hold the doctrine of Evolution ... regard the nebular hypothesis as probable, and in the utter absence of any evidence to prove the act illegal, they extend the method of nature from the present into the past. Here the observed uniformity of nature is their only guide. Within the long range of physical enquiry, they have never discerned in nature the insertion of caprice. Throughout this range the laws of physical and intellectual continuity have run side by side. Having thus determined the elements of their curve in a world of observation and experiment, they prolong that curve into an antecedent world, and accept as probable the unbroken sequence of development from the nebula to the present time.[39]

THE TYNDALLIC 'FANCY SCIENTIFIC'

The concluding stanzas to 'A Tyndallic Ode' that Maxwell reserved for Tait parody Tyndall's efforts to systematically gather together the disparate elements of his thought through the agency of the imagination:

> Here let me pause, these passing facts –
>   These fugitive impressions
> Must be transformed by mental acts
>   To permanent possessions.
> Then summon up your grasp of mind –
>   Your fancy scientific,
> That sights and sounds, with thoughts combined
>   May be of truth prolific.
>
> Go to! prepare your mental bricks,
>   Bring them from every quarter,
> Firm on the sand your basement fix
>   With best asphaltic mortar.
> The pile shall rise to heaven on high
>   To such an elevation
> That the swift whirl with which we fly
>   Shall conquer gravitation.[40]

Tyndall's invocation of the imagination in the 1870 discourse is presented as a desperate *deus ex machina* to save his grand hubristic folly, the speculative system of Scientific Materialism. Figuring this doctrine as an edifice composed of 'passing facts' and 'fugitive impressions' from the lecture demonstrations of the earlier stanzas, Maxwell voices the North British perception that Tyndall's lectures at the Royal Institution served to promote Metropolitan materialism to an influential audience and readership. This offers a decisive reason why he and Tait made them the particular focus of their campaign against popular science, whilst, for instance, ignoring Pepper's even more sensational and popular lectures to lower middle class audiences at the Royal Polytechnic Institution.

Following from the analogy he makes of the Egyptian to the salt pyramids, of one 'illustration of building power to another of a different kind', Tyndall often employs the trope of 'molecular architecture' in the Norwich address. Such minute architecture and the larger structures it engenders can, he insists, be grasped imaginatively, so that, as he declares in a revised version of the Liverpool discourse, 'With accurate experiment and observation to work upon, Imagination becomes the architect of physical theory.'[41] Maxwell's trope construes such imagery sceptically through Descartes's hallmark metaphor of architectural foundations for thought: 'Firm on the sand your basement fix.' Tyndall assembles his cosmology from 'the broken ends of former investigations', elements that Maxwell's poem renders through the prescription made in Genesis 11.3, which describes the construction of the tower of Babel from baked bricks and asphalt mortar, as discrete 'mental bricks'. This model of atomism also recalls the flying bricks, 'Atoms, attracted by some law occult', of James and Horace Smith's parodic Lucretian account of building the new Drury Lane theatre, 'Architectural Atoms', in their hugely popular and much reprinted *Rejected Addresses*: 'I sing how casual bricks, in airy climb, / Encounter'd casual cow-hair, casual lime.'[42]

While the four-stanza version of the '*Tyndallic Ode*', published in *Nature* to the 'Tune' of 'The Brook', affably makes fun of Tyndall's prodigious popular lecturing as a ceaseless babbling, the additional stanzas for Tait harshly accuse their opponent of creating Babel, constructing a huge self-aggrandising system that confounds the language and practice of professional physics. Belonging to the postdiluvian generations of the new Victorian professional science inaugurated by the BAAS, the fashionable lecturer is identified in the poem with the self-promoting descendants of Noah, who in building their tower declare; 'let us make us a name, lest we be scattered abroad upon the face of the whole earth'.[43]

Tyndall argues in his discourse that as the 'fundamental conception' that the mind forms *a priori* (he is thinking of the luminiferous ether) has 'actually forced upon our attention phenomena which no eye had previously seen … but always lands us on the solid shores of fact', then it 'must, we think, be something more than a mere figment of the scientific fancy'.[44] Unsurprisingly Maxwell refuses this inference. His poem pointedly identifies the Tyndallic faculty, by which 'passing facts' and 'fugitive impressions' are 'transformed by mental acts / To permanent possessions', with the trivial 'fancy scientific'. Parodying the molecules and molecular forces that Tyndall sees his scientific imagination to vouchsafe, the 'molecules with fierce desires' described in the second stanza of the '*Ode*' accordingly represent his 'region where things are intellectually discerned' as a place of lewd fantasy. Maxwell considers Tyndall's 'Scientific Imagination' to be a solipsistic failure, whereby the authority attributed to the faculty serves to reify its fancies.

Rather than 'land[ing] us on the solid shores of fact', Tyndall's ether principle corresponds to 'the sand' in Maxwell's poem, upon which the Tyndallic tower, the preposterous 'pile' of Scientific Materialism, rests precariously. This propagating medium, which Tyndall's imagination 'demands', discloses the existence of the atoms and molecules that constitute the 'basement', the foundation of his system. Maxwell was thinking and writing about Descartes's philosophy in early 1873, and his development of this Cartesian metaphor may have been suggested by the *Discourse on Method*, where the 'firm and solid *foundations*' of mathematics are contrasted with 'the moral writings of the ancient pagans', which Descartes likens 'to very proud and magnificent palaces built only on sand and mud'.[45] In Maxwell's poem, the protagonist's theatrical lecture demonstrations give way to – or, more radically, the 'hot embraces' of 'molecules with fierce desires' build deterministically into – the dizzying mental dervishes of a system that defies Newtonian law, repudiates the foundations of professional British science: 'the swift whirl with which we fly / Shall conquer gravitation'. It illustrates 'how', as Maxwell puts it in 1860, 'the unscientific mind has been led from one error to another up to the very pinnacle of absurdity'.[46]

Tait concurs with Maxwell's critique of Tyndall's principle. The main body of 'Imagination in Science', his review essay of Tyndall's book, illustrates the dangers Tait sees to flow from the 'Scientific Imagination' with a recent medical 'communication' to a daily newspaper, 'an account of yellow fever from the imaginative side'. Just as Tyndall is satisfied with the conviction he draws from the imagination that unseen vibrating atoms are the causal agent determining particular wavelengths of light, so similarly:

> Dr Cochrane ... tells us candidly that he states only 'what he believes but does not know', and then takes his flight into the unknown. He imagines 'the yellow fever poison to be composed of living germs in innumerable number, living organisms of inconceivable minuteness, which eat, and drink, and multiply their generations under the sun' ... without paying any attention to facts regarding yellow fever and other diseases which are left untouched by any extant doctrine, he tells us truly that 'the visions of modern science are more wonderful than the visions of Eastern fable'.[47]

Cochrane reads yellow fever as an orientalist tale of excess, a thousand and one febrile nights of microbial high-life. Tyndall's principle of the imagination is seen to provide a rationale for the raciness that Tait identifies with Royal Institution lectures and other popular science in his contemporary essays on 'Sensation and Science'.

The newspaper report on yellow fever highlights press interest in sensational medical stories. Maxwell and Tait's concern over the power of 'bad science' to engage the public imagination finds expression in some verses they wrote about another such health story. The newspaper-driven hysteria that met an 1871 proposal to supplement Edinburgh's water supply by drawing upon St Mary's Loch led Maxwell to write his light poem 'The F.R.S.E.' for Tait, who later added to the manuscript a few lines of his own. The controversy began when *The Scotsman* newspaper received a letter signed by 'A Physician', which claimed that if water from the lake were introduced into the water supply, public health would be undermined by the pernicious effects of the common water insects it contained. Specialist doctors and scientists from Edinburgh and London worked to counter such alarmist warnings. Nevertheless, popular opposition to the scheme only increased as another anonymous newspaper correspondent, who signed himself 'NATURALIST', further inflamed public feelings against the scheme and focused them upon a single minuscule foe, the common water flea *Daphnia pulex*. A contemporary witness to these events, Judge James Colston, records that,

> The opponents of the measure had, to all appearance, secured the public ear. The newspapers, as a rule, were on their side; and no step that could be taken to produce a popular clamour was left unadopted. The walls [of Edinburgh] were placarded with huge representations of that celebrated flea *Daphnia pulex*.[48]

Evidently springing from his amusement at finding the dangerous *Daphnia* hypothesis endorsed by a Fellow of the Royal Society of Edinburgh, Maxwell's poem turns upon a quibble on the letters after the anonymous

Fellow's name (the abbreviation being read here as an acronym) and the biblical Pharisee:

> The F.R.S.E.
> Where Wordsworth's Swan was wont to float
> The man of Science from his boat
> In flasks and phials carefully
> Collects the lively water Flea.
> Then analyses with great pains
> The water which from Flea he drains.
> Water and Fleas! The trout below
> Delights when through his gills they flow
> He, too, the precious mixture drains
> Water ejects and Fleas retains.
> The Edinburgh Pharisee
> Receives both water, fish and Flea
> The water Flea he filters out
> Then, unsuspecting, eats the trout. $\frac{dp}{dt}$
>
> That trout indeed is wondrous bad
> Enough to drive the eater mad
> For 'tis but concentrated Flea
> Flea smaller animalculae
> These feed on spores and deadly germs
> With which their stomachs come to terms
> Think! Edinburgh Pharisee!
> How they will work their will on thee!   G[uthrie Tait].[49]

The opening line of Maxwell's poem introduces St Mary's Loch obliquely through Wordsworth's 'Yarrow Unvisited', which preserves 'The swan on still St. Mary's Lake' as a purely imaginary vision:

> Float double, swan and shadow!
> We will not see them; will not go,
> To-day, nor yet to-morrow;
> ...
> We have a vision of our own;
> Ah! why should we undo it?[50]

The Loch full of killer water fleas is accordingly alluded to with delicate wit as a fiction that refuses empirical testing, a flight of fancy existing only in the mind of the 'Edinburgh Pharisee' and the excitable imagination of a scientifically illiterate public.

THE TYNDALLIC SUBLIME

Maxwell's terse caricature of the Tyndallic 'fancy scientific' can be answered with the personal imaginative vision of Tyndall's poem 'A

Morning on Alp Lusgen'. At home in 'Alp Lusgen', the Swiss chalet that Tyndall and his wife Louisa had built after their marriage in 1876, the persona of the poem meditates on the phenomenal world before him. While the poem was published in 1892, the year before Tyndall's death, a group of untitled manuscript drafts indicate that he was working on it in 1878.[51] Unlike the later part of the poem, the opening lines, which are taken here from the published version, were not greatly changed in the drafting process:

> The sun has cleared the peaks, and quenched the flush
> Of orient crimson with excess of light.
> The tall grass quivers in the rhythmic air
> Without a sound; yet each particular blade
> Trembles in song, had we but ears to hear.
> The hot rays smite us, but a quickening breeze
> Keeps languor far away. Unslumbering,
> The soul enlarged takes in the mighty scene.[52]

Nature is presented as an energy plenum. Its sublimity is theorised scientifically as the full glaring spectrum dissolves the reflective effect of the crimson flush, while its 'hot rays' and the 'quickening breeze' indicate the grand thermodynamic drama of Alpine weather. Gesturing beyond the humanly perceptible range of light radiation, the 'excess of light' parallels the song of each blade of grass that defies our powers of hearing. The dynamism of these blades of grass is concentrated synaesthetically, as having not 'ears to hear' it we are encouraged to imagine their trembling song by analogy with their quivering appearance, much as the 1870 discourse makes the analogy of sound with light, and the scientific imagination 'demands as the origin and cause of a series of ether waves a particle of vibrating matter'. The propagating medium, 'the rhythmic air', allows the imagination to grasp a scientific truth that is not amenable to our limited senses, while the synaesthetic apprehension of the blades of grass likens them to Coleridge's Eolian Harp, an imaginative power that yields 'A light in sound, a sound-like power in light'.[53]

The opening lines of the poem introduce nature as a Carlylean venue for romantic pansemiosis, in which all is radiant and expressive, pushing against and enlarging the limits of the beholder's consciousness in its 'excess of light' and unheard dynamism. Carlyle's *Sartor Resartus*, which pursues the nature and functions of signs through the example and trope of clothing, urges perseverance beyond mediate forms, not only words but other such customs and habits, including science and our conceptions of time and space, and indeed all phenomena, in order to appreciate the noumenal efflorescence of the world, a 'natural supernaturalism'. Sharing this pitch of appreciation for nature, Tyndall's poem resonates with imagery from the chapter on 'Natural Supernaturalism' in *Sartor Resartus*. His 'excess of light' can be matched with Carlyle's image of luminous superabundance, 'Light-sea of celestial wonder', and his trembling grass compared to 'every grass-blade' through which 'the glory of a present God still beams'.[54] The published version of 'Alp Lusgen' also invokes Carlyle directly and elegiacally, as the 'sorrowing shade of him, who preached through life / Obedience to the Highest!'[55]

Tyndall consistently refers to Carlyle, Ralph Waldo Emerson and Fichte as the decisive influences upon the formation and development of his intellectual life. He first read Carlyle in 1843,[56] heard Emerson lecture in January 1848, and in August of that year began to read Fichte, whose thought he studied along with other German philosophy at Marburg.[57] He writes that, in consequence of his time in Germany, his 'whole life' was 'rendered more earnest, resolute, and laborious by the writings of Carlyle. Others also ministered to this result. Emerson kindled me, while Fichte powerfully stirred my moral pulse.' 'These three unscientific men', he declares, 'made me a practical scientific worker.'[58]

Huxley wrote to Charles Kingsley in 1860 that '*Sartor Resartus* led me to know that a deep sense of religion was compatible with the entire absence of theology.'[59] Tyndall similarly finds no conflict between science and Carlylean romanticism, an understanding he tried to convey to Carlyle himself, seeking 'among other things, to remove all prejudice by making clear to him the spirit in which the highest scientific minds pursued their own work'.[60] Carlylean natural supernaturalism allows Tyndall to both sustain his materialist science and maintain a romantic awe of nature. The manuscript notebook that contains both the early drafts of his poem and some recollections of Carlyle is bound together with another, in which Tyndall argues that religion proceeds from the same impulse as poetry, although in the former 'the flowing forces of the soul are thrown into dogmatic forms'. He accordingly refers synonymously to 'poets' and 'the priesthood', whilst distinguishing and condemning the latter for having 'petrified what ought to have remained in solution and based upon external evidence that which sprung solely from the innermost nature of man. It is against this objective rendering of the emotions – this thrusting into the region of facts and knowledge, of conceptions essentially poetic – that science, consciously or unconsciously, wages war.' He then quotes a passage from his 'Apology for the Belfast Address' in which he declares that Genesis '*is a poem, not a scientific treatise*'.[61] 'A Morning on Alp Lusgen' was evidently conceived of as an exemplary expression of the religious impulse, 'the flowing forces of the soul', in which '[t]he soul enlarged takes in the

John Tyndall and 'the Scientific Use of the Imagination' 161

mighty scene' of phenomenal but nonetheless sublime nature. It can be compared to the atheist Shelley's rejoinder to Wordsworth in 'Mont Blanc', which opens with 'The everlasting universe of things' represented as a power that exists through our consciousness, as it 'Flows through the mind, and rolls its rapid waves'.[62] A similar strain of Romantic ideology also informs Tyndall's 1870 discourse, where he declares that without our 'power of imagination' 'the soul of Force would be dislodged from our universe'.

'A Morning on Alp Lusgen' shares with Tyndall's science an historical interest in the origin and development of the natural world. The published version of the poem finds science no more able to answer the question of physical nature's origins than the 'dogmatic forms' of Christian Gnosticism:

> Steep fall the meadows to the vale in slopes
> Of freshest green, scarred by the humming streams,
> And flecked by spaces of primeval pine.
> Unplanted groves! whose pristine seeds, they say,
> Were sown amid the flames of nascent stars –
> How came ye thence and hither? Whence the craft
> Which shook these gentian atoms into form,
> And dyed the flower with an azure deeper far
> Than that of heaven itself on days serene?
> What built these marigolds? What clothed these knolls
> With fiery whortle leaves? What gave the heath
> Its purple bloom – the Alpine rose its glow?
> Shew us the power which fills each tuft of grass
> With sentient swarms? – the art transcending thought,
> Which paints against the canvas of the eye
> These crests sublime and pure, and then transmutes
> The picture into worship? Science dumb –
> Oh babbling Gnostic! cease to beat the air.
> We yearn, and grope, and guess, but cannot know.[63]

Consistent with the nebular hypothesis, the fiery origin of the universe offers a temporal correlate here for the first appearance of the ancient conifers, 'Unplanted groves!' that have no need of an anthropomorphic principle to initiate their being. In contrast to Maxwell's leering and trenchant account of it, Tyndall's radical thesis that atoms articulate themselves into organic forms is posed uncertainly by the poem as a question that it asks directly of the gentian, and then diffuses in the more conventionally poetic references to the other flowers. The Carlylean legacy is sustained here in the rhetoric of nature's phenomenal clothing, the anthropomorphic analogies of crafting, building and dyeing through which it presses its questions about the origin of the natural world, an apparently ingenuous rhetoric of Design that implies that the world was carefully handmade. The 'art transcending thought' registers on the retina as art, a painting, and then as the mysterious object of worship. We are seen to have no other ways of apprehending the noumenal truth of the Alpine scene than 'These crests sublime and pure', mountain peaks that, as though painted on the eye, are like heraldic crests that hold out the promise of a clear lineage back to a distinguished origin. Faced with this ultimate question, Science is struck dumb, while Christianity becomes garrulous, its utterances depicted here as meaningless acoustic vibrations that assault 'the rhythmic air'. An article from 1860, originally entitled 'Physics and Metaphysics', which closes the 1870 volume on *Imagination in Science*, ends on a similar note: 'Whence come we; wither go we? The question dies without an answer – without even an echo – upon the infinite shores of the Unknown.'[64]

The published poem and its earlier drafts bifurcate at this point. While the question of nature's creation and apparent design is seen in the final version to be unanswerable, the early drafts excitedly announce a first principle:

> Ah weary head! the answer is abroad
> Buzzing through all the atmosphere of mind.
> 'Tis Evolution! East, West, North and South
> From droughty sage and spinster shrill we learn
> 'Twas Evolution! When that word has spread
> Its magic to the limits of the world
> Till its reverberation has become
> A lullaby – how sweet 't will be to doze
> Over thy emptied cup of nectar'd sweets
> Divine Philosophy! To doze in peace.[65]

With his 'sky in a bottle' experiment, Tyndall famously debunked the traditional hierarchical trope of an ethereal atmosphere, a heavenly blue sky that transcends the base matter of the earth, by explaining its colour as the effect of light polarised by suspended particles. He claims such imagery for his materialism, most famously in 'the infinite azure of the past'[66] with which the Belfast address closes, but also in his poem's atomist characterisation of the gentian flower with its 'azure deeper far / Than that of heaven itself'. Analogous to the dynamic material medium, 'the rhythmic air', surrounding the earth, the 'atmosphere of mind' is seen to propagate thought, the 'Buzzing' answer to the poet's question.

Darwin's hypothesis, indeed its enchanting 'magic', is envisaged by the poem in physicalist terms of the material word 'Evolution!' that reverberates throughout the 'atmosphere of mind', chiming in with its perpetual creation, a secular version of the Logos. 'In the beginning was the word, in the end the world without end.'⁶⁷ The 'Buzzing' of the Darwinian bee, while still disturbing for some, is presented as a naturalistic music of the spheres, assuring mankind that the garden of nature is being tended and fertilised incessantly, without the need of human, let alone divine, intervention, much as in the Norwich address the 'salt-pyramids' can be trusted to form themselves, the molecules each having 'their inner mechanism, and their store of force'. The poem's metaphor also implies the fertilisation and hence flourishing of human thought, akin to the 'Entrancing truths from bee-touched thoughts' that Tyndall's friend Henrietta Huxley, wife of T. H. Huxley, describes in her poem 'A Wish'.⁶⁸ Once the eschatological resonances of 'that word', are universally recognised, so that there are no longer any discordant notes ringing through 'the atmosphere of mind', buzzing 'Evolution!' will have 'become / A lullaby'. The historical quest of 'Divine Philosophy', the dialectical footing upon which Hegel establishes the historical study of philosophy for the nineteenth century, will find its final fulfilment and rest.

CHAPTER 7

'Molecular Evolution': *Maxwell, Tyndall and Lucretius*

While Tyndall's 'A Morning on Alp Lusgen' offers a useful corrective to the flippant renditions of his cosmology that Maxwell threw to the Red Lions, its encompassing principle of 'Evolution', which the drafts of the poem ground in 'gentian atoms' and 'the flames of nascent stars', is nevertheless described more completely by the phrase 'Molecular Evolution', the title Maxwell gave to two further poems about the Metropolitans' system. The term 'evolution' gained currency not only through the public controversy that followed the publication of Darwin's *Origin of Species* in 1859, but also through a revival of interest in the nebular hypothesis. Paving the way for this expansive usage, the anonymous and much-reprinted *Vestiges of the Natural History of Creation*, which Maxwell read in 1853,[1] facilitated its sweeping speculative narrative by marrying the nebular hypothesis with Lamarckian evolution. Spencer, whom Maxwell describes in the wake of the Belfast address as 'the Apostle of Evolution', had conferred with Tyndall to write an influential essay on 'The Nebular Hypothesis' for *The Westminster Review* in July 1858.[2] Like Tyndall, Spencer argues that 'the doctrine of evolution in its widest sense sets out with that state of matter and motion implied by the nebular hypothesis'.[3] Combining this idea with Darwinism in his Synthetic Philosophy, he furnishes an inclusive evolutionary panorama that, he recalls in 1895, was intended to 'set out with nebular condensation and end with social phenomena':[4] the full historical span of the Metropolitan cosmology familiar from Tyndall's Norwich address and Liverpool discourse.

Suggestively conflating atomist physics and biological evolution, the phrase 'Molecular Evolution' follows directly from Maxwell's account of Tyndall's materialism in '*A Tyndallic Ode*' and the Liverpool address. Taking his cue from Tyndall's characterisation of matter as 'Incipient life', Maxwell represents his molecules anthropomorphically. They accordingly display 'the laws of their existence' by 'clash[ing] together in fierce collision' and engaging in 'fierce' and 'hot embraces'; complementary behaviours that suggest the twin Darwinian exigencies of the competitive struggle for survival and the sexual drive to reproduce. Similarly, Maxwell's use of 'Molecular Evolution' as the alternative title for his 1874 'Song of the Cub', a young Red Lion's impressions of the Belfast address, clearly identifies the phrase with Tyndall's doctrine of Scientific Materialism.

Maxwell's 'Molecular Evolution' poem from 1873, which has historical precedence over the other but only one name, is the focus of the following chapter. Not simply a factional lampoon, its scepticism is diffuse, and its satire lighter and more playful than that of other late poems. Indeed 'Molecular Evolution' recalls the epistemological speculations of the 1850s, with their cautionary principle that 'the whole framework of science' may be nothing more than 'a natural growth on the inner surface of the mind',[5] as it recognises and addresses the efforts of all British Asses and Red Lions, Metropolitans and North Britons alike, to make scientific discoveries and strive for coherent theories of the physical world.

'FORTUITOUS EMBRACES'

'Molecular Evolution' is one of two public statements on Metropolitan materialism that Maxwell made at the Bradford meeting of the BAAS in September 1873. Written for the Red Lions, 'Molecular Evolution' offers a set of unruly observations that complement the more sober views on the subject that Maxwell elaborates in the 'Discourse on Molecules', which was presented to the delegates in their alternate incarnations as British Asses. The lecture furnishes a particular gloss for the phrase 'Molecular Evolution', which renders it preposterous: 'No theory of evolution can be formed to account for the similarity of molecules, for evolution necessarily implies continuous change, and the molecule is incapable of growth or decay, of generation or destruction.'[6] 'Molecular Evolution' parodies the confident system that Tyndall extrapolates from his hypothetical molecular forces with a fluctuant metaphysic that Maxwell develops from similarly arbitrary premises. To do this he invokes Lucretius' great Latin poem on atomism from the first century BC, *De Rerum Natura* (*On the Nature of Things*).

Maxwell's poem opens with the Lucretian account of the creation of the universe:

> At quite uncertain times and places,
> The atoms left their heavenly path,

> And by fortuitous embraces,
> Engendered all that being hath.

The Lucretian story begins with the atoms in 'their heavenly path', raining down in a laminar flow that is finally disrupted by the *clinamen*, in which some atoms 'swerve a little from their equal poise: you just and only just can call it a change in inclination'.[7] The effect of this slight swerving is compounded as the atoms collide with others and rebound to cause further collisions, and through them aggregations that form things, indeed whole worlds.

By tracing the beginnings of the universe to a coalescence of elements, the Lucretian account in the first octave of 'Molecular Evolution' suggests a correlate for the materialist cosmology that Tyndall builds from the nebular hypothesis and his self-organising atoms. Tyndall traces the nebular hypothesis to both Laplace and Kant in his discourse on the scientific imagination. While Laplace's emphasis upon determinism is definitive also for Metropolitan materialism, Kant's earlier formulation of the nebular hypothesis, in which a vast mass of evenly distributed particles form themselves into the universe through polar forces of attraction and repulsion, is consistent with the romantic inspirations and ambitions of Tyndall's scheme. Some lines he cites from Emerson in his essay on 'James Martineau and the Belfast Address' serve to highlight this affinity: 'The journeying atoms, primordial wholes / Firmly draw, firmly drive by their animate poles.'[8]

In writing 'Molecular Evolution', Maxwell evidently has in mind the grand speculative systems of German romantic science that he and his peers associated with Tyndall, who had established a reputation amongst them as a 'Germaniser' through his public championing of Mayer over Joule in the early 1860s. More recently, Tyndall's use of German romantic sources in his discourse had attracted discussion in the press, with the *Pall Mall* attacking him on 20 September 1870 for his use of a quotation from Goethe's *Faust* to illustrate his materialist hypothesis. The *Spectator* came to his defence on 24 September, arguing that, on the contrary, Tyndall had not gone far enough in drawing the parallel between his cosmology and Goethe's. Indeed it made its point with 'A Translation of Goethe's Proemium to Gott und Welt', by J. A. Symonds, which Tyndall evidently agreed offered an apt evocation of his system, as he included it in the second edition of the lecture, the version that Tait reviewed and lent to Maxwell.

Tyndall's discourse on the imagination offered Maxwell ample cues to identify it with the totalising schemes of German romantic science and treat it as an object of fun, much as he does Oken's philosophy in an 1874 letter to Tait.[9] Indeed the system that Oken elaborates in his *Elements of*

Physiophilosophy (1833–45) is especially suggestive as an outlandish parallel to Tyndall's. Following Kant, it begins with polar principles of attraction and repulsion that form the sun and planets, from which proceed successive polarities of light and heat, electricity, magnetism, chemical action and, most momentously, a primal slime from which all forms of life arise in a chain of being that culminates with man. Attributing Tyndall's atoms with polar principles of attraction and repulsion in the Liverpool address, Maxwell similarly 'dowers them with love and with hate' in his poem 'British Association, 1874'.[10] 'Molecular Evolution' toys with such grand mythic schemes, as the poet, allowing himself complete licence for speculation, draws from an atomist premise his apriorist cosmology and theory of discovery. In doing so the poem burlesques the current contest between the North Britons and the Metropolitans for the right to interpret and claim as their own the Lucretian creation story, to seize it as a speculative precursor to (and classical pedigree for) the respective physics they were developing during the late 1860s and early 1870s.

In the following extract from his Belfast address, Tyndall describes the clinamen mechanistically, and Lucretius' atoms as rebuffing gods and the principle of Design. Their molecular evolution is ensured by Spencer's principle of 'the survival of the fittest':

The mechanical shock of the atoms being in his view the all-sufficient cause of things, he combats the notion that the constitution of nature has been in any way determined by intelligent design. The interaction of the atoms throughout infinite time rendered all manner of combinations possible. Of these the fit ones persisted, while the unfit ones disappeared. Not after sage deliberation did the atoms station themselves in their right places, nor did they bargain what motions they should assume ... 'If you will apprehend and keep in mind these things, nature, free at once, and rid of her haughty lords, is seen to do all things spontaneously of herself, without the meddling of the gods.'[11]

Later in the address, Tyndall shapes an imaginary dialogue between Bishop Butler and 'a disciple of Lucretius', who provides a voice for his own views. Tyndall's Lucretian credentials are clearly asserted by the author, as the Bishop forthrightly informs the proxy figure of his identity and creed: 'You are a Lucretian, and from the combination and separation of insensate atoms deduce all terrestrial things, including organic forms and their phenomena.'[12]

The quotation that Tyndall makes at the close of the passage cited above is from H. A. J. Munro's critical edition of Lucretius. Tyndall writes to Hirst on 5 September 1868 that Munro's translation 'fell into my hands a few days ago', a remark that, if taken literally, dates his acquaintance with the text to about two weeks after the Norwich address.[13] First published in 1860, and from 1864 with a supplementary volume of translation and further commentary, Munro's edition was, as Turner observes, decisive in shifting the prevailing Victorian characterisation of Lucretius from poet to scientist.[14] Fleeming Jenkin wrote an influential review essay on Munro's revised second edition (1866) for the *North British Review* in 1868, 'The Atomic Theory of Lucretius', which works to restore the poet's scientific reputation by comparing his ideas with recent advances in physics: 'we may profitably consider what the real tenets of Lucretius were, especially now that men of science are beginning, after a long pause in the inquiry, once more eagerly to attempt some explanation of the ultimate constitution of matter'.[15] While Maxwell knew Lucretius' text from his early studies in Latin and Hamilton's metaphysics class at Edinburgh,[16] the phrase he uses for the clinamen in the first line of his poem echoes those that Jenkin employs in his essay on Lucretius, 'at quite uncertain times and uncertain places', and cites from Munro's translation, 'at quite uncertain times and uncertain points of space'.[17] Maxwell corresponded with Jenkin about his essay,[18] and it is likely that the two discussed the Roman poet through their reading of Munro's edition.

Maxwell was drawn to Lucretius' scientific speculations through his work on the kinetic theory of gases. He begins his 1866 paper 'On the Dynamical Theory of Gases' by tracing precursors to this theory. Less sure of the anticipations of the theory he discerns in Lucretius than those he has traced in the history of modern physics, Maxwell writes to Munro for advice early in 1866: 'With respect to those who flourished since the revival of science I can make out pretty well what they really meant but I am afraid to say anything of Lucretius because his words sometimes seem so appropriate that it is with great regret that one is compelled to cut off a great many marks from him for showing that he did not mean what he has already said so well.'[19] Although Munro's reply has not survived, the account he gives of Lucretius in 'On the Dynamical Theory of Gases' indicates that Maxwell found it encouraging:

The opinion that the observed properties of visible bodies apparently at rest are due to the action of invisible molecules in rapid motion is to be found in Lucretius. In the exposition which he gives of the theories of Democritus as modified by Epicurus, he describes the invisible atoms as all moving downwards with equal velocities, which, at quite uncertain times and places, suffer an imperceptible change, just enough to allow of occasional collisions taking place between the atoms. These atoms he supposes to set small bodies in motion by an action of which we may form some conception by looking at the motes in a sunbeam. The language of Lucretius must of course be interpreted according to the physical ideas of his age, but we need not wonder that it suggested to Le Sage the fundamental conception of his theory of gases, as well as his doctrine of ultramundane corpuscles.[20]

This passage includes Maxwell's earliest use of the phrase 'at quite uncertain times and places', which may have been prompted by Munro's use of it in his lost reply, or in his translation ('at quite uncertain times and uncertain points of space'[21]). Occupying the opening line of 'Molecular Evolution' and occurring insistently at many points in Maxwell's later writings, as well as in Jenkin's essay, the phrase is central to the North British reading of Lucretius, for on it turns their polemical use of his poem as a defence of free will against the deterministic construal of atomism that Tyndall and his peers were promoting.

Maxwell begins some notes he wrote for Thomson 'on the History of the Kinetic Theory of Gases' with Lucretius, explaining that the 'irregularity of the deflexions of the atoms [is] introduced to account for free will &c. This is very important in T[itus] L[ucretius] Carus.' He supports his observation with the following extract from the poem, which is rendered here in Munro's translation:

Wherefore in seeds too you must admit the same, admit that besides blows and weights there is another cause of motions, from which this power of free action has been begotten in us, since we can see that nothing can come from nothing. For weight forbids that all things be done by blows through as it were an outward force; but that the mind itself does not feel an internal necessity in all its actions and is not as it were overmastered and compelled to bear and put up with this, is caused by a minute swerving of first-beginnings at no fixed part of space and no fixed time.[22]

Lucretius' repudiation of force as the ultimate principle governing the physical world, a principle that Tyndall establishes at the molecular level, chimes in with the North British ontology that Maxwell celebrates in his verse rendition of Tait's 1876 lecture: 'The Universe is free from pole to pole, / Free from all forces.'

Maxwell explains the Lucretian derivation of free will from the clinamen in his other Bradford presentation, the evening 'Discourse on Molecules'. The account that Tennyson gives of the clinamen in his 1868 poem 'Lucretius' provides the text for Maxwell's discussion:

In his dream of nature, as Tennyson tells us, he

'Saw the flaring atom-streams
And torrents of her myriad universe,
Ruining along the illimitable inane,

> Fly on to clash together again, and make
> Another and another frame of things
> For ever.'

And it is no wonder that he should have attempted to burst the bonds of Fate by making his atoms deviate from their courses at quite uncertain times and places, thus attributing to them a kind of irrational free will, which on his materialistic theory is the only explanation of that power of voluntary action of which we ourselves are conscious.[23]

Maxwell sees the clinamen to entrench a principle of free will at the most radical level of physical reality.

Maxwell makes a further telling reference to Lucretian chance in the draft version of his *Encyclopaedia Britannica* article on the 'Atom', which P. M. Harman dates to September 1874, a year after the Bradford meeting:

According to [Lucretius'] description of the doctrine of Democritus the atoms are all in motion in a downward stream with an enormous speed ... At quite uncertain times and places these atoms are deflected from their vertical path. This causes them to jostle one another and by their fortuitous concourse they form visible bodies.[24]

The phrase 'fortuitous concourse' is the English translation of the Latin *concursus fortuitus* that Cicero uses to describe the Lucretian clinamen.[25] Maxwell invokes the phrase in the first stanza of 'Molecular Evolution', where the atoms 'by fortuitous embraces, / Engendered all that being hath'. These are chance encounters, akin to those of the bodies in Burns's 'Comin' thro the Rye' and Maxwell's parody of it. Such 'fortuitous embraces' imply an element of discretion that distinguishes them from the Tyndallic 'molecules with fierce desires' in the '*Ode*' and their 'fierce embrace' in the Liverpool address, where lust furnishes a naturalistic metaphor for deterministic materialism.

ATOMS AND ASSOCIATIONS

Seeing space as endless and atoms to be in constant motion, Lucretius hypothesises various patterns that their interactions can take: 'no rest is given to first bodies throughout the unfathomable void, but driven on rather in ceaseless and varied motion they partly, after they have pressed together, rebound leaving great spaces between, while in part they are so dashed away after the stroke as to leave but small spaces between'. The first of these formations, which is seen to 'furnish us with thin air and bright sunlight', is suggestive for Maxwell's theory of gases. Their complementary substance, 'all that form a denser aggregation when brought together', Lucretius describes as 'held fast by their own close-tangled shapes [*Indupedita fuis perplexis ipsa figuris*]'.[26] The etymology of *indupedita* from *pes* entails the sense that, as Don Fowler notes, 'the atoms are tripped up by their own feet',[27] so that they can be seen to gather and cohere centripetally as vortices. Although this discussion occurs early in Book II, before the clinamen is introduced, it also provides a suggestive model for the consequences of this infinitesimal swerving, of vortices engendered within the laminar flow, an idea that chimes in with North British preoccupations and conceptions of the atom during the late 1860s and the 1870s.

The figure of the vortex is familiar from Maxwell's paper 'On Physical Lines of Force', where he develops his 'molecular vortices' model of the electromagnetic field to argue 'that magnetic force is due to the centrifugal force of small vortices, and that these vortices consist of the same matter the vibrations of which constitute light'.[28] William Thomson develops this incipient North British concern with vortices in his model of the atom itself. Building upon an 1858 paper by Helmholtz that Tait translated in 1867, Thomson characterises atoms as interlocking or knotted rings of tubular vortices, which are composed of, and contained by, an infinite etheric fluid of perfectly even density. Such knots are conceived of as continuous loops, they have no loose ends.

Writing to Helmholtz in January 1867, Thomson describes a simple experimental means that Tait devised of producing smoke rings, the dynamics of which he applies to his correspondent's idea of vortex-rings to yield his model of the atom:

If you try it, you will easily make rings of a foot in diameter and an inch or so in section, and be able to follow them and see the constituent rotary motion ... The absolute permanence of the rotation, and the unchangeable relation you have proved between it and the portion of the fluid once acquiring such motion in a perfect fluid, shows that if there is a perfect fluid all through space, constituting the substance of all matter, a vortex-ring would be as permanent as the solid hard atoms assumed by Lucretius and his followers (and predecessors) to account for the permanent properties of bodies (as gold, lead, etc.) and the differences of their characters. Thus, if two vortex-rings were once created in a perfect fluid, passing through one another like links of a chain, they never could come into collision, or break one another, they would form an indestructible atom; every variety of combinations might exist.[29]

Literally moving beyond static diagrams, and building upon his childhood inventiveness in drawing discs for the phenakistoscope, Maxwell used his improved zoetrope to demonstrate and explore cinematographically three of

Helmholtz's vortex rings as they thread through one another: 'H²'s 3 rings do as the 2 rings in his own paper[,] that is[,] those in front expand and go slower[,] those behind contract and when small go faster and thread through the others.'[30] In his *Britannica* article on the 'Atom', he observes that unlike 'The small hard body imagined by Lucretius, and adopted by Newton', the intrinsically elastic nature of Thomson's vortex-ring model 'account[s] for the vibrations of a molecule as revealed by the spectroscope' (i.e., the hypothesis that Tyndall grounds in the authority of the 'Scientific Imagination'). Indeed the vortex-ring 'satisfies more of the conditions than any atom hitherto imagined'.[31] Jenkin writes in his 1868 article that, rather than hard or even elastic particles, 'Lucretius would have . . . much preferred the theory just started by Sir William Thomson . . . that the elasticity of atoms may be due to the motion of their parts – a proposition exemplified by one smoke-ring bounding away from another in virtue of the relative motions of their parts.'[32]

In his Liverpool address, Maxwell contrasts Thomson's mathematically grounded vortex-ring model with another, rather tautologous, speculative model of the atom: 'If a theory of this kind should be found, after conquering the enormous mathematical difficulties of the subject, to represent in any degree the actual properties of molecules, it will stand in a very different scientific position from those theories of molecular action which are formed by investing the molecule with an arbitrary system of central forces invented expressly to account for the observed phenomena.' He clearly has in mind Tyndall's reductionist hypothesis of molecular force. Such forces are, Maxwell observes, not scientific principles but 'occult properties'.[33] Here, no less than in the poem he wrote 'To the Chief Musician upon Nabla' a year later, he is careful to distinguish professional, that is, mathematical, physics from Tyndallic 'dark science'. Tait cites a description of a similarly 'inferior and auxiliary science' in his 1876 BAAS lecture on Force, which, while not mentioning any names, culminates in a notorious phrase from the Belfast address: 'It will endue matter with mysterious qualities and occult powers, and imagines that it discerns in the physical atom, "the promise and potency of all terrestrial life".'[34]

Maxwell warns against such reductionism in his article on the 'Atom', where he argues that 'molecular science . . . forbids the physiologist from imagining that structural details of infinitely small dimensions can furnish an explanation of the infinite variety which exists in the properties and functions of the most minute organisms'.[35] In contrast to Tyndall's Scientific Materialism, which naturalises mechanistic determinism on a cosmological scale, seeing it to apply to all phenomena from atoms and molecules up to the mind that theorises such schemes, the statistical theory championed by Maxwell and his peers recognises that molecules have different properties from the bodies they compose, and that their mechanistic behaviours can interact in unforeseen ways, a principle that has its emblem in the clinamen. Indeed, even in the early stages of his researches into gases, writing to Thomson in 1857 about Saturn's rings, Maxwell evidently conceived of such phenomena in Lucretian terms: 'The general case of a fortuitous concourse of atoms each having its own orbit & excentricity is a subject above my powers at present, but if you can give me any hint as to the point of attack I will go at it.'[36] By the early 1870s Maxwell had found several ways of addressing this problem.

Through the extrapolations he draws from his statistical theory, which are most boldly illustrated by his hypothetical 'Demon', Maxwell finds the instabilities and variability commonly identified with the workings of complex organisms and human moral character entrenched at the submicroscopic level of atoms and molecules. In an essay on 'Science and Free Will' he delivered in February 1873 to the Eranus Club, a group of his fellow Apostles from the 1850s, Maxwell notes that 'It has been well pointed out by Professor Balfour Stewart, that physical stability is the characteristic of those systems from the contemplation of which determinists draw their arguments, and physical instability that of those living bodies, and moral instability that of those developable souls, which furnish to consciousness the conviction of free will.'[37] As Thomson recognises with his appellation of 'Maxwell's Demon', by extending this principle of instability to the behaviours of atoms and molecules, the mechanistic relations of physics come to assume an anthropomorphic character, just as they do in Maxwell's poem 'In Memory of Edward Wilson, *Who repented of what was in his mind to write after section*'.

'In Memory of Edward Wilson' clearly had currency for Maxwell and his peers in the early 1870s. In the second of his 'Lectures to Women on Physical Science', which was first published in November 1874, Maxwell has a bright female student observe of her boorish professor:

> The only song you deigned to praise
>   Was "Gin a body meet a body",
> "And even there," you said, "collision
>   Was not described with due precision."[38]

His revival and revision of the undergraduate parody of Burns may have been prompted by Sylvester's 1866 paper 'On the motion of a free rigid body acted on by no external forces', which Maxwell discusses in August 1869 in a report for the Royal Society.[39] Edward Wilson is presumably the Irish astronomer and physicist (1851–1908), who accompanied Huggins to Algeria in 1870 to witness an eclipse. He may have rashly admitted to having distracting thoughts during a Mathematics and Physics session at the BAAS, or was simply a young colleague who could be teased in this way, and jokingly set Tripos problems as punishment '*after section*'.

The counterpart to the yielding human body of Burns's 'Comin' thro the Rye', the correspondingly rigid and inorganic body of 'In Memory of Edward Wilson' does what it wants. 'Flyin' through the air' freely and having no understanding of the problems it gives rise to and how they could be solved ('But what the waur am I?'), the Rigid Body is depicted as carefree and careless. The world of elementary mechanics is characterised here by phenomenal promiscuity and abandon. A suggestive model for the clashing atoms that cannot be measured individually and are consequently theorised statistically in Maxwell's dynamical theory of gas, the rigid bodies of the poem move irregularly, they 'meet' and 'hit' one another, with impacts that cannot be measured, while, furthermore, 'How they travel afterwards / We do not always see.' As an amalgam of mechanism and agency that defies Tyndallic determinism, the Rigid Body is like the molecular bodies in Maxwell's allegory of the 'Demon', which can act in ways that affront and modulate the second law of thermodynamics.

The fundamental phenomenon of the clinamen, in which the individual atom unexpectedly deviates from its path and so hits another, is also suggestively modelled by the simple physical bodies of 'In Memory of Edward Wilson': 'Gin a body meet a body / Altogether free.' The analogy that Maxwell's parody makes of simple physical dynamics to human sexual relations corresponds to that in the first stanza of 'Molecular Evolution', which quietly observes that Lucretius' account of the successive creation of worlds by the clinamen parallels the following account he gives of sexual passion in Book IV:

> Usque adeo cupide in Veneris compagibus haerent,
> membra voluptatis dum vi labefacta liquescunt.
> Tandem ubi se erupit nervis collecta cupido,
> Parva fit ardoris violent! Pausa parumper.
> Inde redit rabies eadem et furor ille revisit,
>
> [... so eagerly in Venus' toils
> They cling, while melt their very limbs, o'ercome
> By violence of delight. But when at length
> The gathered passion from their limbs hath burst,
> There followeth for a space a little pause

'Molecular Evolution': *Maxwell, Tyndall and Lucretius*

> In their impassioned ardour. Then once more
> The madness doth return ...]⁴⁰

The second quatrain of Maxwell's poem extends the imagery of the 'fortuitous embraces' by describing the existing order and its rupturing with words that translate key terms from Lucretius' account of sexual passion, *haerent* (to hold fast, cling) and *erupit* (bursting forth):

> And though they seem to cling together,
>   And form "associations" here,
> Yet, soon or late, they burst their tether,
>   And through the depths of space career.

As in the Lucretian account of passion, where the lovers strive futilely to fuse into one, the atoms in their crucial 'fortuitous embraces' only 'seem to cling together' and after a climactic bursting separate again, thereby furnishing the conditions for a new clinamen, another bout of embraces and the instauration of a new order. The confidently enunciated ontology of the stanza's first quatrain is undermined, deemed merely apparent here.

The use of the conditional verb 'seem' in the first line of the poem's second quatrain announces an epistemological shift from the categorical statement of the clinamen in the first quatrain to a sceptical perspective in the second. The Lucretian cosmology is now presented as a matter of perception rather than objective fact, its contingent formation the consequence of mental as much as physical '"associations"'. The quotation marks encasing this word make it into a citation that refers directly to the British Association, but which also invokes 'the association of ideas'.⁴¹ Although the phrase was coined by Locke, it is more usually identified with Hume, for whom mental impressions and other simple ideas are like contingent particles that the mind draws together through principles of resemblance, contiguity, and cause and effect. Indeed Hume provides a precedent for the analogy that Maxwell makes suggestively in the first stanza, and confirms in the second, between a scientific ontology and a perceptual psychology, as he describes the workings of the associationist principles in the mind by analogy with the Newtonian law of gravitation: 'Here is a kind of ATTRACTION, which in the mental world will be found to have as extraordinary effects as in the natural, and to shew itself in as many and as various forms.'⁴²

The shifting and ambiguous account of the Lucretian cosmology in the first stanza of 'Molecular Evolution' not only accommodates different readings but also encourages them. The foregrounding of the word '"associations"' serves to facilitate and demonstrate the Humean mental function it cites, as the various contexts the poem establishes, including its implied audience of Red Lions, encourage the reader to make mental associations with such things as the Lucretian clinamen, sexual intimacy and the BAAS, as well as Hume's doctrine itself. Observing in his Liverpool address that 'the atomic theory' is 'a branch of physics which not very long ago would have been considered rather a branch of metaphysics',[43] Maxwell finds in the Lucretian clinamen a punning principle of freedom of association that recalls and renews his earlier preoccupations at Edinburgh and Cambridge with the relations between metaphysics and scientific knowledge.

'A SWIFT METAMORPHOSIS'

The reverberant hermeneutical provocations of the first stanza of 'Molecular Evolution' – as it gestures toward cosmological and human relations, science and mythology, ontology and epistemology – are enhanced and extended momentously by the second, which announces that its atomist story is a figure for the transformation that the members of the British Association undergo after the day's meeting:

> So we who sat, oppressed with science,
>   As British asses, wise and grave,
> Are now transformed to wild Red Lions,
>   As round our prey we ramp and rave.
> Thus, by a swift metamorphosis,
>   Wisdom turns wit, and science joke,
> Nonsense is incense to our noses,
>   For when Red Lions speak, they smoke.

Parodying Tyndallic reductionism, the poem explains the transmutation of 'British asses' into 'Red Lions' as a mock instance of evolution that suggests a fable, or one of Ovid's *Metamorphoses*. The second stanza fully mobilises the pivotal word of the first, '"associations"', which focuses the poem's audaciously encompassing analogy between the most radical and the most recent and sophisticated points of 'Molecular Evolution'. It economically parodies the sweeping developmentalist narrative of Tyndall's Scientific Materialism, sketching a scheme that, beginning with the original associations of atoms clinging together as molecules, finds its acme in the poem's audience, much as Tyndall does in the Norwich address, where, referring to biological evolution, he remarks that it is 'a long way from the Iguanodon and his contemporaries, to the President and Members of the British Association'.[44]

The pun that 'Molecular Evolution' makes on '"associations"', as it short-circuits the graduated stages of Tyndall's reductive argument by analogy in a single equivocal term, itself demonstrates the witty type of mental associationism that Maxwell identifies with his generic principle of 'Nonsense'. Hume describes associationism, his 'uniting principle among ideas', as 'a gentle force'. This phrase could be used also for the *inclination* of the Lucretian clinamen, which reciprocally suggests a *laissez-faire* model for the association of ideas, the likes of which the Scots philosopher alludes to early in his discussion of the principle: 'Were ideas entirely loose and unconnected, chance alone wou'd join them.'[45] For Hume this implies the production of pure unmeaning nonsense, whereas in Maxwell's Lucretian analogy the principle of chance, represented by the clinamen, is an engine for creating new regimes of order.

Gilles Deleuze also finds fruitful affinities between the Lucretian clinamen and Humean associationism. His 1972 essay 'Hume' uses the same set of locutions and tropes as the 1969 'Lucretius and the Simulacrum' to describe 'a world in which terms are veritable atoms and relations veritable external passages; a world in which the conjunction "and" dethrones the interiority of the verb "is"; a harlequin world of multicoloured patterns and non-totalisable fragments where communication takes place through external relations'. 'Hume's originality', Deleuze emphasises, '... comes from the force with which he asserts that *relations are external to their terms*.'[46] While this cognitive autonomy troubles Maxwell's early poems and essays with worrying possibilities that reality could be methodically hallucinated by us, it is presented in 'Molecular Evolution' as a liberating facility for creating new knowledge. Lucretian atomism offers both Deleuze and Maxwell a model that recognises dynamic pluralism in thought and nature without surrendering it to entropic randomness and meaninglessness, empty nonsense. For Maxwell it also distinguishes associationism from materialist psychology, the mechanistic mental combinations of the 'fancy scientific' described in the '*Tyndallic Ode*', crude aggregations of 'mental bricks', which, having been 'Fetch[ed] ... from every quarter', suggest a merely entrepreneurial collocation of ideas. Contrasted with both 'rigid reason' and 'too, too solid sense', the 'combinations' of 'Nonsense' in 'Molecular Evolution' are formed imaginatively, by analogy with the Lucretian clinamen, as exercises in free will that can yield wisdom and truth: 'What combinations of ideas / Nonsense alone can wisely form!'

The second stanza's transformation of the 'British Asses' into 'wild Red Lions' is an allegory for the 'swift metamorphosis' of sense into nonsense, a fable that points a moral about scientific thought and discovery that the remainder of the poem consolidates. This is grounded in the parity that the professional theories of the British Asses and the Nonsense of the Red Lions have through their analogy with Lucretian physics, as complementary forms of mental '"associations"', different 'combinations of ideas'. The established order, the even laminar flow of atoms, is the cosmos *as we know it*, which is identified with the sense of the British Asses and destined in its turn for revolution as the atoms 'burst their tether / And through the depths of space career'. The Red Lions' anarchic gesture of 'Nonsense', the free mental '"associations"' of a new clinamen, liberates ideas from the 'tether' by which the British Asses hold them in the theories they canonize at their annual meetings. Tyndall's inflexible atoms are identified with such asinine regulation and regularity in 'British Association 1874', where they are described as 'Like spherical small British Asses in infinitesimal state'.[47]

While, as Chapter 3 observed, Maxwell identifies 'science' with analogy, its transformation into 'joke' turns it on its head, casts it in its reciprocal form, ostensibly confusing or conflating it in the preposterous manner of the pun. By being placed in quotation marks the word '"associations"' is singled out for proliferent punning, while the various 'combinations of ideas' that the word forms, as it concentrates the poem's analogy of the Lucretian clinamen to the BAAS, demonstrates the reflexive relationship of puns and analogies that Maxwell introduced in 'Analogies in Nature'.

Such carnivalesque Nonsense is likened in the poem to 'incense'. This not only suggests the exhilarating and intoxicating communion of laughter, but also an evanescent essence, a correlative for the elusive and teasing hybrid meanings and associations that arise in the mind at play, as it makes and contemplates puns and other forms of wit. Indeed, the *pungency* of incense suggests a quibble on the semantic concentration and sharp wit of the pun. 'Wisdom', good sense drawn from broader experience and thought, is similarly distilled here, it 'turns wit', a process of paring down that is enacted by the diction, as the two softened syllables of 'Wisdom' are shortened and pointed in the curtailed form of 'wit'.

If, as the venerable cliché has it, humour is all about timing, then wisdom is honed to incisive 'wit' by history. The Lucretian laminar flow furnishes a trope for an entropy of ideas, totalising theories that furnish the heat death of scientific thought, the end of history, a parallel to the promised end, indeed dead-end, of the utopia. The clinamen, however, provides Maxwell with an allegory for the historically contingent nature of scientific thought, as an unforeseen independent element emerges from the complacent established order to cut through and devastate it, relegate it to history.[48] Disavowing an entropic model of mental processes, Maxwell declares in the Liverpool address that the mind resembles a tree. He suggests that thought is like the shooting roots and branches, focused and resolute forms that are also lithe and exploratory, by which the tree reaches into an unfolding temporal world:

But the mind of man is not, like Fourier's heated body, continually settling down into an ultimate state of quiet uniformity, the character of which we can already predict; it is rather like a tree, shooting out branches which adapt themselves to the new aspects of the sky towards which they climb, and roots which contort themselves among the strange strata of the earth into which they delve. To us who breathe only the spirit of our own age, and know only the characteristics of contemporary thought, it is as impossible to predict the general tone of the science of the future as it is to anticipate the particular discoveries which it will make.

Physical research is continually revealing to us new features of natural processes, and we are thus compelled to search for new forms of thought appropriate to these features.[49]

Appreciated as analogies for one another by *De Rerum* and Maxwell's poem after it, the clinamen and sexual passion trace their respective '"associations"' to a principle of *inclination*, a subtle but irreducible exercise of free will that holds the possibility of 'new forms of thought', a semantics in which word-atoms are newly *inflected*, as through puns and analogies. Championing plurality, the clinamen contests the naturalism of the Metropolitans, the deterministic lust of the Tyndallic atoms, indeed it frustrates all such idealist ambitions for grand systems. Like the echoes from the mountain-side in 'Reflex Musings', the clinamen marks the incursion of matter upon the totalising purity of thought, as the rational Newtonian mechanics of the stream of atoms in their Euclidean plane, an order analogous to the Pythagorean cosmology that the young Maxwell was drawn to, is disrupted subtly but devastatingly by the swerve of the individual atom.

HUMEAN DELIRIUM

Attributed with a ritual significance, 'Nonsense is incense to our noses, / For when Red Lions speak, they smoke.' Such utterances suggest fire-breathing, but are also so much hot air. As smokers, the Lions are in a sense sucklings:

> Hail, Nonsense! dry nurse of Red Lions,
>   From thee the wise their wisdom learn,
> From thee they cull those truths of science,
>   Which into thee again they turn.
> What combinations of ideas,
>   Nonsense alone can wisely form!
> What sage has half the power that she has,
>   To take the towers of Truth by storm?

The reference to the dry nurse is, as Campbell notes, taken from Horace. It occurs in the fourth stanza of Ode 1.22, where it describes a wolf: 'nor does the land of Juba, dry nurse of lions, give it birth' ('*nec Iubae tellus generat, leonum arida nutrix*').[50] One of the better known of the Horation Odes, the reference would have been easily recognised by Maxwell's audience at Bradford, not least by North British insiders, many of whom would have remembered it from both their school Latin studies and the *Scottish Student's Songbook*, where it is set in the style of a church hymn. The parallel that Maxwell's poem draws with the desert land of Juba represents Nonsense as parched and empty, the source of the miniature siroccos of smoke that give voice to it at the close of the second stanza. This expansive emptiness furnishes a nursery for Nonsense, a capacious field for free thought, full of the paradoxical promise of Biblical deserts that come into bloom epiphanically.

Following upon Hume, Kant observes in his *Critique of Judgement* that 'All the richness of imagination in its lawless freedom produces nothing but nonsense.'[51] This idea of nonsense is for Deleuze 'something essential' that 'Kant owes ... to Hume', the recognition that 'we bathe in delirium'. The Humean 'physics of mind', simple ideas present to the mind as so many atoms, are seen to be matched with 'a logic of relations' by which the imagination can go beyond Nature 'in all the directions of a delirium that forms a counter-Nature, allowing for the fusion of anything at all'.[52] Such Humean and Kantian strains of thought converge in Sir William Hamilton's idea of the subconscious, from which the young Maxwell develops his conception of dreams as 'wandering fancies, by some lawless force entwined', a principle of nonsense entrenched deep in the mind. This mental physics of 'Empty bubbles, floating upwards' is like the unforced languid drift of smoke emitted by the Red Lions, the 'Nonsense [that] is incense to our noses', or indeed the shoots of the tree in the analogy from the Liverpool address, its roots securely and intricately grounded in a mysterious substrate, like the rocky subconscious in 'Recollections of Dreamland'.

The Horatian oxymoron of the dry nurse, '*arida nutrix*', is formed by analogy with the Latin for nursemaid, *assa nutrix*, the punning implication here being that the BAAS fulfils the role of the wet nurse, or indeed of a nanny, the 'tether' that the Red Lions break free from like so many naughty children. Maxwell's poem articulates and orchestrates the oxymoronic nature of the Lucretian clinamen, the instauration of order as disorder, into his principle of Nonsense, which in turn corresponds to Horace's description of Epicurean thought in Ode 1.34 as 'crazy wisdom' ('*insanientis ... sapientiae*').[53] Whewell presents a comparable, if rather more sober, conception of scientific discovery as the generation of sense from nonsense, as with the supervention of an appropriate hypothesis: 'A mass of facts which before

'Molecular Evolution': *Maxwell, Tyndall and Lucretius*

seemed incoherent and unmeaning, assume, on a sudden, the aspect of connexion and intelligible order.'[54] Maxwell's Nonsense 'can take the towers of Truth by storm', with all the epiphanic suddenness of laughter, of recognising puns or other witty 'combinations of ideas', such as bold analogies, 'the semantic impertinence of the creative metaphor'.[55] Immanent to objective reality, 'the towers of Truth' have their counterfeit in the preposterous materialist structure of 'mental bricks' described in the closing stanzas of '*A Tyndallic Ode*'.

'Molecular Evolution' culminates with a model of the 'combinations of ideas, / Nonsense alone can wisely form' that, offering another cheeky counterpoint to the Metropolitans' cosmology, follows the analogy of Laplace's nebular hypothesis:

> Yield, then, ye rules of rigid reason!
>   Dissolve, thou too, too solid sense!
> Melt into nonsense for a season,
>   Then in some nobler form condense.
> Soon, all too soon, the chilly morning,
>   This flow of soul will crystallize,
> Then those who Nonsense now are scorning,
>   May learn, too late, where wisdom lies.

With the release of the 'tether', the yielding of 'rigid reason' and dissolving of 'too, too solid sense', these mental capacities 'Melt into nonsense', the gaseous state instanced earlier by incense, smoke and the thin desert air. Analogous to his hypothetical 'Demon', which vouchsafes clinamen-like perversity as the prerogative of individual molecules of gas, Maxwell's principle of Nonsense allows atoms of thought to behave in free and surprising ways. Through the exercise of such freedom, the radical elements of reason and sense 'in some nobler form condense'. Having first melted into the dissipated dynamism of Nonsense, they emerge as a fresh distillation, furnishing a new universe of understanding, a new paradigm of scientific knowledge. The Red Lions, who when they 'speak, they smoke', are like the Homeric poet that Horace congratulates in his *Art of Poetry*: '*His* plan is not to turn fire to smoke, but smoke to light.'[56] Gaseous Nonsense, newly ennobled, is accordingly described as 'This flow of soul', imagery that recalls 'Recollections of Dreamland', where dreams are likened to bubbles of gas 'floating upwards through the current of the mind', emissaries from the subconscious depths 'where Self in secret lies'. 'This flow of soul will crystallize', assume the form of the great guarantor of knowledge with which Maxwell ends his Apostles essay on 'Analogies in Nature', the mineral crystal that marks the coalescence of mind and matter, of mathematics and natural form, 'bursting in upon us, and sparkling in the rigidity of mathematical necessity'.

The free play of the Red Lions, the radical thought experiments of Nonsense, gives way to 'the chilly morning' in which the 'flow of soul' of the Red Lions dinners crystallises, through complementary analytical modes of thought and scientific procedures, into formal papers for the BAAS meetings. The Red Lions, which (in a quibble on the clichéd 'rant and rave') 'ramp and rave', are figured as rampant in the manner of the heraldic lion, a balance to the more formal and stabilising British Ass, so that together they furnish the armorial bearings for British science.[57] Once the creative discoveries of the Red Lions are developed and presented soberly in 'the chilly morning' of the BAAS meeting, then, 'too late', 'those who Nonsense now are scorning' will learn the importance of the poem's teasingly oxymoronic accounts of its relations to truth, which are focused in the pun on the poem's final word, the concluding definition of Nonsense as the place 'where wisdom lies'.

CHAPTER 8

James Joseph Sylvester: the romance of space

James Joseph Sylvester not only wrote and translated poetry but also theorised such efforts in his book *The Laws of Verse or Principles of Versification Exemplified in Metrical Translations*, which was published by Longmans in 1870, the year he turned fifty-six. While the main title signals a scientific intent, the longer alternative title announces that the book includes some poetry, not for its own sake but to illustrate the author's theories. Although this arrangement contradicts modern, post-classical, tastes in poetry, it accords with mathematics, where instances are subordinated to the formal principles they demonstrate. Axioms, theorems and other such constructions are accordingly esteemed over the particular equations and cases that exemplify them, the symbolic forms of algebra over the individual values they can represent. The title page of *The Laws of Verse* continues by declaring that this work is published *Together with an Annotated Reprint of the Inaugural Presidential Address to the Mathematical and Physical Section of the British Association at Exeter*, which Sylvester had delivered at the 1869 meeting. Furthermore, this intriguing combination of texts is enhanced with an appendix, 'On the Incorrect Description of Kant's Doctrine of Space and Time Common in English Writers'.

Sylvester's volume is not, as it may seem, a random grab-bag of texts, but an idiosyncratic omnibus that draws together his mathematics, philosophy, prosody and poetry through a common concern with *space*. The Exeter address reaches its climax in his excited declaration that an unfettered principle of space constitutes the ground of mathematics:

if I were asked to name, in one word, the pole-star round which the mathematical firmament revolves, the central idea which pervades as a hidden spirit the whole corpus of mathematical doctrine, I should point to Continuity as contained in our notions of space, and say, it is this, it is this! Space is the *Grand Continuum* from which, as from an inexhaustible reservoir, all the fertilizing ideas of modern analysis are derived; and as Brindley, the engineer, once allowed before a parliamentary committee that, in his opinion, rivers were made to feed navigable canals, I feel almost tempted to say that one principal reason for the existence of space, or at least one principal function which it discharges, is that of feeding mathematical invention.

The preface to *The Laws of Verse* recalls and builds upon this passage:

> The great law of Continuity (continuity of sound and continuity of mental impression) has been my guiding star throughout. Some of my readers may chance to remember that I laid much stress upon this principle in the inaugural address delivered by me as President of the Mathematical and Physical Section of the British Association for the Advancement of Science, at Exeter. I went, on that occasion, so far as to say that Mathematics under its existent aspect might be defined as the Science of Continuity. I have thought, then, that this would be a not unsuitable occasion for republishing the address in a separate form.[1]

Both passages identify mathematics itself with the mysterious and apparently unending spaces of the heavens, where *Continuity* prevails as the polestar. The pervasive nature of such space describes Sylvester's mathematical concern with the locally continuous, the continuity of each point between given intervals, an idea that is analogous to, but distinct from, causal continuity, the principle of the unity of nature, which is similarly foundational for the physical sciences. Space is accordingly presented in all its grand Continuity as a propagating medium for mathematics, rather as the similarly ubiquitous luminiferous ether was seen to be for light. Described by analogy with the more substantive medium of water, space is imaged here as fertilising, as facilitating creative thought and transporting the mind, much as the canal system does its vehicles.

Mathematics is for Sylvester a creative activity, which consists in not merely the analysis of concepts but their construction in intuition. Kant identifies this capacity with geometry in his *Critique of Pure Reason*: 'The mathematics of space (geometry) is based upon th[e] successive synthesis of the productive imagination in the generation of figures.'[2] Space and time are for Kant the immediate *a priori* condition of our mental representations, which is to say that such objects appear to consciousness spontaneously in these Forms of Sense, as a phenomenal manifold that is synthesised through the imagination and subjected to the mediate forms of the legislative Understanding. As the science of our *pure* intuition of space, geometry is not predicated upon such applications of the Forms of Sense to empirical intuitions, our sense impressions, but is accordingly seen to have its own synthetic *a priori* ground. This leads to what Michael Friedman describes as 'The standard modern complaint against Kant', that he fails to distinguish 'between *pure* and *applied* geometry', the study of the *a priori* formal relations of an axiomatic system, which makes no appeal to experience, and that concerned with the truth value of such a system in relation to an empirical object world.[3]

Acknowledging the distinction between pure and applied geometry, Sylvester demonstrates his understanding, or rather his fruitful misunderstanding, of Kant's concept of space in the Exeter address:

> Like his master Gauss, Riemann refuses to accept Kant's doctrine of space being a form of the understanding, and regards it as possessed of physical and objective reality.[4]

Sylvester corrected this mistake in revising the address for publication in the BAAS *Report* and *Nature*, both of which accordingly refer to 'Kant's doctrine of space and time being forms of intuition'. This crucial amendment was suggested by his friend Ingleby, who, in the footnote on Kant that appears with it, Sylvester credits with having first 'pointed out to me' the 'very common, not to say universal' misunderstanding of the Kantian Forms by 'English writers, even such authorised ones as Whewell, [G. H.] Lewes, or Herbert Spencer, to refer to Kant's doctrine as affirming space "to be a form of thought", or "of the understanding"'.[5] Lewes responded to this provocation in the issue of *Nature* for 6 January 1870, the first installment of the crowded correspondence that Sylvester participated in, and reprinted as his appendix to *The Laws of Verse*. The controversy continued until 10 February, when Lockyer declared that 'This correspondence must now cease'.[6]

It is curious that Sylvester should provide a strenuous defence of Kant's 'doctrine of space' as an adjunct to a sentence that rejects its validity in order to more clearly affirm, with Gauss and Riemann, the 'physical and objective reality' of space. He evidently wishes to retain both conceptions. This implication is confirmed and clarified by the 'Probationary Lecture' that Sylvester gave in applying for the Chair of Geometry at Gresham College London in 1854, in which he makes a similar distinction between pure and applied geometry. The lecture allocates to the metaphysician a concern with 'the nature of space as it exists in itself, or in relation to the human mind', and to the geometer 'space as an objective reality, and to view it in its relation to matter, and as the substratum or the condition necessary to the existence of our conception of form'. This last clause, however, reintroduces the pure intuition of space, but refers it to the object world, as a noumenal reality. Sylvester then proceeds to specify the nature of 'our conception of form', the ground of apodeictic certainty for the geometer: 'The first property which strikes the mind in dwelling upon the idea of space is its infinitude, its capacity of boundless extension.'[7] 'Space is', as Kant puts it, 'represented as an infinite *given* magnitude',[8] within which particular spaces can be delimited and the concepts of geometry constructed analytically. This is the fundamental condition of Sylvester's principle of Continuity, 'the *Grand Continuum*' of space that he represents metaphorically in the Exeter address as 'an inexhaustible reservoir'.

Sylvester elaborates upon the relation he sees to exist between space as the substratum for both objective reality and mental intuitions in the Exeter address:

mathematical analysis is constantly invoking the aid of new principles, new ideas, and new methods, not capable of being defined by any form of words, but springing direct from the inherent powers and activity of the human mind, and from continually renewed introspection of that inner world of thought of which the phenomena are as varied and require as close attention to discern as those of the outer physical world (to which the inner one in each individual man may, I think, be conceived to stand in somewhat the same general relation of correspondence as a shadow to the object from which it is projected ...).[9]

In this idiosyncratic form of metaphysical realism, the 'inner' and the 'outer' ideas of mathematical space are each able to match the other in variety and complexity, and indeed coincide with their counterpart as knowledge. Functioning in the synthetic *a priori* manner of Kant's pure intuitions, the mind's 'inner' phenomena are imaged in Sylvester's adaption of Plato's parable of the cave[10] as the shadow that can describe reflectively the ontologically prior and objective noumenal truth of space itself. Reminiscent of Rowan Hamilton's comments on Newton, this view of the mind as able to grasp the nature of reality *a priori* is seen by Sylvester in the 'Probationary Lecture' to be demonstrated by Plato himself, for as he 'revelled in a world of his own creation ... he was writing the grammar of the language in which it would be demonstrated in after ages that the pages of the universe are written'.[11]

Sylvester deduces the basic forms of geometry from the pure intuition of space, and hence the dimensions they inhabit, in his 'Probationary Lecture'. The simple superficies of the plane are produced by dividing infinite space, its 'ideal separation into two parts, distinct but contiguous', while the 'continuation of this process, that is to say, the dichotomy of superficies in its turn into distinct contiguous parts, gives birth to the notion of an infinite line; a surface limits space, a line limits a surface, ... now again, conceive a line to separate into two parts and we arrive at the notion of a point, the lowest term in the scale of geometrical being'.[12] This deduction demonstrates Kant's doctrine that 'geometrical propositions are one and all apodeictic, that is, are bound up with the consciousness of their necessity; for instance, that space has only three dimensions'.[13] Kant exemplifies his doctrine with proofs from Euclid's *Elements of Geometry*.[14] This, however, marks another point at which Sylvester requires Kantian space to be brought into line with Gauss and Riemann.

Sylvester describes Euclid's propositions and the axioms they are derived from pathologically in the Exeter address 'as a sort of morbid secretion, to be compared only with the pearl said to be generated in the diseased oyster'. Furthermore, he regards the *Elements* as a museum piece to be kept well away from children: 'I should rejoice to see ... Euclid honourably shelved or buried "deeper than e'er plummet sounded" out of the schoolboy's reach.'[15] He asks that schools renounce the powers that the *Elements* hold over them, just as Shakespeare's Prospero does his book in the final scene of *The Tempest*: 'And deeper than did ever plummet sound / I'll drown my book.'[16]

Kant's doctrine of space is not, however, purely and simply Euclidean, for it attributes a temporal dimension to acts of pure intuition: 'I cannot represent to myself a line, however small,' he writes, 'without drawing it in thought, that is gradually generating all its parts from a point.'[17] A fundamental means of articulating the infinite intuition of space, this principle grounds Continuity in calculus. The concerns with motion and the lengths of variable lines that Kant describes belong not to Euclidean geometry, but to the investigations made by such natural philosophers as Johannes Kepler and Galileo Galilei into the physics of planetary orbits and other moving bodies. This required a mathematics that could gauge changes in velocity and movements along curvilinear trajectories. Differential calculus developed to allow the measurement of such varying rates of change, as by positing infinitesimal increments of time and distance, variable quantities that approach but just stop short of parity with zero, it can simulate the continuity of motion and curves.

Isaac Newton's announcement in 1671 of his method of 'fluxions' gives him historical precedence in the invention of calculus, while G. W. Leibnitz in Germany is generally acknowledged to have developed his own more systematic version between 1673 and 1676. Curves are generated by the continuous motion of a point in Newton's differential calculus, the conception that Kant applies in his description of the synthetic *a priori* representation of a line. Having taken the Mathematical Tripos in the 1830s, Sylvester was well schooled in continental 'analysis', the principles of which he describes in the Exeter address as deductions from the infinite magnitude of space, 'as from an inexhaustible reservoir'. Such imagery makes the conceptual basis of Newton's calculus in 'fluents' or 'flowing quantities' concrete and elemental. Sylvester finds in the motionable medium of water, with its protean curvilinear forms, a suggestive analogy for the subtle and dynamic conceptions of space, the continuity, that the calculus traces and indeed facilitates. Furthermore, such imagery gestures beyond the Euclidean geometry that Newton based his physics upon to a non-Euclidean geometry of curved spaces.

After making his deduction of the plane, line and point in the 'Probationary Lecture', Sylvester goes on to observe that curves and spheres question and complicate the Euclidean notion of space: 'To say that every solid has length, breadth and depth, and therefore that space has three dimensions, is to convey a very inadequate explanation of the fact to be accounted for; for if we consider a spherical or any other body bounded by a continuous surface without angles or edges, we see nothing to indicate the existence in it of three or any other specific number of directions of measurement.'[18] Sylvester is following Gauss here. Rather than assuming that curved surfaces are embedded in three-dimensional Euclidean space and so constituted by the qualities of this medium, a consequence of measuring them from the 'outside' through their deviations from the plane, Gauss proposed that they be treated discretely and their intrinsic properties studied in detail. In his 1827 memoir, 'General investigations of curved surfaces', he establishes a measure of curvature that, rather than requiring reference to the whole of a curve or surface, was inherent to small parts. By focusing upon the local properties of curves and surfaces and applying to them the methods of calculus, Gauss's differential geometry yielded metrics that, as such subsequent thinkers as Riemann realised, may not coincide with Euclidean coordinates, and could accordingly facilitate conceptions of non-Euclidean spaces. The observation that Sylvester makes about the sphere gives his 1854 'Probationary Lecture' to the Gresham Committee an affinity with the qualifying lecture that Riemann gave at Göttingen in the same year, 'On the Hypotheses which Lie at the Bases of Geometry'. Indeed his early remarks offer a prepossession of the non-Euclidean geometry that Riemann's lecture expounds and that Sylvester is quick to champion after its text is finally published in 1868.

PROJECTIVE GEOMETRY

Sylvester comes to non-Euclidean geometry through projective geometry, which brought into question Euclid's propositions by theorising geometrical forms not metrically, in terms of distance, but formally through invariant properties, qualities of Continuity. The origins of projective geometry can be traced to the theory of conic sections. Begun by Menaechmus, a contemporary of Plato, and developed by Aristaeus, Euclid, Archimedes and Apollonius of Perga, this theory conceives of the cone as the projection of a circle, which forms the base of the figure, from a point, its vertex. In this way the figures of the ellipse, the parabola and the hyperbola can each be understood as a section of a cone. Renaissance theories and practices of perspective drawing provided another forerunner to projective geometry. Producing their effects by projecting lines and three-dimensional figures on a plane, and so departing from Euclidean principles of fixed angles, these methods further stimulated projective geometry by presenting metric relations as relative rather than absolute, with such objects as human figures and buildings retaining their forms despite the alterations in size and shape they endure in being cast perspectivally. Michel Serres observes that the body 'sees in projective space', which he distinguishes from the Euclidean 'space of work', the practical metrical relations 'of the mason, the surveyor, or the architect'.[19] Preparing the way for non-Euclidean geometry, projective geometry encourages ideas of space as multiple rather than homogeneous.

The modern foundations of projective, or descriptive, geometry are laid in the seventeenth century by Gérard Desargues and Blaise Pascal, whose theorems describe particular geometrical properties that remain unaltered by projection, while it is established for the nineteenth century by Gaspard Monge, who devised a way of representing the essential geometrical attributes of three-dimensional bodies in two-dimensional drawings. Michel Chasles, the last great indigenous representative of the school that Monge started and a pioneering historian of mathematics, writes in 1837 that the first object of projective geometry is, as its name indicates, 'to represent all bodies of a definite form on a plane area, and thus to transform into plane constructions graphical operations which it would be impossible to execute in space'. 'The second', he explains, 'is to deduce from this representation of the bodies their mathematical relationships resulting from their forms and their relative positions',[20] a task that benefitted greatly from the use of calculus. These foundational principles proceed from Monge's ambition to establish a geometry that could, as Fauvel and Gray observe, 'match the generality of algebra'.[21] This goal was effectively realised by Monge's student Jean-Victor Poncelet, who developed his teacher's ideas in 1813–14 whilst a prisoner of war from Napoleon's disastrous Russian campaign. His pressing duties as a military engineer, which he resumed after his release, prevented him from publishing his results until 1822.

Following from the two objects of projective geometry that Chasles describes, and facilitating the generalisation that Monge pushed for, Poncelet's central principle of Continuity describes the preservation of geometrical properties that occurs as their figures undergo continuous transformations: 'Is it not evident that when', he argues, '... we subject certain parts of the original figure to some continuous movement, that the properties and relations that were found for the first system will be applicable to the successive stages of this system, if only the particular modifications are taken into account ... for instance, certain quantities may have vanished or changed their sense or sign.'[22] A simple example, which both E. T. Bell and Joan L. Richards outline, is provided by the figure of two intersecting circles that gradually move apart from one another. A straight line that passes through the two points at which the circles intersect will retain its set of relations to each circle (as, e.g., equidistant from the centre of each and at right angles to the line that would pass through these central points) as they move outwards, so that it passes through the tangential point at which they meet, before they finally separate. As the circles move apart, the straight line remains the common axis between them, while the various points of intersection established earlier are retained through the principle of Continuity as 'imaginary' or 'ideal' points.[23] As Richards observes, Poncelet's continuous transformations 'involved establishing connections among propositions previously considered separate'. The example she gives of the circles accordingly connects two Euclidean propositions that had been separated for millennia, as the demonstration that 'A circle does not cut a circle at more points than two' proceeds transformatively to show that 'A circle does not touch a circle at more points than one.' Grounded in concrete forms of representation, such connections nevertheless facilitate generality. They give geometry the means of 'incorporating algebra's general powers'.[24]

Poncelet's 1822 treatise argues that the discrete propositions of the Euclidean catalogue must give way to a systematic development of geometry based upon principles of projection, Continuity, and 'imaginary' or artificial points (by which 'new' geometries can be constructed).[25] Sylvester champions such principles in the Exeter address, where he defends mathematics against recent public utterances by Huxley that identify mathematics with Euclid, its staple text in British schools, and so characterise it as dogmatic and deductive, fundamentally unprogressive. In doing so Sylvester is lobbying not only Huxley but his other peers at the Exeter meeting, which appointed a committee to report on the state of geometry teaching in schools.[26] Abridged and published in *Nature* as 'A Plea for the Mathematician', Sylvester's address argues that Euclid should be supplanted by a school curriculum that sees 'projection, correlation, and motion accepted as aids to geometry – the mind of the student quickened and elevated and his faith awakened by early initiation into the ruling ideas of polarity, continuity, infinity, and familiarisation with the doctrine of the imaginary and inconceivable'.[27]

Amongst the various references he makes in his address to mathematical developments after Poncelet, Sylvester discusses an important area of his own researches, invariants. He coined the term in 1851, to designate a new field of algebra he invented and explored with Cayley. George Boole and Gotthold Eisenstein had each recorded isolated examples of invariants during the early 1840s, while Cayley established methods that could systematically yield all members of such classes in his 1845 paper *On the Theory of Linear Transformations*.

Bell gives some simple examples that, by indicating the sorts of applications that invariants offered to geometry, allow their broad character to be understood without recourse to the esoteric idioms of algebra. Invariants lend algebraic expression to immutable patterns and order, such basic forms as the points in geometrical figures of intersecting lines and curves that are retained throughout various transformations, as, for instance, when the plane upon which they are projected is distorted, as in the case of crumpling the paper or stretching a sheet of rubber they are inscribed upon, resulting in their original curves and straight lines being altered in shape and measurement. Bell also gives the example of the knot in a rope or string, the form of which remains invariant throughout the transformations it can be subjected to in being loosened and tightened at various points.[28] An invariant is, as Sylvester explains, 'a function of the coefficients of the form, as remains absolutely unaltered when instead of the given form any linear equivalent thereto is substituted'.[29] The work of projective geometry, of tracing persistent relations in the projections of geometrical figures as they pass through continuous processes of transformation, is expressed, supplemented and developed by algebraic invariants.

'TRANSCENDENTAL SPACE'

Cayley's 1859 'Sixth Memoir upon Quantics' concludes that Euclidean or 'metrical' geometry is a special case of projective or descriptive geometry: 'Metrical geometry is thus part of descriptive geometry, and descriptive geometry is *all* geometry.'[30] Projective space is prior to metrical considerations, it is not constituted by distance, the concept that, as Riemann demonstrates, distinguishes Euclidean from non-Euclidean geometries. Effectively inverting the prevailing assumption that the space of projective geometry is Euclidean, Cayley's Memoir establishes the means by which others, most importantly the German mathematician Felix Klein, but also Sylvester, would find that projective space can also generate non-Euclidean metrics.[31]

Sylvester's preoccupations with space, and his expansive conception of it as a '*Grand Continuum*', follow from Gauss and Cayley's recognition that metrical properties do not define space, but rather depend upon the space in which they subsist. He sketches some grounding ideas from Gauss and Cayley in his Exeter address:

I may mention that Baron Sartorius von Waltershausen (a member of the Association) in his biography of Gauss ('Gauss zu Gedächtniss'), published shortly after his death, relates that this great man was used to say that he had laid aside several questions which he had treated analytically, and hoped to apply to them geometrical methods in a future state of existence, when his conceptions of space should have become amplified and extended; for as we can conceive beings (like infinitely attenuated bookworms in an infinitely thin sheet of paper) which possess only the notion of space of two dimensions, so we may imagine beings capable of realising space of four or a greater number of dimensions. Our Cayley, the central luminary, the Darwin of the English school of mathematicians, started and elaborated at an early age, and with happy consequences, the same bold hypothesis.

Sylvester adds a footnote to this passage, in which he develops the analogy between our experience and conceptions of space and those of the two-dimensional bookworms: 'the laws of motion accepted as a fact suffice to prove in a general way that the space we live in is a flat or level space (a "homoloid"), our existence therein being assimilable to the life of the bookworm in a flat page'.[32] While we can imagine spaces of up to three dimensions, 'we may', as Sylvester puts it, only 'imagine beings capable of realising space of four or a greater number of dimensions', we cannot directly imagine such spaces themselves. The mathematician's vision is metaphorical here, not actual and observational. For empirical confirmation of further dimensions Gauss accordingly looks forward to a geometer's heaven, 'a future state of existence', which will allow experience of other forms of space that would accordingly enable 'his conceptions of space' to 'become amplified and extended'.

In his Habilitation lecture, Riemann provides the means of describing mathematically conceptions of space that exceed the three dimensions that can be represented graphically by Euclidean and projective geometries. One year after it was published, and three years after the author's death, Sylvester offers a summary account of the lecture in the Exeter address, where he observes

that other kinds of space might be conceived to exist subject to laws different from those which govern the actual space in which we are immersed, and that there is no evidence of these laws extending to the ultimate infinitesimal elements of which space is composed.[33]

In the first section of his lecture, Riemann develops a philosophical and mathematical 'notion of a multiply extended magnitude',[34] which furnishes the means of extending the representation of spaces beyond the Euclidean trinity to the 'space of four or a greater number of dimensions' that Sylvester alludes to in his passage on Gauss. This is Riemann's idea of the manifold, a broadly applicable concept that can be understood as a set of points or objects, each of which can be specified by a numerical value. Their relations to one another determine them as either 'a *continuous* or *discrete* manifoldness'. The first of these, consisting of collections of points between which there is 'a continuous path from one to another', applies to multiply extended magnitudes. The fundamental principle here is one of local Continuity, akin to that recognised earlier by Kant's description of the line as it proceeds from the point, 'a simply extended manifoldness, whose true character is that in it a continuous progress from a point is possible only on two sides, forwards or backwards'.[35]

Riemann gives two examples of continuous manifolds, 'the positions of perceived objects and colours'.[36] In a spatial manifold, the numbers that designate each of its points are determined by the measurement of distance, which identifies their relative positions. Helmholtz describes the less familiar application of the idea to colour in an 1870 review essay he wrote for *The Academy*, where he translates Riemann's term not as 'manifolds' or 'manifoldnesses' (as William Kingdom Clifford does) but as 'varieties':

We have also other objects besides the situation of a point which may be defined by three measurable quantities and which we are able to change by infinitesimal degrees, changing at the same time their co-ordinates in the same way. Riemann comprehends them under the name of varieties (*Mannichfaltigkeiten*) of three dimensions. Thus every colour may be represented, according to Thomas Young and Maxwell, as a mixture of three measurable quantities of three primary colours. Space, therefore, and the system of colours may be called 'varieties' of three dimensions.[37]

The respective ratios of each primary colour as they constitute an instance of colour yield three numbers, the coordinates of a three-dimensional manifold. Represented by such quantities, manifolds can extend to n dimensions and their relations can be studied through mathematical formulae. Sylvester finds in Riemann's principle the means of reinvigorating traditional appreciations of poetry as mathematical ratio. The title 'Studies in Monochrome', which he gives to a group of his poems in his 1876 collection *Fliegende Blätter*, signals his recognition that the manifold encompasses the ratios of poetry as of colour, so that it can be conceived of spatially through his principle of Continuity. Riemann maintains that, despite their abstract forms, manifolds 'are decomposable into relations which, taken separately, are capable of geometric representation'.[38]

Types of magnitudes, such as colour and musical tone, are distinguished by empirical propositions, 'matters of fact' or 'hypotheses', which in the case of space are also necessitated by the abnegation of the axiom's universal purview, a consequence of Gauss's local metrics that will be returned to later. While Riemann's references to 'matters of fact' are rather cryptic, Sylvester nevertheless makes them the ground for the modern idea of mathematics he asserts against Huxley's Euclidean characterisation of it: 'the ever to be lamented Riemann has written a thesis to show that the basis of our conception of space is purely empirical, and our knowledge of its laws the result of observation'.[39]

Having introduced his principle of the manifold, Riemann turns his attention to the relations of measurement they allow. Noting that 'Measure ... requires a means of using one magnitude as the standard for another', and building '*on the assumption that lines have a length independent of position, and consequently that every line may be measured by every other*', Riemann determines the '*Measure-relations of which a manifoldness of* n *dimensions is capable*'. The grounding principle that the lengths of lines are independent of place, which is assumed in Euclid and warrants the supposition of manifolds of *n* dimensions and their mathematical expression, is found by Riemann to entail that the space in which they subsist must have a constant curvature: 'The common character of these continua whose curvature is constant may be also expressed thus, that figures may be moved in them without stretching ... whence it follows that in aggregates with constant curvature figures may have any arbitrary position given them.'[40]

'That space is an unbounded three-fold manifoldness', Riemann acknowledges, 'is an assumption which is developed by every conception of the outer world', constantly confirmed by experience as the 'region of real perception' and the place in which 'the possible positions of a sought object are constructed'. Such unboundedness does not, however, mean that space extends infinitely. Indeed, 'if we assume independence of bodies from position, and therefore ascribe to space constant curvature, it must necessarily be finite provided this curvature has ever so small a positive value'. Such 'an unbounded surface of a constant curvature' yields a model of the Riemannian plane that, he observes, 'would take the form of a sphere, and consequently be finite'.[41]

Helmholtz uses the figure of the sphere to elaborate some consequences of the Riemannian plane for Euclidean geometry in his 1870 review essay on the Habilitation lecture and other recent non-Euclidean texts. A point in Riemann's plane corresponds to a point on the sphere and straight lines to circles, so that, as Helmholtz observes, if we imagine beings living on this plane:

The shortest lines which the inhabitants of a spherical surface could draw would be arcs of great circles [i.e., circles defined by the intersection with a plane that passes through the centre of the sphere]. The axiom, that there is only one shortest line between two points, would not be true without exception; for between two points diametrically opposed, they would find an infinite number of shortest lines, all of equal length.

Helmholtz also observes that on the Riemannian plane, contra Euclid: 'The sum of the angles of a triangle would be greater than two right angles, and the difference would grow with the area of the triangle.'[42] Most momentously, Riemann excludes from his geometry Euclid's fifth postulate, which asserts that, given a point and a straight line, only one straight line can pass through the point and be parallel to the first line, however long each of them stretches. If we imagine that the earth is a sphere representing the Riemannian plane, one line can be seen to pass through both poles as a great circle while the equator describes another, so that the two will intersect at right angles. Such lines are finite in length and will always intersect, they can never be parallel to one another. Riemann's geometry demonstrates that space is not homogeneous, so that axioms cannot necessarily be predicated of all spaces. This local appreciation of space is the meaning of his thesis, that it is not (Euclidean) axioms but 'Hypotheses which lie at the foundation of geometry'.

The manifold allowed Riemann a means of further specifying space as local and intrinsic, and so build upon Gauss, drawing out the non-Euclidean implications of his teacher's metrics. So, for example, a curved surface can not only be examined independently of the Euclidean space it is traditionally assumed to inhabit, as it is in Gauss, but, as was noted earlier, appreciated as constituting its own distinct space, that is, as a different metric that yields a different geometry. Riemann's theory of space as a manifold of n dimensions opens the possibility of specifying a vast range of localised spatial behaviours and geometries, so that, as he puts it early in his lecture, 'a multiply extended magnitude is capable of different measure-relations, and consequently that space [i.e., our familiar three-dimensional space] is only a particular case of a triply extended magnitude'.[43] Sylvester alludes to the possibility that we may be living in another such triply extended magnitude, as he adds to his description of the 'homoloid', 'our level space of three dimensions', the mischievous possibility that this space could be taking a non-Euclidean turn: 'but what if the page should be undergoing a process of gradual bending into a curved form?'[44] Riemann describes mathematically three-dimensional spaces that are curved positively in the manner of a ball and negatively in forms approximating that of a saddle, while flat planes are characterised in these terms as cases of 'zero curvature'.

How are we who are hemmed in by the three-dimensional experience of space to join Gauss in that 'future state of existence, when his conceptions of space become amplified and extended'? Like other versions of heaven, this state is reached through faith: 'In philosophy, as in aesthetic,' Sylvester affirms, 'the highest knowledge comes by faith.' Hence his grand declaration, familiar from Chapter 6: 'If Gauss, Cayley, Riemann, Schalfli, Salmon, Clifford, Krönecker, have an inner assurance of the reality of transcendental space, I strive to bring my faculties of mental vision into accordance with theirs.'[45] This follows the criteria of his Platonic parable, as the 'shadow' of his mental representations, his 'inner world of thought', endeavours to comprehend the 'outer physical world', 'the reality of transcendental space'.

Ingleby responded to Sylvester's statement in January 1870 with the first of his letters to *Nature* headed 'Transcendent Space', the source of Salmon's denial that he had any such 'inner assurance' of the fourth dimension. A month later, in the second letter, which introduces itself as the 'epilogue' to the controversy over Kantian space he had inadvertently triggered with his correction of Sylvester, Ingleby clarifies the meaning of his phrase with the observation that 'Hamilton's speculations [i.e., quaternions] had borne a very remarkable relation to Transcendent Space of Four Dimensions'.[46] Having earlier alerted Sylvester to his misreading of Kant's doctrine of space, Ingleby was sensitive to the connotations of his phrase 'transcendental space', as it shades the Kantian Form of Sense into a noumenal non-Euclidean conception of space, which he accordingly distinguishes as 'Transcendent Space'.[47] This implication becomes explicit in the correspondence that Ingleby initiated with Clifford in *Nature* in February 1873 over Kant and the curvature of space. Ingleby finds such non-Euclidean notions refuted by Kant's Form of Sense, which he sees to simply yield experience of Euclidean space. Clifford observes conversely that, as 'The transcendental object is *unreasonable*, or evades the processes of human thought', our incapacity to imagine noumenal space argues for its consistency with non-Euclidean space.[48] Ingleby considers this line of argument to be based upon a misreading of Kant that Clifford gets from Rowan Hamilton, a diagnosis that he could also have applied to Sylvester's 'transcendental space'.[49]

Science is for Sylvester the struggle of the finite human mind to grasp the transcendental, the sublime. He describes this aspiration in his 'Probationary Lecture', as he does also in the passage on Continuity from the Exeter address, with the celestial tropes of the heavens and infinite space: 'Sciences, true sciences, spring from celestial seeds sown in mortal soil, they outgrow the restrictions which human shortsightedness seeks to impose upon them, and spread themselves outwards and upwards to the heavens from whence they derive their birth.'[50] The commonsense conviction in the sufficiency of Euclidean three-dimensional space is for Sylvester a case of such 'human shortsightedness'. While Continuity is figured as the *pneuma* that breathes life into the body of mathematics, 'the central idea which pervades as a hidden spirit the whole corpus of mathematical doctrine', and extends throughout the infinite heavenly expanse of space, its fulfilment in non-Euclidean geometry is accordingly seen to require a superior Kantian subject for its appreciation. This is one who is able to have pure intuitions of space beyond the three dimensions described by Euclidean geometry and mortal experience, intuitions that anticipate an afterlife of pure mathematical spirit, Gauss's 'future state of existence'. Sylvester looks to 'beings capable of realising space of four or a greater number of dimensions', rather as in the 1820s Rowan Hamilton, elaborating a similar strain of idealism, conceives of 'an order of beings of pure and passionless intellect, to whom Science in all its fullness of beauty is unveiled'. Hamilton tells Wordsworth that, as Eliza records, '*he* believed Mathematics to be a connecting link between men and beings of a higher nature; the circle and triangle he believed to have a real existence in their minds and in the nature of things, and not to be a mere creation or arbitrary symbol proceeding from human invention'.[51]

'STILL-MORE HEAVEN-REACHING THEORY'

Sylvester's 'transcendental space', like his reference to Gauss's 'future state of existence', implies a continuum between earth and heaven. Figured as 'the pole-star', Continuity is an emissary from the celestial realms that furnishes bearings for the terrestrial world beneath it. The audacious extrapolation of 'transcendental space' that Sylvester makes from his principle of Continuity has a telling counterpart in Stewart and Tait's *The Unseen Universe*, which similarly finds causal Continuity to entail a transcendent reality. According to Knott, Tait's 'favourite theme was the Law of Continuity'.[52] Stewart and Tait argue that Continuity means that we can trace the conditioned from one form to progressively higher forms that logically culminate in the Unconditioned, thereby assuring us of its existence: 'Now, what the principle of Continuity demands is an endless development of the conditioned', an 'argument [that] has led us to regard the production of the visible universe as brought about by an intelligent agency residing in the unseen'.[53]

Of all the celestial imagery of Continuity he uses in the Exeter address to describe his mathematical work toward 'a future still-more heaven-reaching theory', Sylvester became notorious amongst his peers at the Exeter meeting for the following striking analogy: 'Were it not unbecoming to dilate on one's personal experience, I could tell a story of almost romantic interest about my own latest researches in a field where Geometry, Algebra, and the Theory of Numbers melt in a surprising manner into one another, like sunset tints or the colours of the dying dolphin.'[54] Huxley cites the passage in a letter to Tyndall from June 1870, referring to Sylvester as 'The genius which sighs for new worlds to conquer beyond that surprising region in which "geometry, algebra, and the theory of numbers melt into one another".'[55] Nor did such imagery go unnoticed a few months later in Maxwell's Liverpool address, where he declares that his predecessors' respective addresses at Norwich and Exeter transported him to contrasting realms of atomism. Tyndall's diabolical microscopic world, where molecules 'clash together in fierce collision, or grapple in yet more fierce embrace', is familiar from earlier chapters. Following this account, Maxwell presents Sylvester conversely as having delivered him to a heavenly sphere, which he in turn renders with an extract from Tennyson's 'Lucretius':

I have been guided by Prof. Sylvester towards those serene heights

> 'Where never creeps a cloud, or moves a wind,
> Nor ever falls the least white star of snow,
> Nor ever lowest roll of thunder moans,
> Nor sound of human sorrow mounts, to mar
> Their sacred everlasting calm.'[56]

Sylvester's science is identified with Tennyson's description of the divinities' domicile, 'The lucid interspace of world and world' that 'The Gods . . .

haunt'.[57] The extract from 'Lucretius' recognises Sylvester's twin passions for mathematics and poetry, and indeed the principle of space that he sees to link the two, his '*Grand Continuum*'.

The thematic imagery that Sylvester uses for 'The great law of Continuity' instances the second application he sees it to have for poetry, 'continuity of mental impression'. This is described in *The Laws of Verse* as the distinctive capacity of poetry to sustain patterns of imagery, and is demonstrated in the volume by a group of purportedly anonymous poems. Indeed it is precisely their preponderance of astronomical and celestial imagery, along with stylistic affinities, that draws them together with poems that Sylvester puts his name to.[58] Such imagery gathers together his personal preoccupations with not only science, but also women.

Figures of the heavens are often associated in Sylvester's verse with the glamour of women. The first of his poems in *The Laws of Verse*, 'The Evening Star', describes the beauty of an unassuming woman it addresses with the conceit of a celestial counterpart, 'Venus seated in the sky / Amidst the starry choir'. Other verses that use such tropes of a 'star of beauty' include 'Eyes, like stars in heaven that glow', and the intriguingly entitled 'To an Ink-spot upon a Lady's Cheek', in which the 'Presumptuous spot!' on 'this matchless cheek' is likened to clouds and sun-spots:

> Pencilled with clouds, the azure vault
> But beams more heavenly pure,
> And spots upon the sun exalt
> The brightness they obscure.[59]

Similarly, the heroine of a later poem, 'The Lily Fair of Jasmin Dean: A Reminiscence', is described 'as Venus in the air' and 'heavenly-graced', 'Mirror of heaven's own loveliness' and, finally, 'The loved Pale Star of Jasmin Dean'.[60]

Women are the stars that ornament Sylvester's private cosmology, 'the illimitable sphere of mathematical discovery and invention'. Reflections of such heavenly beauty, they are seated Venuses poised to view the Promethean heroics of its scientists, such as the eponymous mathematician and astronomer of 'Kepler's Apostrophe', who 'Shall flash out to a meteor's blaze / And stream along the skies'. Indeed his heroic fate is figured astrologically, written in the very stars that he further enlightens with his science: 'Me, heaven's bright galaxy shall greet / Theirs by primordial choice.'[61] The mathematical tradition that begins with Pythagoras is envisaged similarly in another of his verses, rising, like Satan's flight in Milton's *Paradise Lost*, as a 'pyramid of flame',[62] while in the poem 'Indifference' the 'hateful name' announced by the title is opposed by the divine creative powers that 'Lent to the orbs of eve their fire'. Dedicated 'To all that's great and high aspires', and so 'From Heaven draws down Promethean fires', the poem 'Remonstrance' assures its reader that 'The truthful course pursues and knows / By Heaven-imparted light.'[63]

Sylvester describes himself in similarly grandiose terms in a note to a couplet he wrote in Latin in 1865, 'Urbi et orbi' and published in 1895 with an English translation, as 'To Oxford and the World', with the explanatory subtitle '*When I had demonstrated the theorem of Newton concerning the imaginary roots of equations which had remained undemonstrated for a long course of years.*' Newton describes how such imaginary roots can be discovered in his *Arithmetica Universalis* (1707), but neglects to offer a justification for this practice. Sylvester provides a limited proof of the theorem in a paper he submitted to the Royal Society in 1864, 'An Inquiry into Newton's Rule for the Discovery of Imaginary Roots'.[64] On 11 June 1865 he wrote to Hirst announcing that he had 'obtained *a simple and completely general proof* of Newton's Theorem', which he presented in a paper to the London Mathematical Society eight days later.[65] On 28 June it received notice in *The Times* under the title 'A Mathematical Discovery', while that evening Sylvester gave a public lecture on it at King's College. He notes that a character in a novel published later in the year cites an extract from the lecture, a reference to Newton's Rule as 'the beautiful child of Newtons [sic] youth'.[66] Sydney Lanier, one of Sylvester's colleagues while he was professor of mathematics at Johns Hopkins University from 1876 to 1883, wrote a verse tribute in 1880 to 'the so[a]ring-genius'd Sylvester / That, earlier, loosed the knot great Newton tied'. Sylvester describes his accomplishment in 'Urbi et orbi': 'Sprung from Descartes, which Newton to a higher level bore, / See now the fount and source lie open of that lore.' Not only is this brief poem encumbered with a disproportionately long subtitle, it is also furnished with a note to rival it, another couplet which, drawing upon his thematic imagery of heaven and fire, revises Pope's epitaph for Newton: 'Descartes' and Newton's law lay hid in night: / Heaven touched my heart with fire, and all was light.'[67]

Such celestial tropes are well exercised in a poem from 1886 that Sylvester addresses to the Savilian Professor of Astronomy at Oxford, Charles Prichard. Known to each other from their undergraduate days together at St John's College, Cambridge, the two men became good friends after Sylvester took up the Savilian Chair of Geometry in 1884. The poem was published in *Nature* in April 1886 under the informative title '*Sonnet To the Savilian Professor of Astronomy in the University of Oxford, Author of a*

Memoir on the Proper Motion of Forty Stars in the Pleiades, On receiving the Gold Medal of the Royal Astronomical Society for his Investigation of the Relative Brightness of the Fixed Stars'.[68] It was recited at the award ceremony.[69] An apotheosis of the astronomer, larded with classically inspired poeticisms, it begins with the line 'Pritchard! thy praise is lifted to the skies', and ends 'Whilst yet with watery ray yon Pleiads shine / Or strew with sands of gold their hair divine, / Thy praise shall flourish in immortal song.' The final line may involve a Pythagorean conceit, in which the light waves of the 'watery ray' are likened to vibrations of sound, the Pleiades' contribution to the music of the spheres. Alternatively, in a more direct and less apologetic reading, the final line assumes that posterity will recognise and retain Sylvester's poem as 'immortal song'.

Like most of his peers, Sylvester's formal acquaintance with poetry came from his classical studies at school and university, a legacy that combines with science to furnish his ambiguous imagery of the heavens, a conflation of mythology and astronomy that recalls Rowan Hamilton's 'Ode to the Moon under Total Eclipse'. In the 'Sonnet to the Savilian Professor', for instance, Pritchard's scientific investigations of the Pleiades, a group of stars in the constellation Taurus, are glossed through their origin in Greek mythology, where they are the seven daughters of Atlas and Pleione, and accordingly seen to 'strew with sands of gold their hair divine'. While such personifications are in scientific terms primitive and in verse mere poeticism, in Sylvester's poems they often suggest objective correlatives for the poet's wonder at the residual mystery of space and its phenomena, feelings that are also associated, through shared patterns of imagery, with female beauty. They describe the fascinating sphinx-like objects of inquiry that beautify space. Such feelings are attributed to Pritchard, 'Who in the starry fields find'st pure delight, / Noting each ray that gilds the brow of night.' Each such ray is further described in a draft version as an emissary 'From depths unfathomed, baffling Man's surmise',[70] a phrase that suggests Sylvester's open mathematical domain of 'transcendental space'. A curious poem that dates from a few months before his sonnet to Pritchard, 'To a Missing Member of a Family Group of Terms in an Algebraical Formula', the subject of the following section, playfully casts a mathematical value as a Pleiade missing in the immensity of space.

LOST IN SPACE

Sylvester is apt to quote poetry, both his own and that of others, in his mathematical papers and lectures, a further testimony to the Continuity he sees not only to characterise each of these activities separately but to run between them. One notable paper is structured as a play, in which he deploys a series of verse quotations as mathematical puns. 'A Constructive theory of Partitions, arranged in three Acts, an Interact and an Exodion' marks plot developments by heading sections with quotations from Shakespeare's *Twelfth Night* ('seeming parted, / But yet a union in partition'), Chaucer's *The Canterbury Tales* and Spenser's *The Faerie Queene*.[71] Sylvester places his poem 'To a Missing Member' in the middle of his inaugural lecture as Savilian Professor, 'On the Method of Reciprocants as Containing an Exhaustive Theory of the Singularities of Curves', which he presented belatedly in December 1885. He decided to use his inaugural lecture to speak on his most recent field of research and what would be his last great contribution to mathematics, 'reciprocants' or differential invariants, which he introduces in a subsequent undergraduate lecture as 'A new world of Algebraical forms, susceptible of important geometrical applications'.[72] Research in this area was part of the long work Sylvester shared with Cayley of making geometry algebraic, for as he explains in his inaugural lecture, 'every pure reciprocant corresponds to, and indicates, some singularity or characteristic feature of a curve, and *vice versâ* every such singularity of a general nature and of a descriptive (although not necessarily of a projective) kind, points to a pure reciprocant'.[73] His work on reciprocants refers back to the newly prominent study of curves he discusses in the Exeter address, an algebraic means of tracing the organisation of space that he sees to be the larger task of his mathematics.

In a table he gives of some rudimentary forms discussed in the inaugural lecture, 'THE H RECIPROCANTIVE PROTOMORPH', Sylvester lists a set of twelve terms that are functions of values he designates as U and W, and notes a further such form, 'a Thirteenth (a banished Judas), equally *a priori* entitled to admission to the group, but which does not make its appearance amongst them, namely, b^4d'. Sylvester explains and develops this term in figures that are not so much mathematical as anthropomorphic and poetic:

Still, in the case before us, this unexpected absence of a member of the family, whose appearance might have been looked for, made an impression on my mind, and even went to the extent of acting on my emotions. I began to think of it as a sort of lost Pleiad in an Algebraical Constellation, and in the end, brooding over the subject, my feelings found vent, or sought relief, in a rhymed effusion, a *jeu de sottise*, which, not without some apprehension of appearing singular or extravagant, I will venture to rehearse. It will at least serve as an interlude, and give some relief to the strain upon your attention before I proceed to make my final remarks on the general theory.

To a Missing Member

Of a Family Group of Terms in an Algebraical Formula.

> Lone and discarded one! divorced by fate,
> Far from thy wished-for fellows – whither art flown?
> Where lingerest thou in thy bereaved estate,
> Like some lost star, or buried meteor stone?
> Thou mindst me much of that presumptuous one
> Who loth, aught less than greatest, to be great,
> From Heaven's immensity fell headlong down
> To live forlorn, self-centred, desolate:
> Or who, new Heraklid, hard exile bore,
> Now buoyed by hope, now stretched on rack of fear,
> Till throned Astræa, wafting to his ear
> Words of dim portent through the Atlantic roar
> Bade him "the sanctuary of the Muse revere
> And strew with flame the dust of Isis's shore."[74]

The formal continuity of this eccentric sonnet is enhanced accordingly as the rhyme scheme (*abab baba cddcdc*) is muffled by pararhymes; a graduated effect of differential invariants that suggests the curved forms that concern reciprocants.

An indeterminate value that nonetheless has geometrical applications, the thirteenth reciprocant accordingly formulates mathematically the condition of being lost in space. Electra, the eldest of the seven daughters of Atlas and Pleione, is known as the lost Pleiad, with whom Sylvester identifies the algebraic term in his prefatory remarks to the poem. The protagonist is 'Like some lost star' that has abandoned, or been exiled from, the Pythagorean cosmos, 'the mathematical firmament' he figures in the Exeter address as revolving around 'the pole-star' of Continuity, the heaven of harmonious mathematical relations. Identified in the lecture with Judas, the missing term is similarly compared in the poem to the other great traitor in Christian mythology, Satan, who is of course also known as Lucifer, the morning star. It is not only this implication of exile, but also that of a (mathematically) derivative nature, that the poem delivers through its description of the term as a 'new Heraklid', one of the many descendants of Herakles who repeatedly tried to return to their ancestral homeland.

Help comes for the forlorn term in the form of the sympathetic goddess Astræa, the 'Starry Maid', who is herself an exile. The last of the gods to leave the earth after the Golden Age, when they dwelt among humankind, she withdrew to the heavens due to the wickedness of men. She was allocated the divine portfolio of Justice in recognition of her moral scruples, and transformed into the constellation Virgo, while her scales of justice became the nearby Libra. In a surprising parallel to Baudelaire's magisterial poem 'Le Cygne', Sylvester's 'To a Missing Member' sustains and vivifies the theme of exile by similarly rolling it through a series of parallel images of forlornness: 'Lone and discarded one!', the 'lost star', the 'buried meteor stone', Lucifer, a 'new Heraklid' and Astræa furnish a strain of imagery, a train of analogy, a 'continuity of mental impression', through which the mind can grasp its thematic 'invariant'.

Like Astræa and many of the figures that his poem likens to the missing term, Sylvester was also forced, through injustice, to be something of an exile for most of his life. Having sat the examination in 1837, Sylvester emerged from the Mathematical Tripos as second wrangler, as would Maxwell, William Thomson, Clifford and J. J. Thomson. His Jewish faith, however, precluded him from subscribing to the Thirty-Nine Articles of the Anglican faith. This disqualified him from graduating and taking a Smith's prize, and barred him from a fellowship and further opportunities: 'I feel what irreparable loss of facilities for domestic and foreign study', he declares in his Commemoration Day Address to Johns Hopkins in 1877, 'for full mental development and the growth of productive power, I have suffered, what opportunities for usefulness been cut off from, under the effect of this oppressive monopoly, this baneful system of protection of such old standing and inveterate tenacity of existence'.[75]

Sylvester places his faith elsewhere. Observing that Horace's *Odes* is 'the English public-schoolboy's own book ... (just as much as Euclid unfortunately is or was in another direction)',[76] he recognises Euclidean geometry as the establishment mathematical religion of Anglican boys' schools and Oxbridge, its propositions a suggestive parallel to the Thirty-Nine Articles. Justifying his idea of 'transcendental space' with his tenet that 'the highest knowledge comes by faith', Sylvester opposes Euclid with his modern continental creed of analysis, projective and non-Euclidean geometries, and his invariants.

The original figure Sylvester uses in his inaugural lecture for the thirteenth term, 'a banished Judas', suggests that a personal identification moderates his levity in the passage and the poem it prefaces. Because, as he puts it, 'he professed the faith in which the founder of Christianity was educated', being, as Geoffrey Cantor observes, 'among the very few Jews to study at Cambridge before 1856', Sylvester was not only precluded from graduating, but subject to prejudice and aggravation from his fellow students. One of his contemporaries reports that, having been taunted about his religion by another undergraduate at a College dinner, he 'lost his temper and threw a plate or dish at the offending person's head'.[77] A precedent and premise for Sylvester's identification of his mathematical figure with the stigmatised Jew can be traced to the opening paragraphs of the Exeter address, where, having recounted witnessing a Royal Institution audience enthralled by a lecture on the Talmud, he parallels (and personalises) Judaism and Mathematics in the following remarkable declaration: 'Now, as I believe that even Mathematics are not much more repugnant than the Talmud to the common apprehension of mankind, and I really love my subject, I shall not quite despair of rousing and retaining your attention for a short time.'[78] Sylvester can be credited with some sincerity when he writes in his inaugural lecture of being affected by the plight of the missing term: 'this unexpected absence ... made an impression on my mind, and even went to the extent of acting on my emotions', and so prompted him to write his poem; 'my feelings found vent, or sought relief, in a rhymed effusion'.

Sylvester embarked upon a peripatetic working life after Cambridge, becoming professor of natural philosophy at University College London in 1838, then an unwelcome foreign professor of mathematics at the University of Virginia from November 1841 to February 1842. For the next ten years he took private pupils in mathematics, amongst them Florence Nightingale, and worked as an actuary and an executive manager for an insurance company from 1844. Beginning legal studies in 1846 with a view to further advancement in his actuarial position, he was called to the Bar in 1850. He resumed university work in 1855, as professor of mathematics at the Royal Military Academy in Woolwich, London, a position he held until 1870. He was belatedly awarded his Cambridge BA and MA in 1872, the year after the religious tests were repealed. In July 1877 he returned to the United States to take the foundation chair of mathematics at Johns Hopkins, where he stayed until he was appointed to the Savilian professorship, a move that occasioned his inaugural lecture and is alluded to in the poem he embedded in it.

That Sylvester wrote 'To a Missing Member' expressly for his lecture is clear from its final lines, as the sestet builds dramatically to further disclose his personal identification with its protagonist. '[W]afting to his ear / Words of dim portent through the Atlantic roar', Astræa is attributed with correcting the original injustice that exiled Sylvester from the Oxbridge system, bidding 'him "the sanctuary of the Muses revere / And strew with flame the dust of Isis' shore"'. Dusty Oxford is to be disturbed and enlivened by a resurgent flame. Sylvester observes in the Exeter address that 'No mathematician now-a-days sets any store on the discovery of isolated theorems, except as affording hints of an unsuspected new sphere of thought, like meteorites detached from some undiscovered planetary orb of speculation.'[79] Like such meteorites, the meaning of the missing member will become clear as it is brought into relation with its family of terms. No longer to be compared to a 'buried meteor stone', Sylvester will, like Kepler in his earlier poem, 'flash out to a meteor's blaze'.

As indicated earlier, such fiery imagery is used in a romantic, almost Blakean, manner in poems from *The Laws of Verse* to designate creative energy and genius. This is clearly exemplified by the vow that opens 'Kepler's Apostrophe':

> Yes! on the annals of my race,
>   In characters of flame,
> Which time shall dim not nor deface,
>   I'll stamp my deathless name.

The autobiographical impetus behind this poem is obscured by the anonymous attribution it receives in *The Laws of Verse*, much as it is by levity in 'To a Missing Member'. Kepler's academic career was, like Sylvester's, badly disrupted by his religion. His Lutheran Protestantism led to his expulsion from universities in Graz, Prague and Linz. Sylvester combines mathematical prowess with religious nonconformity in his defiant imagery of fire, which, as was noted earlier, makes at least one reference to Milton's Satan. The final stanzas of 'Kepler's Apostrophe' affirm the scientist's resilience, its last lines that 'steadfast faith unkindly used / Hardens to stubborn pride'. Sylvester's mathematical and aesthetic conviction in an overarching principle of Continuity may also have been informed by an abiding personal imperative to persevere with his mathematical career and prevail over various obstacles and disruptions, a catalogue of exclusion and exile that is rendered hyperbolically in 'To a Missing Member' and resolved triumphantly with his appointment as the Savilian Professor. In 'Kepler's Apostrophe' the astronomer anticipates his apotheosis with 'some lovelier star' who will 'Ascend with me Fame's fiery car / And claim celestial place',[80] a model Sylvestrian romance that draws together the two types of heavenly imagery he establishes in the 'anonymous' poems from *The Laws of Verse*, a team that anticipates the return to 'Isis' shore' of Astræa and the missing algebraic/Hebraic term in the later poem.

CHAPTER 9

James Joseph Sylvester: the calculus of forms

The parity that Sylvester gives to his mathematical and poetical work is nicely displayed by each of them. He not only makes them companionate by citing poetry in his lectures on mathematics, but intimate through a shared terminology. The idiom in which he discusses and renovates prosody in *The Laws of Verse* is esoteric, indeed much of it is drawn from his mathematical coinages. Like its classical quantitative and modern accentual counterparts, Sylvester's prosody focuses upon the 'Rhythmic' aspect of verse, which he sees to be composed of three elements: 'Metric is concerned with the discontinuous, Synectic with the continuous, aspect of Art', while 'Between the two lies Chromatic, which comprises the study of the qualities, affinities, and colorific properties of sound.'[1] This announces the principal innovation of his prosody. Metric, whether it be accentual syllabic or quantitative, divides lines of verse into segments, simple arithmetical and geometrical ratios defined by fixed beats or periods. Synectic, however, appreciates the sound patterns of poetry as tonal modulations, as subtly variable and continuous in the manner of curves, the motion of the point traced by differential calculus.

'My chief business', Sylvester declares, 'is with Synectic.' Describing unmediated relations of cause to effect, the word synectic was, as Sylvester explains, adopted by the French mathematician Augustin Cauchy in a specialist meaning that designates the mathematical principle of Continuity: 'For the benefit of my non-mathematical readers, I may say that "Synectique" is a word used, and I believe for the first time, by Cauchy in his Theory of Functions, the true and very insufficiently acknowledged foundation and origin of Riemann's great doctrine of Continuity.'[2] 'A function of a complex variable which is continuous, one-valued, and has a derived function when the variable moves in a certain region of the plane,' explains Benjamin Williamson in 1888, 'is called by Cauchy *synectic* in this region.'[3] Similarly, his colleague at Johns Hopkins, the mathematician, philosopher and fellow enthusiast for Riemann's thought, Charles Sanders

Peirce, acknowledging the term's etymology from the Greek word for the continuous, and perhaps encouraged by Sylvester's idiosyncratic extension of its application, writes in 1893 that he 'proposed to make *synechism* mean the tendency to regard everything as continuous'.[4] As Sylvester applies the term to prosody, synectic describes the conditions that define the locally continuous in particular sequences of verse.

Sylvester writes that 'Synectic, which leads down from the Alps of Cauchy and Riemann to the flowery plains of Milton and Byron', provided his 'crossing-point' between 'scholarship and science'. Mathematics, once again represented as sublime and reaching heavenward, dominates Sylvester's cultural landscape. Synectic is implicitly likened to Alpine water-flows that irrigate 'the flowery plains' of poetry below, a fluid Principle that derives conceptually and rhetorically from the representation of space in the Exeter address, 'the Grand Continuum from which, as from an inexhaustible reservoir, all the fertilizing ideas of modern analysis are derived'. The 'reservoir' of space that curves trace and circumscribe is paralleled in poetry by Synectic, which is seen in similarly fluid terms to 'run into three channels – *Anastomosis, Symptosis*, and between them the main flood of Phonetic Syzygy [*sic*]'.[5]

'Syzygy' is for Sylvester the crux of poetry and hence the focus of *The Laws of Verse*: 'It is to Syzygy that I have called most attention in my annotations, and it was with this principle exclusively that at the outset I intended to deal.'[6] The original use of the term in Greek prosody refers to a dipody, literally a combination of two feet. The earliest definition of *syzygia* recorded in the *OED*, by Edward Phillips in 1658, identifies it with the sound of two feet clapping: 'Among Grammarians, the coupling or clapping of different Feet together.'[7] Sylvester expands this idea of two metrical feet combined in one measure, a pair brought together in a single sound, into a fluid principle of phonetic amalgams and tonal relations between word sounds.

While both classical and modern scansion focus upon the vowel sound, Sylvester's syzygy foregrounds the neglected contribution that consonants make to the music of lyric poetry as they mediate such sounds. Accentual stress and indeed quantity are accordingly subordinated to synectic, to principles of Continuity. Syzygy describes the continuous 'colorific properties of sound' encompassed by the Chromatic moment. 'There is', Sylvester writes in *The Laws of Verse*, 'quite as much room for the exposition of a method of distributing sound as of laying on colour.'[8] In his 1880 book *The Science of English Verse*, Lanier explains that 'Such a succession of consonant-colors has been called Phonetic Syzygy (syzygy, from *sunsugia*, yoking together) by Professor Sylvester, in his *Laws of Verse*,' and he congratulates his colleague for the originality and usefulness of his principle; 'the term seems so happy as to be a genuine contribution to the nomenclature of the science of English verse'.[9]

Early in *The Laws of Verse*, Sylvester acknowledges the influence of Edgar Allan Poe's essay 'The Rationale of Verse',[10] which argues that existing schemes of scansion, especially accentual metrics, are crude and deforming impositions upon the natural music of poetry. Verse is defined on the first page of Poe's study, primarily through relations of metrical sequence and other formal patterns, both as and by arithmetical ratio: 'the subject is exceedingly simple; one tenth of it, possibly, may be called ethical; nine tenths, however, appertain to the mathematics'. Poe goes on to contrast traditional systems of scansion, the perseverance of which he attributes to an ungrounded respect for authority, with a scientific study of verse: 'English Prosodists have blindly followed the pedants ... in place of facts and deduction from fact, or from natural law, were built systems of feet, metres, rhythms, rules.' He accordingly writes of the need for poetry to be 'scanned by the true *laws* (not the supposititious *rules*) of verse',[11] a formulation that describes the reforming thesis, and may also have suggested the title, of Sylvester's study.

In place of existing prosodies Poe proposes his own system of quantitative measures, a set of relative values that he designates numerically and places directly under the syllables they describe. As in classical prosody, the long syllable furnishes the basic unit, beneath which Poe accordingly places the number 1, while the short syllable, being half this length, is represented by a 2, a quarter with a 4, and so on. This parallels musical scansion, the modern sequence of semibreve, minim, crotchet, quaver and so on, whilst of course allowing a far greater and finer range of quantitative values to be discerned and named, such as the sixth that he discusses and exemplifies in his essay.[12] Sylvester's principle of syzygy similarly brings forward the musical flow of the line in poetry. This is the basis of Lanier's appreciation and development of it in *The Science of English Verse*, which, he maintains, 'gives an account of the true relations between music and verse'. Indeed Lanier argues that music and poetry share the same laws and hence the same measure, and applies musical notation directly to the explanation of verse.[13] Sylvester draws his own parallels between the measures of music and those of verse, as for instance in describing 'the office of rhyme in marking off a line as a sort of compound foot (like a bar in music)'.[14] Lanier wrote to Sylvester in 1878 thanking him 'for the pleasure I have drawn from your *Laws of Verse*', saying that he will quote from it 'several times' in his lectures on Shakespeare.[15] The

Science of English Verse suggests the completion of a sequence, as it follows upon what its preface describes as 'the defiant metrical outburst of Poe in his *Rationale of Verse*, and the keen though professedly disconnected glimpses of Professor Sylvester in his *Laws of Verse*'. Recognising it as 'a branch of poetic science', Lanier devotes a chapter of his book to 'Phonetic Syzygy'.[16]

The enlarged type that Sylvester uses to introduce 'Phonetic Syzygy' in *The Laws of Verse* suggests not only his enthusiasm for the principle, but its unwieldy nature, the reason why this 'main flood' needs to be flanked on either side by the conditioning principles of Anastomosis and Symptosis. Anastomosis offers a connection, a mouth, between two vessels or channels. It organises and directs syzygy: 'Anastomosis gives to verse its cohesion, syzygy its flow and consistency.'[17] The use of the term, and indeed the currency it acquired amongst his peers, is illustrated in a correspondence in *Nature* in 1871 on scientific neologisms, in which Ingleby objects to Wallace's coinage 'prolificness' on the grounds that 'the syllables *ic* and *nes* will not inosculate. To use Mr. Sylvester's phraseology, there is not a perfect anastomosis.' He accordingly suggests *'prolificence'*.[18] Symptosis, which Sylvester coins from the Greek *Symptoma*, for an accident, chance event or disease, suggests a disruption to this flow: 'rhymes, assonances (including alliterations, so called), and clashes (this last comprising as well agreeable reiterations, or congruences, as unpleasant ones, i.e. jangles or jars)'.[19] These *Principles of Versification* are, as the subtitle of the book promises, *Exemplified in Metrical Translations*, primarily through Sylvester's fastidious translation of Horace's *Ode to Maecenas*.

Sylvester's regard for Horace's *Odes* as a happy alternative to Euclid's *Elements* surfaces quietly in his notes to the translation of the *Ode to Maecenas* that opens *The Laws*, where he recalls working on the poem's first line: 'I hesitated, and chopped and changed a long time, as my printers can too well attest, between the two readings, "Birth of Tyrrhenian [regal line!]" and "Tyrrhenian, birth of [regal line!]"; and yet it is as certain as any proposition in Euclid can be that the former is the proper order of words.' What makes this order apodeictic is, however, as he goes on to explain, its Continuity, the mathematical quality that he opposes to Euclid in the Exeter address. Sylvester has two objections to his second formulation of the line from Horace:

> 1st, of the *b* in 'birth' following the *n* in 'Tyrrhenian,' contrary to the laws of Anastomosis; and, 2nd, of the number and measure of 'Tyrrhe' being to the number and measure of 'Tyrrhenian, birth of re', as 1:3; whereas, in the contrary order, the corresponding ratio is as 2:3, – which latter, by the principles of Symptosis (here applying to the clash or congruence of the open ē sounds) is preferable, especially at the opening of the piece, as being less suggestive of subdivision of measure.[20]

What Lanier usefully refers to as the 'Junction of terminal and initial consonants'[21] in the case of Sylvester's preferred rendering of the phrase, describes a syzygistic relation ('Tyrrhenia*n* *r*egal'), while in the alternative sequence it baulks Anastomosis, disrupts the synectic flow of phonetic sound ('Tyrrhenia*n*, *b*irth'). The unit of sound he focuses upon in his second argument, 'Tyrrhe', is a syzygy, as, mediated by its softened central group of breathy consonants, the two vowels are drawn into a coherent flow of sound, an extended metrical foot in which the diatonic effect of accent is correspondingly diminished.

While stress usually lengthens the vowel-sound and hence the foot it constitutes in traditional metrics, it also clearly marks its separation, and so threatens the synectic flow. Consequently, for Sylvester, '*Metrik*, ... including the study of accents, pauses, and interruptions, may be referred to the antagonistic principle of discontinuity.'[22] He accordingly conceives of the main stress in 'Tyrrhe' as one of the poem's 'open ē sounds', as assonant rather than accentual. Sylvester's placement of the syzygistic foot 'Tyrrhe' within the variant translations he gives of the first line from Horace yields different measures. 'Birth of Tyrrhenian re[gal line!]' does indeed engender a more regal metrical line, as the accentual stresses and the synthesising flow of the syzygistic foot are dwelt upon and allowed a fuller measure than its more hurried instance at the abrupt start of the alternative, 'Tyrrhenian, birth of re[gal line!]'. The former is judged preferable by the criteria of Symptosis, which appreciates the way that the close proximity of 'the open ē sounds' (in 'Tyrrhenian' and 'regal') facilitates an assonant effect of congruence. This chiming is, however, dissipated in the alternative form, as the syllables are separated from one another at either end of the musical phrase and on opposite sides of the caesura that the comma effects. As more distant echoes of one another, their effect is dissonant rather than assonant. They tend to break rather than enhance the synectic flow.

Symptosis focuses upon relations of sound elements, including those of alliteration, assonance and rhyme, which in being drawn together through their chiming with one another correspondingly extricate themselves from the main synectic flow in which they are embedded, so that they serve to both enhance and punctuate it. Thus, 'Alliteration and rhyme ... are under one point of view only extreme cases, or as we say in algebra, limiting forms of phonetic syzygy.' Sylvester clarifies the functionalist nature of symptosis, 'the theory of clashes and beats', by posing and then addressing a rhetorical question:

how, I say, can a theory dealing with discreet [*sic*] matter of this kind, come under the head of Synectic: but the answer is easy, for if the elements with which it deals, its *matter*, is discontinuous, not so is the object to which it tends (its *form*); just, for instance, as in an iron shield or curtain or a trial target, the bolts and screws and rivets are separate, but serve to consolidate and bring into conjunction the plates, and to give cohesion and unity to the structure.

Symptosis strengthens the synectic form of poetry as rivets do a metal structure, being at once distinct from and integral to it. Sylvester's tropes of industrial construction, which makes poetry analogous to the contemporary technology of iron bridges, railway trains and grand stations, as well as the ocean liners of the following century, anticipates the scientistic functionalist aesthetics of Le Corbusier and Amadée Ozenfant's *L'Esprit Nouveau* in the 1920s. Poetry is for Sylvester similarly geared metaphorically to the functionalist integrity of metal: 'it is to Synectic, and to its main branch Syzygy, that we must attend in order to secure that coherence, compactness, and ring of true metal, without which no versification deserves the name of poetry.'[23]

PLAYING 'SYZYGIES'

Sylvester's 1876 volume *Fliegende Blätter* is introduced on its half-title page as *Rosalind, and other Poems*, with both being allocated the subtitle *Supplement to the Laws of Verse*. It accordingly follows its predecessor in being 'not a treatise on Prosody' but 'an attempt to illustrate' its ideas about versification 'by examples', principally that named by its alternative title, which like the *Ode to Maecenas* is placed first in its book. His own composition, 'To Rosalind', allows Sylvester much greater scope than the earlier translation to demonstrate his principle of syzygy. Accompanied by a substantial introduction and commentary, there is something of the stridency of the manifesto in the over-determined instance of the 'continuity of sound' that 'To Rosalind' embodies, as each of its legion lines is made to rhyme with the addressee's name.[24]

Published before he left for Johns Hopkins, the version of 'To Rosalind' printed in *Fliegende Blätter* runs to 268 lines, but grew in length through subsequent revisions. Daniel Coit Gilman, the first President of Johns Hopkins, records that the versions Sylvester read 'to many unwilling hearers' while he was at the university were considerably longer:

he became possessed by a sort of monomania for rhyme, and soon after he came among us his friends were confidentially treated to a long series of lines, every one of which ended with a syllable pronounced both *īnd* and *ĭnd*. Rosalind was the theme... This extraordinary composition, a veritable *tour de force*, reached four or five hundred verses, each closing with the three monotonous letters or their vocal equivalents ... An early manuscript copy is in the archives of the University, and I will give a few lines from it I am afraid to give more.[25]

While this account effectively foregrounds the workings of symptosis, the rhymes that the poem makes on the long and short sounds of the final vowel in Rosalind are seen by Sylvester to participate in more subtle aspects of synectic flow, principally syzygy.

Having made his case for syzygy in *The Laws of Verse*, Sylvester identifies himself with the principle a year later by signing his translation of the poem 'The Ballad of Sir John de Courcy' with the pseudonym 'Syzygeticus'. Coming, as Lanier noted, from the Greek word for a yoking together, syzygy has also been used in astronomy since the seventeenth century to describe the conjunction of heavenly bodies along one line, a simple instance of mathematical Continuity and a clear analogy to dipody in poetry. Sylvester finds a further, mathematical, analogy for the term in an 1850 paper on conics: 'the members of any group of functions, more than two in number, whose nullity is implied in the relation of double contact ... must be in syzygy. Thus PQ, PQR, QR, must form a syzygy.'[26] With his enlargement of this term Sylvester once again asserts that poetry and mathematics are united by comparable concerns with formal relations between quantities. Carroll demonstrates this Continuity conversely by applying Sylvester's mathematical relation of syzygy to relations between words in his game 'Syzygies'.

Carroll devised his word-puzzle game of Syzygies in the late 1870s as a sequel to his popular 'Word-Links' or 'Doublets', in which by changing one letter at a time a word undergoes a set of transformations into another prescribed final term, as for example; 'HEAD / heal / teal / tell / tall / TAIL'. Syzygies similarly specifies the first and the last word in a chain that the player needs to complete. Here, however, each word must be linked to the next by a shared sequence of letters, which is then superseded by another such sequence, until the final term is realised. Carroll gives the following chain as a solution to the problem '*Introduce* Walrus to Carpenter':

> WALRUS
> (rus)
> peruse
> (per)
> harper
> (arpe)
> CARPENTER[27]

Such sequences of letter-clusters correspond directly to the mathematical terms of Sylvester's syzygy. Belonging to only two adjacent words, their 'nullity is implied in the relation of double contact', so that they proceed in the pattern: 'PQ, PQR, QR'.

Sylvester returns his distinctive mathematical principal of syzygy to prosody in his renovated application of the term to describe gradual transformations in word-sounds. He glosses such subtle relations in a note to a line from 'To Rosalind', '*Dainty dame treads brodequin'd*', where he observes that 'the word *treads* ... collects into a focus ... all the leading sounds of the environment – it *remembers* the dropped *t* sound of "dainty" and *prepares* the *rd* sound of "brodequin'd"'. Another example from the poem, a line that describes spray issuing from Niagara Falls, 'Tone-spray tossing on the wind',[28] similarly pivots upon its verb. The letters *to, s* and *n* of 'Tone-spray' are gathered up and moderated in 'tossing', while the *n* sound perseveres, carried along with the flattened *o* in their permutations 'on' and 'wind', and the alliterative *t* persists mildly in the spirant 'the'. In a further, more contained, example, the mathematician Sylvester forms a szyzygy with his poetic alter ego, Syzygeticus.

The uses to which Carroll and Sylvester put their mathematical principle of syzygies highlight their eccentricity in relation to prevailing mid-Victorian literary culture. Both defer good-humouredly to more established and 'professional' practitioners of literature, recognising that their own writing proceeds from certain premises and practices that they import from mathematics, displaced habits of problem solving and exercises in permutations and combinations, which allow it to be construed as 'nonsense'. Carroll offers his readers hints for making chains of syzygies 'in the form of a soliloquy, supposed to be indulged in by the possessor of what Tennyson would call "a second-rate sensitive mind", while solving the problem "*Turn* CAMEL *into* DROMEDARY"'.[29] Acknowledging that their respective exercises are incommensurable, Carroll makes fun of both his soliloquy and Tennysonian high-mindedness: 'If any of my readers should fail, in attempting a similar soliloquy, let her say to herself, "It is not that my mind is not *sensitive*: it is that it is not *second-rate!*"'[30] The form of his soliloquy may be judged inferior by extrinsic criteria of sensibility and aesthetics, but not ostensibly on utilitarian and intellectual grounds, for it is directed toward solving a problem, albeit one that, like Lear's limericks, is likely to be tautologous.

Fliegende Blätter ('Flying Leaves'), which appears to have borrowed its title from a popular German comic paper of the time, is 'Dedicated to Frederick Locker, Esq., Laureate of the Lighter Muse, in Token of Appreciation and Regard.' The well-connected Locker, who changed his name to Locker-Lampson in 1885, included among his close friends

Tennyson, Eliot, Trollope, Dickens, the British royal family, the Brownings and Franz Liszt. As a poet, he was best known for his popular *London Lyrics*, which was first published in 1857 and regularly revised for its successive editions. His poem 'On "A Portrait of a Lady". *Vide Royal Academy Catalogue*. By the Painter', which provides the cues for Sylvester's 'To Rosalind', was first included in the 1868 issue of the collection.[31] Referring implicitly to Locker and his poem, Sylvester explains in his preface to 'To Rosalind' that it 'is to be understood as the out-come of a "wit-combat", into which its author, Marsyas-like, had the temerity to enter with a veritable Apollo in wit and song'.[32] Akin to Carroll's syzygistic alternatives to Tennyson, and in contrast to the courtly art forms embodied by Locker's 'On "A Portrait of a Lady"', 'To Rosalind' springs from the lowly and transparently contrived form of the riddle and other word-games, including his sport of continuous rhyming, as through his restless revisions of the poem he progressively increases his record score.

Sylvester's approach to writing his poem, which in devising its syzygies and new rhymes is at once playful and workman-like, directly affronts the romantic myth of the inspired and solitary poetic genius, the figure of the artist who in Locker's poem 'stand[s] aloof'.[33] He thanks Ingleby and other friends for supplying him with particular rhymes for his poem, evidently approaching poetry as a communal practice, similar to his mathematical work with Cayley and his other correspondents in England and on the Continent.[34] Such cooperation is complemented by competition, much as it is in mathematics,[35] so that it is also a 'wit-combat', a jovial contest as well as a shared 'puzzling out'. Carroll provides an algebraic formula for scoring his game, with variables representing such things as the numbers of links and waste letters, and the 'greatest No. of letters in an end-Syzygy'.[36] Correspondingly, Sylvester focuses mainly upon the simple criterion of length and his witty conceits and particular syzygies. Both mathematicians permit themselves some leeway, with Sylvester allowing the punning eye-rhyme that conflates the long and short i-sounds, and Carroll, in Rule 6 of his game, that 'The letters "i" and "y" may be treated as if identical. Thus "busy(usy) using" would be a lawful Syzygy.'[37]

UT PICTURA POESIS

As the title of Sylvester's poem indicates, it draws its dominating formal conceit from the verses that Orlando addresses to Rosalind in Shakespeare's

As You Like It. Each line of Orlando's poem reiterates his obsession with his love by closing with a rhyme on her name:

> *Enter* ROSALIND [*with a paper, reading.*]
> *Rosalind*: From the east to western Inde
>   No jewel is like Rosalind.
>   Her worth being mounted on the wind,
>   Through all the world bears Rosalind.
>   All the pictures fairest lin'd
>   Are but black to Rosalind.
>   Let no face be kept in mind
>   But the fair of Rosalind.

Sylvester's poem evidently tries to fulfill the extravagant pledge to surpass Orlando's efforts that Touchstone makes to Rosalind: 'I'll rhyme you so, eight years together; dinners and suppers and sleeping-hours excepted.'[38] Such continuous rhyming is an obsessive form of containment, a nonsense effect that Lear uses in a poem he wrote for his sister Ann when he was thirteen, consisting of 110 lines of birthday salutations that rhyme with the word 'relation'.[39] As Chapter 1 noted, such solipsistic and inward-looking uses of rhyme have their *reductio ad absurdum* and paradigm in the enclosing repetition of the final word in the first and last lines of Lear's limericks.

'To Rosalind' receives a two-page preface, which begins with an account of the poem's origins:

A lady ('a born actress' distinguished for her successful impersonation of Shakespeare's 'Rosalind') was supposed by the author of the annexed lines, with or without reason, to have sat in part for the portrait of Cecilia (can one live near the rose without imbibing some of its fragrance?), as delineated in a tale written several years ago and destined to be read with delight for many years to come.

The model for the protagonist is identified in some lines that a friend addressed to Sylvester ('Whose verse "with thanks" is oft declined') after reading his poem: 'Algebraic master-mind, / Helen Faucit's grace refined.'[40] Rosalind was one of the two most often acted and popular parts played by Faucit (1817–98). The actor was celebrated for her idealised feminine construal of the role, which broke with traditions of the time and corresponds to the image that Sylvester pays hyperbolical homage to with his poem. Indeed, as G. H. Lewes remarks at the time of her 1865 revival of Rosalind, Faucit is 'accustomed to be[ing] smeared with fulsome undiscriminating praise'.[41]

Holding within it the further recursive masquerade of Ganymede, the role of Rosalind provides an apt label for the otherwise unnamed protean protagonist of Sylvester's poem. A mere signifier, 'Rosalind' nonetheless offers a nominal ground for the various transformations that the persona undergoes in the poem, just as it does for the sustained rhyme scheme that emblematises these changes so insistently. A semi-algebraic case of Miss X, the anonymous heroine is cast as a series of projections, as she performs Rosalind, is painted and sketched as Cecilia and is cast descriptively and formally in the poem's continuous lines of verse and rhyme. Indeed the fictitious and arbitrary nature of her main designation is emphasised by these rhymes on 'Rosalind', which pivoting upon pararhymes of long and short *i* sounds, fail to specify precisely how it is pronounced.[42] From 'Rosalind' Sylvester's protagonist is changed by verse and painting into Cecilia, whose name is in the poem drawn from Celia, the friend of the Shakespearean character. The reference in Orlando's poem to 'All the pictures fairest lin'd' offers a link to her transformation in the painting of Cecilia, and the 'tale' in which it was originally 'delineated', Locker's poem 'On "A Portrait of a Lady"'.

Heading the 'Studies in Monochrome' section of *Fliegende Blätter*, 'To Rosalind' sustains the Horatian *ut pictura poesis* conceit of Locker's poem. The first two stanzas of 'On "A Portrait of a Lady"' are dedicated to replicating the eponymous Portrait, albeit ambiguously, by picturing either the Lady directly or the painting of her, it is not clear which. These stanzas introduce her synecdoche, and the poem's central motif, 'A rose with a passionate heart [that] is twined / In her crown of golden hair.' The poem turns about the conceit of 'the bloom of bygone days' that these admirers find the portrait to have captured. The identity of the painter and his relation to his subject, questions that intrigue the viewers of the picture in the third stanza, are addressed directly by the stanzas that follow it. These are spoken in the first person by the painter-persona, who is of course also the poet, and hence presents an encompassing allegorical figure of the artist. We learn from this fifth and final stanza that the persona had placed the rose in his beloved's hair, and that, as the closing words reveal, it was returned to him, to be preserved through his love and art: 'She gave me that rose, it is fragrant yet, / And its home is near my heart.'[43] Sylvester alludes directly to this poem, which itself comes to be enfolded by the symbol it develops, with the references he makes to living 'near the rose' and 'imbibing some of its fragrance'.

'To Rosalind' opens by introducing its portrait of Cecilia as a work of art that, contrasting directly with Locker's 'Portrait of a Lady', is not in oil, but in pencil, taking the form not of a love lyric, but of a riddling conceit:

> Friendship's not like love purblind –
> Spelled with sympathetic mind
> In Cecilia's name, I find

> Pencilled in with art refined,
> (Richest ore is deepest mined) –

The first two lines, huddled together as a couplet prior to being engulfed by the poem's gargantuan rhyme scheme, can be read as addressed directly to the friends to whom Sylvester sent the privately printed volume as his 'Swan-Song before leaving England' for Johns Hopkins.[44] They accordingly acknowledge that these readers cannot be expected to suspend their critical judgement, to be entirely indulgent in the manner of the Shakespearean lover. After a long parenthetical interjection, in which the poet again identifies himself apologetically with 'Marsyas stript and skinned', the riddle is completed and resolved with a tortuous pun on the name Cecilia:

> Be not my presumption fined
> That would peer athwart the blind:
> Is not well the clue divined,
> Or the hap was't undesigned
> In the words thus underlined
> (Precious ore in knotty rind),[45]
> *Celia* with a *sigh* combined?

Sylvester defers to the high art of Locker's poem and its subject matter, the oil painting of the lady, with a riddle that is little more than a name written in pencil.

'In Cecilia's name, I find / Pencilled in with art refined', '*Celia* with a *sigh* combined'. These lines are cued, and brought together across the intervening divide, by the parallel parenthetical lines that refer respectively to 'Richest ore' and 'precious ore', a trope that provides a clue to the riddle. Gilman glosses the conceit with arithmetical directness: 'Celia + ci = Cecilia.' The 'ci' sound is extracted syzygistically from 'Celia', like precious metal from ore. Sylvester elaborates the solution to his riddle more fully or indulgently in one of his notes to the poem:

> *Celia with a sigh combined?*
> *Do* spells *dough* our loaves are made with,
> *Bo* spells *beaux* by ladies played with,
> *Ci* spells *sigh* Cecilia's said with.[46]

Celia is supplemented with a punning sigh to become Cecilia. Sighing is the lover's defining activity in *As You Like It*, so that, for instance, the vignette in Jaques's 'All the world's a stage' speech has 'the lover, / Sighing like furnace'.[47] 'Spelled with sympathetic mind' by 'love purblind', rather as Rosalind is in Orlando's verses, the name of her friend 'Celia' is informed at its heart with the lover's sigh to become 'Cecilia', the beloved of the portrait painting.

With the precarious pun on the additional letters and the breath expended in uttering the fuller name of Cecilia, Sylvester asserts the dignity of the riddle and the pun against the high arts identified with Locker's poem, much as Shakespeare's Touchstone does in the face of Orlando's courtly poetry. A defence of the quasi-nonsense form of the pun, akin to Maxwell's in his essay on 'Analogies in Nature', it can be traced to the poem's opening gambit, the couplet that turns about the word 'Spelled'. In order for 'the Cecilia notion', the foundational pun that he describes as the poem's '*Primum Mobile*', to be credited it must be, letter by letter, 'Spelled with sympathetic mind'. The critical judgement of the friend needs after all to be suspended and superseded by the lover's 'sympathetic mind', which, as Orlando's poem demonstrates, allows him to be enchanted by the beloved's name, the incantation of its sequent letters that punningly enacts the spell. In this way, as the poem reflects after answering the riddle, the 'five letters loose align'd' of Celia become 'Magic set and re-entwined'.[48]

Like that of Locker's poem, the poet-persona of 'To Rosalind' is figured as the lover, and in both cases the love object is identified with a work of art. Indeed Sylvester's aestheticist conceit anticipates Wilde's *The Picture of Dorian Gray*, the eponymous hero of which recognises and loves the actress Sibyl Vane only as the incarnation of Shakespearean heroines (indeed he first speaks with her after she has played Rosalind).[49] The powers of spelling and enchantment that are associated with the lover and poet in Sylvester's poem suggest Theseus' well-known speech in the final scene of *A Midsummer Night's Dream*, which declares that 'The lunatic, the lover, and the poet / Are of imagination all compact.'[50] This convergence is enacted in the section of *As You Like It* that Sylvester draws upon, where the lover's besotted repetition of the beloved's name lends Orlando's verses a foolish and facile quality, which Touchstone easily exploits in his parody of them.

Sylvester prefaces 'An Inquiry into Newton's Rule for the Discovery of Imaginary Roots', the first public notice of his celebrated, but at this stage partial, proof of Newton's theorem, with an epigraph from Theseus' speech. Depriving them of their immediate context, he quotes only the last two lines of the following:

> And as imagination bodies forth
> The forms of things unknown, the poet's pen
> Turns them to shapes and gives to airy nothing
> A local habitation and a name.

Sylvester observes in his essay that while Newton's rule exists 'without proof or indication of the method by which he arrived at it, or the evidence upon which it rests',[51] it perseveres as a 'marvellous and mysterious rule'. The epigraph implicitly identifies Newton's single ungrounded theorem with the raw speculative workings of imagination, which Sylvester, poet-like, finally locates and substantiates. Conversely, his quibble in 'To Rosalind' on '*Ci*' and 'sigh' proceeds from the Shakespearean conceit of poetic conjuring, as the spell of 'Cecilia' gives shape to the 'airy nothing' of breath.

'STUDIES IN MONOCHROME'

Akin to a mathematical variable, x or y, the unnamed subject of 'To Rosalind' is known only through her transformations, Sylvester's projections of her as the pictorial and literary figures of the lady in Locker's poem, the portrait of Cecilia, and the role of Rosalind. More radically, she exists as a vocal projection, in the sonorous properties of shaped breath ('*Celia* with a *sigh* combined') through which synectic properties persist as invariants. Such formal qualities have their semantic counterpart in the relations of persistent pulchritude that pulse through the protagonist's various incarnations, generated by the remorseless motor of the rhyme scheme: 'Sum of worth in woman-kind! / Whose dear praises I could grind'.[52]

Maxwell's 1874 poem 'To the Committee of the Cayley Portrait Fund', which Sylvester not only knew, but offered 'corrections' for,[53] demonstrates concisely how Locker's *ut pictura poesis* conceit can be enhanced by projective and non-Euclidean geometries. It concludes with a conceit that presents portrait painting as akin to projective geometry and the algebraic methods by which Cayley and Sylvester extended such explorations of space. This representation also describes a set of conventions by which a figure of at least three dimensions is rendered in only two, as the rich character of its subject is projected upon the canvas:

> In two dimensions, we the form may trace
> Of him whose soul, too large for vulgar space,
> In n dimensions flourished unrestricted.[54]

The 'form' that the portrait defines, the irregular outlines of the face, is suggestive for non-Euclidean preoccupations with curvature, while the work of the portraitist, like that of Gauss and Riemann, is accordingly seen to be one of local definition, of intrinsic measurement and relations, an effort to recognise and register specific and unique qualities.

Having read an article by Sylvester in 1866 that brought back to him his early interests in Cartesian Ovals, Maxwell begins a letter to its author with another genial allusion to their shared interests in mathematics and poetry, as he carefully refers to 'the R[oyal] S[ociety] Laureat (Chasles, not Tennyson)'.[55] With his abiding delight in such geometrical forms, Maxwell would have followed with interest the rising mathematical fortunes of curves over the period from the 1840s to the 1870s. Sylvester marks their recent progress in the Exeter address. Persevering with his apotheosis of Continuity beyond the extract made at the start of Chapter 8, he assumes that his peers in Section A understand that a concern with curves is no longer the province of certain practical arts (as it was for the young Maxwell attending Hay's lectures in the mid 1840s):

Everybody knows what a wonderful influence geometry has exercised in the hands of Cauchy, Puiseux, Riemann, and his followers Clebsch, Gordan, and others, over the very form and presentment of the modern calculus, and how it has come to pass that the tracing of curves, which was at one time to be regarded as a puerile amusement, or at best useful only to the architect or decorator, is now entitled to take rank as a high philosophical exercise, inasmuch as every new curve or surface, or other circumscription of space, is capable of being regarded as the embodiment of some specific organised system of Continuity.[56]

The intrinsic geometry of surfaces pioneered by Gauss and Riemann is summed up here in 'the tracing of curves', a standard phrase that Maxwell references in his poem, using its attentive verb to envisage the portrait of Cayley diagrammatically, as 'the form [we] may trace'. Of course, Sylvester sees verse to participate in such Continuity, as '[an]other circumscription of space', 'a high philosophical exercise', that accordingly facilitates the parallel he draws in a letter to Hirst of a mathematical 'linear function to a *foot* in scanning'.[57] Concordant with the *ut pictura poesis* conceit of Locker's 'On "A Portrait of a Lady"', Sylvester and Maxwell find in poetry another set of lines by which their forms can be projected. Conventional references to *lines* of poetry as *inflected* (by, for example, rhyme, assonance and accentual stress) are vivified by this conception of verse, which draws it into analogies with contemporary mathematical understandings of localised space, 'every new curve or surface'. Extending its original mathematical application to describe syzygistic flows of syllables and their conditions, Sylvester's principle of synectic appreciates poetry as 'the embodiment of some specific organised system of Continuity'.

Sylvester opens *Fliegende Blätter* by reiterating this premise of his prosody, the analogy he sees verse and mathematics to have to one another through the principle of Continuity. 'Studies in Monochrome', the heading he gives to the first group of poems in the collection, is completed by his assertion of authorship, 'By a Professor of Advanced Mathematics', conjoint phrases that he defends with the following footnote:

> The incongruity between advanced mathematics and verse composition is more apparent than real – Mathematic commencing as a practical art, thence passing into the form of a science, having again emerged into an art of a higher order – a fine art – plastic in the hands of the Mathematician, obedient to and taking shape from his will, and almost admitting of the free play of fancy upon it, thanks to the deeper principles evolved in the new Geometry, the higher Algebra, and the calculus of the Continuous Riemonn [*sic*].[58]

The title 'Studies in Monochrome' that Sylvester glosses with this note alludes to the only examples of continuous manifolds that Riemann gives in his Habilitation lecture, 'the positions of perceived objects and colours', with the doubly extended manifolds of the picture plane and monochromatic tonal values. Each furnishes an analogy for musical tone in poetry, the two-dimensional manifold of sharp and flat pitches. Noted earlier for developing this analogy in his discussion of syzygy as a 'succession of consonant-colours', Lanier illustrates it with 'two perfect lines of Tennyson[,] whose physical beauty depends on their suave syzygy of M-colours, aided by a delicious distribution of vowel-colours: "Or moan of doves in immemorial elms / And murmur of innumerable bees."'[59] Sylvester's 'Studies' are not bleached of such polychromatic elements, but rather foreground the two-dimensional tonal sequences of their fixed rhyme schemes. His synaesthetic analogy is the converse of that made by James McNeill Whistler in the titles he gave to many of his tonalist paintings, including *The Little White Girl: Symphony in White* No. 2 (1864) and *Harmony in Yellow and Gold: The Gold Girl–Connie Gilchrist* (c. 1876–7), which render monochromatic hues in musical terms. The words that furnish the monochrome rhymes in Sylvester's 'Studies' are accordingly appreciated as local and differential, the continuous extension of a point. This mathematically suggestive characterisation of his continuous rhyme schemes is further focused by his description of these poems as '*Monochord*' exercises, the monochord being the simple single-stringed sounding-box by which the mathematical intervals between musical tones are measured. He accordingly likens his poems to the virtuoso violinist 'Paganini's performance on a single string'.[60]

'To Rosalind', the most extravagant of the 'Studies in Monochrome', not only perseveres further with its single rhyme than Sylvester's other poems, it also, as its author observes, 'with the exception of the line at the beginning and the two at the end', extends in 'one single sentence' that runs uninterrupted throughout the main body of the poem. Laocoon-like the reader is left to wrestle with 'the Anaconda-like sinuosities of this portentous period'.[61] The poem is projected on the two-dimensional plane of the page as a continuous line consisting of numerous subordinate lines, each of which bears a formal 'invariant' relation to the others through the rhyme scheme.

THE HYPERSYZYGETIC CANONICO-MEIO-CATALECTICIZANT

Sylvester grounds the appreciation of rhyme and other 'continuity in verse ... on the physical fact of the permanence of the auditory impressions'. Analogous to the persistence of vision effect harnessed by the thaumatrope, it allows sound images to be retained and their similarities to be recognised reflexively by the mind: 'Its existence ... is proved directly by the bare fact of the sensuous pleasure (the so-called tickling of the ear) derived from rhyme.'[62] How then does the poet arrive at such satisfying ratios? In his preface to the poem, Sylvester claims that the inexorable rhymes of 'To Rosalind' proceed by a mathematical operation, to which he assigns a monstrous agglutinative coinage:

The rhymes, which might be indefinitely extended, have been deduced by a somewhat indirect and circuitous process from a new form of Hypersyzygetic Canonico-meio-catalecticizant, varied and enlivened by an occasional use of a few carefully selected Pippians and Quippians, borrowed from the inexhaustible laboratory of his esteemed friend, Professor Cayley.[63]

Writing in an 1888 essay for *Nature*, Sylvester rightly claims to have contributed to mathematics more neologisms than any of his peers: 'Perhaps I may without immodesty lay claim to the appellation of the Mathematical Adam, as I believe that I have given more names (passed into general circulation) to the creatures of the mathematical reason than all the other mathematicians of the age combined.'[64] The formidable formula he enunciates in the extract suggests a patented technique for stamping out rhymes in industrial quantities, so that it is perhaps understandable that he is circumspect about explaining this 'somewhat indirect and circuitous process'. Like the secret of life discovered by the eponymous hero of Shelley's *Frankenstein*, Sylvester's process appears to be lost to science and literature. Nevertheless, the apparently over-determined nature of the compound term provides an opportunity to trace the dual usages of some of his coinages, as they move between mathematics, music and prosody. Having clarified the meaning of these component terms, the operation that they each contribute to the greater term, a mechanism for proliferating continuous chains of rhyme, can then be discussed more fruitfully. Each such constituent of the coalition, and indeed their interactions within it, can be grasped at least in part through their derivations in music and prosody. Some aspects of their original meanings effectively leap-frog their mathematical applications in the new term, while others are significantly mediated by them.

To deal first with Sylvester's outrider terms, the Pippian (or 'Caleyan') is defined by Tony Crilly as 'A curve dual to a given cubic curve'.[65] Sylvester evidently found in this parallelistic form a suggestive analogy for certain continuous relations of rhyme. His use of this type of curve, and the other that Cayley designates as Quippians,[66] demonstrates the formal parallel he sees to exist between rhyme and the 'high philosophical exercise' of tracing curves. As was noted earlier, the 'one single sentence' of 'To Rosalind' is for Sylvester crowded with curves, 'Anaconda-like sinuosities'. However, inflected by 'the continual recurrence of the same rhyme', the long line of the poem needs to be protected from monotony, making it, as Sylvester puts it, 'absolutely necessary to secure variety by all other means at the disposal of the composer'.[67]

There is a certain sinuosity also to Sylvester's coinages, as his compound 'Canonico-meio-catalecticizant' serves to demonstrate. In his momentous 1852 paper 'On the Principles of the Calculus of Forms', in which he establishes the theory of invariants, Sylvester explains that the 'Meicatalecticizant' is a type of invariant that usually, 'for the sake of brevity, I denominate the catalecticant'.[68] Writing to Hirst ten years later, he argues for the utility of the term 'catalecticant' through the relation it bears to the principle he attaches to it in the later compound: 'This sort of invariant is so important and stands in such a close relation to the Canonizant that we cannot afford to let it go unnamed and as this name has been used by Cayley as well as myself it may as well remain.'[69]

Like syzygy, Sylvester's catalecticant is lifted from prosody and developed into a mathematical principle before being returned to prosody in a new application: 'I took the Idea of the name from the Iambicus Trimeter Catalecticus.' This imperious phrase describes three iambic feet, the last of which lacks a syllable. Sylvester describes the mathematical Catalecticant through the coinage's source in prosody:

Usually a function of an even degree of x, y may be expressed as a sum of powers of linear functions of x, y with an outstanding term (The product of the linear functions be a certain covariant of such a product), and I compared in my mind the presence of each linear function to a *foot* in scanning; so that for ex. $(x, y)^6$ would be represented by 3½ feet, the ½ foot corresponding to the outstanding term; but the Catalecticant being zero, this mode of representation no longer remains possible and in some way or another (*how* at this moment my memory fails me) the *scanning* loses as it were half a foot.

This paragraph 'shows', as Parshall observes, 'the sort of mental gymnastics th[e] naming process could involve' for Sylvester.[70]

While the Catalecticant is drawn from prosody, the Canonizant derives from musicology. Sylvester played the piano and was a keen tenor, who had taken singing lessons from the French composer Charles Gounod.[71] Finding close parallels to exist between music and mathematics, he poses the rhetorical question: 'May not Music be described as the Mathematic of sense, Mathematic as the Music of the reason?' A canon is a formal structure in which the melody in one part of a composition is replicated in its entirety in another, albeit with some subtle variation, as of temporal distance, tone, inversion or measure. The term's mathematical application derives from this principle of irreducible and correspondent form. In one of the earliest references to it, from Sylvester's 1851 'Sketch of a Memoir on Elimination, Transformation, and Canonical Forms', the mathematical principle's simplicity and symmetry is accordingly seen to warrant the parallel with the aesthetic qualities of the musical canon: 'I now proceed to the consideration of the more peculiar branch of my inquiry, which is as to the mode of reducing Algebraical Functions to their simplest and most symmetrical, or as my admirable friend M. Hermite well proposes to call them, their Canonical forms.'[72] The Canonizant can be understood accordingly as the agent that reduces a quantic, an algebraic form, to its canonical form.

The Catalecticant and the Canonizant are both determinants, a term originating with Cayley in the 1840s that describes a number calculated from 'a group of quantities arranged in square order'.[73] In 1850, Sylvester gave the name 'matrix' to this form. Maxwell provides a description of it in 'To the Committee of the Cayley Portrait Fund', although he does not follow Cayley and Sylvester in carefully distinguishing between determinants and their matrices. Having resolved that to honour Cayley 'The symbols he hath formed shall sound his praise', Maxwell has the Determinants lead the procession:

> First, ye Determinants! in ordered row
> And massive column ranged, before him go,
> To form a phalanx for his safe protection.
> Ye solemn powers of roots of minus one!
> Around his head your ceaseless cycles run,
> Primordial increate spirits of direction.[74]

This stanza was written in collaboration with Sylvester, who revised its final lines. Writing in *The Laws of Verse*, four years before helping Maxwell make the matrix a subject for verse, Sylvester translated a stanza form favoured by Horace into this quadrilateral regime:

> As regards metre, let us denote a spondee, the first epitrite, a dactyl, and a trochee, by A, B, C, D respectively: then the construction of the Alcaic stanza, *as commonly practised by Horace* . . . will be represented by the scheme (or as, say, in determinants, the square matrix) –
>
> $$\begin{matrix} A & B & C & C \\ A & B & C & C \\ A & B & D & D \\ C & C & D & D \end{matrix}$$
>
> which (as is apparent) has a pure algebraical or tactical deep-seated harmony of its own. Denoting the lines of the symbols by single letters, and reading the square upwards, the scheme assumes the type L M N N, which is homœomorphic with the upper two lines of the square.[75]

Constituted by its classical metre, the Alcaic stanza corresponds to what Sylvester describes in his definition of Determinants as 'a group of quantities', which he has accordingly 'arranged in square order', as a matrix. The summary alphabetical symbols, from L to N, that he allocates to each line of poetry correspond to the product of each line of a mathematical matrix. Sylvester's analysis of verse here is like his work in geometry, dedicated to finding deeper algebraic principles in its phenomena and representations, a task that he sees 'the theory of determinants' to facilitate most radically and completely: 'It is an algebra upon algebra; a calculus which enables us to combine and foretell the results of algebraical operations, in the same way as algebra itself enables us to dispense with the performance of the special operations of arithmetic. All analysis must ultimately clothe itself under this form.'[76] The resultant of the square matrix into which Sylvester casts the Alcaic stanza is accordingly seen to be a unique and 'pure algebraical or tactical deep-seated harmony', that is, its determinant. It is for Sylvester 'evidence of the strong mathematical bias of Horace's mind, wherein perhaps is to be sought the secret of the peculiar incisive power and diamond-like glitter of his verse. Had Athens been Cambridge', he proclaims, Horace would have been a senior wrangler in the Mathematical Tripos.[77]

The metrical version of the catalectic determinant he gives in his letter to Hirst, and, conversely, his matrical rendition of the Alcaic stanza in *The Laws of Verse*, demonstrate that Sylvester theorises poetry and mathematics alike as, in the phrase he uses for his important early paper, a 'Calculus of Forms'.

As the Alcaic matrix demonstrates, Sylvester's prosody shares with calculus a defining concern with differential and integral quantities. The basic mathematical idea of functions, of unique correspondent relationships between the elements of two sets, is intrinsically suggestive for verse. Indeed, as Sylvester's Alcaic matrix demonstrates, the parallelistic relations that largely define verse can be understood as a set of linear functions, while in his earlier metrical transcription of the Catalecticant 'a *foot* in scanning' furnishes the equivalent of a letter in the sequence 'L M N N' in the matrix, that is, a 'linear function'.

Sylvester's paper 'On the Principles of the Calculus of Forms', the first part of which is entitled 'Generation of Forms', begins by outlining the workings of determinants in generic terms of 'forms or form-systems':

> The primary object of the Calculus of Forms is the determination of the properties of Rational Integral Homogeneous Functions or systems of functions: this is effected by means of transformation; but to effect such transformation experience has shown that forms or form-systems must be contemplated not merely as they are in themselves, but with reference to the ensemble of forms capable of being derived from them, and which constitute as it were an unseen atmosphere around them.[78]

Also grounded in 'The great law of Continuity', rhyme and other constituent effects of poetry similarly occur 'by means of transformation' within the rules of their linguistic and generic 'form-systems'. Transposed in this way, Sylvester's conception of the 'form-systems' yields a broadly structuralist model of language, in which each speech act depends implicitly for its meaning upon its relation to the other components of the system. The Continuity that facilitates the transformations that constitute poetry is like that of projective geometry, where 'imaginary' or 'ideal' points are implicitly retained in their subsequent transformations. Rhyme functions only insofar as the particular word before us summons the ghosts of related word-sounds, the imaginary points at which earlier lines ended. Syzygies also depend upon this principle of Continuity, so that, for example, in Carroll's game of syzygies, 'Walrus' persists through its transformations into 'Carpenter'. Particular language uses have an 'unseen atmosphere around them', which is especially charged in the case of the heightened formal applications that language is put to by poetry.

The 'Canonico-meio-catalecticizant' draws together the 'Canonizant' and 'Catalecticant' into a single noun of action. The interactive and transformational possibilities of such 'cognate determinants' are stated in the definition that Sylvester gives in 1851 of 'a "Matrix" as a rectangular array of terms, out of which different systems of determinants may be engendered, as from the womb of a common parent; these cognate determinants being by no means isolated in their relations to one another, but subject to certain simple laws of mutual dependence and simultaneous deperition'. He describes this formal pattern in 'An essay on Canonical Forms' (1851), where 'the canonical form becomes catalectic by one or more of the linear roots disappearing', while reciprocally, 'this catalectic form . . . gives place to a singular form', as 'the canonical form reappears catalectically'.[79] In the analogous case of the 'Canonico-meio-catalecticizant', the Catalecticant describes the moment of deperition, the wasting away of the final half-foot of a line of poetry, where the rhyme needs to be placed. Its sibling determinant, the Canonical agent, accordingly draws upon the continuous form-system of language to fill this lack with a symmetrical value.

In its application to prosody, the Canonizant suggests the ideal schematic form of a specific rhyme that accordingly provides the pattern from which its varied instances can be deduced, a generic value in the algebraic formula of the 'Hypersyzygetic Canonico-meio-catalecticizant'. The nature of this schema as a determinant implies the full complement of combinations and permutations that the matrix makes available, and hence the possibility of a continuous rhyme, a constant replication with formally satisfying and interesting modulations, in the manner of the canon in music. The final syllable of 'To Rosalind' furnishes an example of a 'canonical' form or keynote, not least because, as an equivocal note '(Eye! cheat ear! with ĭnd for īnd)', it serves to highlight the schematic nature of the principle.

The function designated by Sylvester's cumbersome coinage effectively describes what he refers to in a note to his later poem *Spring's Début: A Town Idyll*, another exercise in monochromatic rhyme, as 'certain laws of resolution and substitution'.[80] As was suggested earlier, the value that his compound principle furnishes is like that of the missing term of a function (such as the term Sylvester eulogises in 'To a Missing Member'), namely a specific example or embodiment of 'the form, which', as he explains in an 1853 memoir on 'Syzygistic Relations', 'resides in the function as the soul in the body. A form is always common to an infinity of functions.'[81] Following the broad pattern of his 'Calculus of Forms', just as 'the ensemble of forms capable of being derived from' the mathematical 'forms or form-systems . . . constitute as it were an unseen atmosphere around them', the continuous principle of the 'Hypersyzygetic Canonico-meio-catalecticizant' is haunted by all the rhymes that can be 'deduced' by it.

The final determination of rhyme in its context, in harmonious relation to the instances that go before and after it, is the work of syzygy. Sylvester accordingly further compounds his unwieldy term with the adjective 'Hypersyzygetic', which in this context suggests a ramping up, an amplified quality, of syzygy. Each particular incarnation of the schematic 'Canonico' form that completes the vacant term of the 'meio-catalecticizant' to sustain the rhyme requires syzygy, for as rhyme is usually organised around a regular vowel sound, its crucial complementary requirement of difference or irregularity depends upon variations in at least some of the consonants that are fused with it from line to line. The 'Hypersyzygetic' function accordingly facilitates the smooth flow of the rhymes, enhancing synectic fluidity to overcome their naturally symptotic tendency.

Sylvester's most original and momentous contribution to prosody, syzygy is a measure of subtle and varying rates of change, the great *raison d'être* of calculus. It is a 'Calculus of Forms' to supersede what was described earlier as the traditional arithmetic of form, accentual and classical quantitative metrics. While these divide verse into fixed and arbitrary units, regular segments of lines, restricting it to simple Euclidean geometry and space, Sylvester's prosody brings to the fore the continuous principles of invariance and local metrics, which he images as free to bend and curve. He writes in his note to *Spring's Début* of 'my view of English Rhythm as containing two distinct schemes or strands of time and accent in general coincident, but occasionally and subject to certain laws of resolution and substitution, free to open out into loops and waves or to shift lengthways, so as to render sensible the fact of their independent although usually blended existence.'[82] His description of the 'Anaconda-like sinuosities' of 'To Rosalind' undermines the traditional unit of the line, much as in *The Laws of Verse* syzygy is seen to eviscerate metrical line-segments, by smoothing them into a subtle synectic flow, so that the new unit is determined symptotically according to 'the office of rhyme in marking off a line as a sort of compound foot (like a bar in music)'.[83] In *Fliegende Blätter*, the calculus of the 'Hypersyzygetic Canonico-meio-catalecticizant' is pledged to extend the synectic flow by modulating a base word-sound in a single continuous rhyme that rolls through the length of the poem, as 'To Rosalind' is intended to demonstrate.

Sylvester observes that the generative principle of the 'Hypersyzygetic Canonico-meio-catalecticizant' gives to 'the continual recurrence of the same rhyme' a likeness to 'the regularly recurring dash and plash of the waves on the seashore',[84] an effect that is described schematically, and meant to be enhanced, by the parallel curves of his Pippians. Through this 'Hypersyzygetic' mechanism, the disciplined distortions enacted by each rhyme in 'To Rosalind' are seen to generate those that follow it, much as Clifford, developing upon Riemann, sees to occur with the curvature of space: 'this property of being curved or distorted is continually being passed on from one portion of space to another after the manner of a wave'.[85]

No longer the conspicuous principle of symptosis he describes in *The Laws of Verse*, rhyme is the main agent for 'continuity of sound' in 'To Rosalind'. Indeed, Sylvester records that Locker, his 'rancorous opponent' in their 'wit-combat', wrote to him that the poem's powerful will to rhyme threatened to dissolve the entire English language: 'So powerful a will must in time disintegrate the dictionary like water on a lump of sugar, and make every final syllable flow into the channel of Ind; in fact, / "Language all is Sylvest'rined / In the light of Rosalind."' Sylvester, however, appears to have received these remarks soberly as testimony to his achievement, observing for the benefit, if not the further wonderment, of his readers that 'When the above was written the poem extended to only two-thirds of its present length.'[86]

AIRS AND GRACES

Sylvester's game of syzygies is brought into focus by the following 'key' he gives for 'To Rosalind', a dramatically curtailed version of nine lines:

> In Cecilia's name I find
> (Deem not thou the guess unkind)
> *Celia* with a *sigh* combined,
> Whose five letters loose align'd
> Gems new set and recombined,
> Fairest O! of lily-kind
> Shall disclose to every wind
> With each mortal thing untwinned,
> Thy sweet name, dear Rosalind!

This version allows the structure and conceits of its cumbersome original to be more easily appreciated. The riddle with which the poem opens is accordingly represented schematically here, complete with the parenthetical apology that punctuates it. The opening volley in Sylvester's 'wit-combat', the 'Cecilia notion' is the original source of the poem's synectic movement, which he refers to in his prefatory notes to the poem as 'the wished-for impetuous flow of the verse'.[87] The word-sound 'Celia' is brought into syzygy as 'Cecilia', a relation that allows the shorter form's 'knotty rind' of disparate vowel sounds and soft consonants to form a pattern through the 'Precious ore' of the central '*Ci*', the core of the name. The 'five letters loose align'd' of Celia are 'Gems in fashion new assigned', 'new set' around the stressed '*Ci*' sigh, the setting of 'Richest ore' that brings them into relation as a new continuous measure, as it informs them pneumatically, and articulates them synectically, in a coherent pattern of word-sounds.

The 'Cecilia notion', the crucial sigh of breath that informs the name on the page, inflates it into a word-sound, is the poem's *'Primum Mobile'*, much as the *pneuma*, the breath or soul, breathed into Adam may be said to be for not only him but also all succeeding generations of humanity. The '*Ci*' of Cecilia is later recast as the interjectional vowel in the line 'Fairest O! of lily-kind', around which the protagonist's other designation of Rosalind is shaped. This line begins what Sylvester calls 'the invocation part' and 'the *Heart* of the poem', a long chanting recitation that catalogues its protagonist's charms:

> Fairest, O! of lily-kind,
> Perfect pearl and priceless find!
> Pure as poet's milk-white hind,
> Spirit! from all dross refined,
> Hearts to ravish Heav'n-designed;
> Fresh as rills that sparkling wind
> On their way the sea to find

The first line is, Sylvester writes, 'a great favourite with the author, which starts and gives the key-note to the Invocation'. It accordingly furnishes the synecdoche for this section in the curtailed 'key' version of the poem, while 'the key-note' itself can be traced more precisely to the interjective 'O!' at the heart of the line. Sylvester's notes for the line focus upon this sound, which he associates with purity of expression and a magic show:

How characteristic the recovered purity of our interjectional sounds O! and Ah! as contrasted with the Irish and German Och! and Ach! of the refinement of the English race – seizing, so to say, the soul of the sounds and throwing the husk away. Do I not remember, and did I not in a letter written to a lady at the time record the wonderful O! that issued from the breasts around me at the close of Joachim's magic renderings (I forget of what piece), superseding the ordinary vulgar forms of applause, betokening the sense of relief of the audience from an enjoyment rendered also painful from its intensity and seeming to say in audible language 'It is *too* beautiful'?[88]

Sylvester's approving observations about these radical expressive sounds are shared and more fully vindicated by an abrupt new work, that like 'To Rosalind' was completed in 1876 and similarly constitutes the longest poem of another English poet of even greater obscurity at this time, the Jesuit priest Gerard Manley Hopkins. 'The Wreck of the *Deutschland*' records and commemorates five German nuns exiled from their homeland who died when their ship was wrecked off the English Coast in 1875, and focuses in particular upon a Christ-like tall nun that Hopkins had read about in reports of the accident from *The Times* newspaper. The poem frequently reiterates the exclamatory 'O' of the tall nun's definitive expiration: '"O Christ, Christ, come quickly"', a gesture that returns breath back to the Creator who first breathed life into her, and which accordingly looks to the Trinity for its response and meaning: 'Breathe, arch and original Breath. / ... / Breathe, body of lovely Death.'[89] As wildly different as their verses are, both the Jesuit priest and the Jewish mathematician find the font of poetry in the interjectory exhalation.

Corresponding to Sylvester's description of the 'Ah!' and 'O!' as 'the soul of the sounds and throwing the husk away', the pneumatic 'Fairest O!' can be seen to represent its subject, the protagonist of the poem, precisely as 'Spirit! from all dross refined'. In the final couplet of the poem the involuntary exhalation of the 'O!' sound is shaped by the word and the fragrant figure of the Rose, the latter object of inhalation and syzygetic exhalation being a gesture back to Locker's poem: 'Rose smells sweet and soft spells *lind*, / Soft, smooth, sweet, spell Rosalind.' The name is completed by this exhalation, as Cecilia was by the 'sigh' that supplements and informs 'Celia', and the meaning of it underlined by the German word *lind*, which Sylvester translates as 'Soft, mild'. He is careful to justify the name Rosalind by explaining its component parts, much as he does in constructing and defending his mathematical neologisms. Contrasted with the railway's 'stertorous cars engíned', a hyperbolic image of noisily expelled air, the name of Rosalind is in the last lines of the long poem 'To those tones of Orpheus twinned' and declared to be 'Soft as notes of Jenny Lind', the Swedish opera singer.[90] Each inflected aspiration of the lover's sigh and the Invocation's 'Fairest O!' encapsulates, swiftly and directly, the poem's intemperate testimonies to its protagonist's graces. The names 'Cecilia' and 'Rosalind' are linked in this way as variant forms. The shaping of breath into these sounds echoes the protagonist's protean subjectivity, as it assumes the forms of the painted Cecilia, the performed Rosalind and the mysterious unnamed heroine who ostensibly dwells within and informs these names.

Like Hopkins, Sylvester stresses that his poetry is written to be read aloud, actively informed by breath as utterance. In preparation for the final couplet of his poem, its antepenultimate line 'Thy swéet name, deár Rosalind' is, as Sylvester explains in his commentary, artificially loaded with emphasis and breath, so that it is effectively made to function in the manner of Hopkins's sprung rhythm, and to require some engineering:

The accents are intended to indicate that the words in the line are not to be taken trippingly off the tongue, but to be as it were dropped from the lips one by one like olives bubbling out of the mouth of a narrow-necked bottle. To support so vast and heavy a superincumbent mass of sound as this last line of the period has to bear, need is that it should be made of extra breadth and strength. Obvious truism as it is, people are still not generally alive to, or do not practically realise the fact that Poetry is addressed immediately to the *Ear* and only secondarily to the *Eye* as a guide to the ear.[91]

Hopkins similarly writes that his poetry should be read with the 'ears, as if the paper were declaiming it at you'.[92] Albeit coming to their analyses from different directions, with strong interests and complementary aptitudes in science and poetry, both Sylvester and Hopkins treat words in poetry as physical objects with their own mass and momentum, which accordingly enter into dynamic relations with one another and resonate in the air.[93] This is integral to the ontology of stress and instress with which Hopkins approaches poetry,[94] and key to the hydrodynamic and other mechanistic tropes that Sylvester habitually uses to describe the workings of poetry in both *The Laws of Verse* and *Fliegende Blätter*.

Depending upon breath to impel its sounds and so realise 'the wished-for impetuous flow of the verse', poetry, for Sylvester, is organised formally as a respiratory infrastructure. In *The Laws of Verse*, 'the main flood of Phonetic Syzygy' is accordingly flanked by two 'channels', Anastomosis and Symptosis, complementary principles that facilitate and temper the synectic flow. Anastomosis is, Sylvester explains, 'necessarily familiar to writers of words for song music and all judicious singing masters, a great part of whose business it is to teach the art of keeping back the breath'. As the application of form to a fluid medium, it accordingly gives poetry an affinity with plumbing: 'Anastomosis regards the junction of words, the laying of them duly alongside one another (like drainage pipes set end to end, or the capillary terminations of the veins and arteries) so as to provide for the easy transmission and flow of breath (unless a suspension is desired for some cause, or is unavoidable) from one into the other.'[95]

In a note to *The Laws of Verse*, Sylvester explains that the synectic triad facilitates a Continuity of both breath and sound, and expresses an intention to demonstrate this, which he later fulfils in 'To Rosalind': 'I may seek some other opportunity of stating the result of my speculations in this direction, and especially the principle of twofold – i.e. respiratory and sonorific – continuity in verse.'[96] In this conception, Continuity occupies and articulates physical space as breath and acoustic vibration. Measured by its regular rhyme and modulated chromatically by its syzygies, the long line of 'To Rosalind' is endowed with a conspicuous materiality through its vocalisation, its inflected expirations of air. It is a type of spatial Continuity that we enact in reading it aloud, a 'respiratory and sonorific' line, an *air*.

CHAPTER 10

Science on Parnassus

Anxious to refute Huxley's account of mathematics as 'that study which knows nothing of observation, nothing of experiment, nothing of induction, nothing of causation', Sylvester describes Eisenstein's 1844 discovery of an invariant as the fortuitous consequence of mathematical fieldwork; 'the accidental observation by Eisenstein, some score or more years ago, of a single invariant (the Quadrinvariant of a Binary Quartic) which he met with in the course of certain researches just as accidently and unexpectedly as M. Du Chaillu might meet a gorilla in the country of the Fantees ... Fortunately he pounced down upon his prey and preserved it for the contemplation and study of future mathematicians.'[1] This conciliatory gesture to Huxley and his science casts Eisenstein as an energetic Dupin, his fictitious contemporary in Poe's 'The Murders in the Rue Morgue', who, in applying his analytical skills, discovers a murderous escaped orangutan. Having made a pact with Cayley to furnish their 'New Algebra' with its own taxonomy and hence a new nomenclature,[2] Sylvester prides himself on his ability to give names to such 'creatures of the mathematical reason' as Eisenstein's invariant.

While the unity of poetry and mathematics is figured as a prelapsarian state in Rowan Hamilton's early poem 'To Poetry', it abides as a foundational assumption in the work of Sylvester, 'the Mathematical Adam'. Many of his coinages, such as the terms 'syzygy' and 'catalectic', move freely between prosodic and mathematical applications, and in so doing recapitulate the larger cultural changes that, as Chapter 9 noted, he sees mathematics to have undergone over the course of its history; 'commencing as a practical art, thence passing into the form of a science having again emerged into an art of a higher order – a fine art – plastic in the hands of the Mathematician'. Consistent with his naturalistic trope of a mathematical menagerie, and corresponding description of Cayley as 'the Darwin of the English school of mathematicians', Sylvester introduces an evolutionary mechanism to explain what he sees as the restless Parnassian reach of mathematics to 'a future still more heaven-reaching theory': 'This higher art is again passing into science embodied in the sublimer theories of Physics and Elasticity, and so may we anticipate a continual round of palingenesis, or series of Evolutions, in an ever more and more refined and ethereal medium of operation.'[3]

Sylvester's model of mathematical regeneration, his 'series of Evolutions', is drawn not from Darwin but from Oken, whose developmentalist doctrines are based upon palingenesis, the principle of biological recapitulation. Following a suggestion from Owen, Oken's principal British advocate, Sylvester accounts for mathematical man's place in nature with a teleological construal of this developmental principle:

> There is an old adage, 'purus mathématicus, purus asinus'. On the other hand, I once heard the great Richard Owen say, when we were opposite neighbours in Lincoln's Inn Fields (doves nestling among hawks), that he would like to see *Homo mathematicus* constituted into a distinct subclass, thereby suggesting to my mind sensation, perception, reflection, abstraction, as the successive stages or phases of protoplasm on its way to being made perfect in Mathematicised Man. Would it sound too extravagant to speak of perception as a quintessence of sensation, language (i.e. communicable thought) of perception, mathematic of language? We should then have four terms differentiating from inorganic matter and from each other – the Vegetable, Animal, Rational, and Supersensual modes of existence.[4]

This recollection, which is made in a note to the Exeter address, evidently refers to the period from late 1844 until 1855 when Sylvester lived and worked as an actuary at 26 Lincoln's Inn Fields, across the square from Owen's medical practice.[5] Oken's system is, as Chapter 7 noted, propelled in the post-Kantian fashion by polar opposites. Life emerges in this cosmology as the various polarities of electricity, magnetism and chemical powers yield a primal slime, the 'protoplasm', as Sylvester refers to it, from which higher forms are believed to come into being. Sylvester follows Oken by seeing such higher forms as recapitulating all earlier, lower, forms.

Oken substantiated his evolutionary speculations experimentally by dissecting human embryos, from which he discerned formal homologies between the stages that the incipient human being passes through and a historical sequence in which progressively more complex animals were seen to develop, a chain of being from infusoria to insects, fish, reptiles, birds and finally mammals. Following Owen's suggestion, Sylvester translates Oken's hypothesis into cognitive terms, as if to bring post-Kantian science back to Kantian theory of mind. Indeed, as Owen observes in his entry on Oken for the eighth edition of the *Encyclopedia Britannica*, his subject's early work, *Grundriss der Naturphilosophie* (1802), 'extended to physical science the philosophical principles which Kant had applied to mental and moral science', so that, as he quotes Oken, 'the animal classes are virtually nothing else than a representation of the sense-organs, and that they must be arranged in accordance with them'. Oken accordingly names five forms of animals:

The *Dermatozoa*, or Invertebrata; 2. the *Glossozoa*, or Fishes, as being those animals in which a true tongue makes, for the first time, its appearance; 3. the *Rhinozoa*, or Reptiles, wherein the nose opens for the first time into the mouth and inhales air; 4. the *Otozoa*, or Birds, in which the ear for the first time opens externally; and 5. the *Ophthalmozoa*, or mammals, in which the organs of sense are present and complete, the eyes being movable and covered with two lids.[6]

Sylvester begins at the last point, with sensation, to extend the scheme to describe distinctively human forms of development. Consistent with his criticism, in another note to the Exeter address, of his peers' conflation of the Kantian Forms of Sense with the Categories of the Understanding, Sylvester begins by distinguishing sensation from perception as the first and second stages in his hierarchy of cognitive functions.

Mathematics is placed at the top of Sylvester's hierarchy and described akin to poetry as the quintessence of language, for as he puts it in his Commemoration address, it aims 'to condense the Maximum of meaning into the Minimum of language'.[7] In a wholesale Pythagorean parallel to Pater's famous dictum, '*All art constantly aspires towards the condition of music*',[8] Sylvester sees poetry to constantly aspire toward the celestial condition of mathematics. As Chapters 8 and 9 observed, he treats poetry as a science: 'I affirm that every science becomes more perfect, approaches more closely to its own ideal, in proportion as it imitates or imbibes the mathematical form and spirit.' Subsumed under his grand mathematical principle of Continuity, poetry's synectic qualities reproduce the graduated series of projective geometry and differential calculus. Its patterns of Alcaic stanzas find their refined formal expressions in square matrices, its subtle word sounds in syzygies, and the rolling sustained rhyme forms of the various 'Studies in Monochrome' in such determinants as the Catalecticant and the Canonizant. Indeed, as the derivations of such coinages indicate, Sylvester regards mathematics as 'the Music of the reason'. Furthermore, completing the ethereal evolutionary scheme outlined in the Exeter address, in his essay 'On Newton's Rule' Sylvester anticipates the quintessence of language and the perfection of '*Homo mathematicus*' in the final synthesis of mathematics and music:

Music the dream, Mathematic the working life – each to receive its consummation from the other when the human intelligence, elevated to its perfect type, shall shine forth glorified in some future Mozart-Dirichlet or Beethoven-Gauss – a union already not indistinctly foreshadowed in the genius and labours of a Helmholtz![9]

Like Sylvester, the Austrian physicist Ludwig Boltzmann, writing in an 1888 *Festschrift* for Kirchhoff, conceives of mathematics, and indeed mathematical physics, as akin to music, a matter of style. J. J. Thomson and Max Planck concur with this aestheticist appreciation of their sciences, as in their respective centenary essays on Maxwell they each quote approvingly parts of the following passage from Boltzmann:

> Just as the musician recognizes Mozart, Beethoven or Schubert from the first few bars, so does a mathematician recognize his Cauchy, Gauss, Jacobi or Helmholtz from the first few pages. Perfect elegance of expression belongs to the French, the greatest dramatic vigour to the English, above all to Maxwell. Who does not know his dynamical theory of gases? First, majestically, the Distribution of Velocities develops, then from one side the Equations of Motion in a Central Field; even higher sweeps the chaos of formulae; suddenly are heard the four words: 'Put $n = 5$'. The evil spirit V (the relative velocity of two molecules) vanishes and the dominating figure in the bass is suddenly silent; that which had seemed insuperable being overcome as if by a magic stroke. There is no time to say why this or why that substitution was made; he who cannot sense this should lay the book aside, for Maxwell is no writer of programme music obliged to set the explanation over the score. Result after result is given by the pliant formulae till, as unexpected climax, comes the Heat Equilibrium of a heavy gas; the curtain then drops.[10]

Boltzmann finds in Maxwell's mathematical expositions of physics an operatic splendour, more literal forms of which he enjoyed as a subscriber to the Vienna Opera.[11] Of course, Maxwell similarly appreciates equations and physical theories as aesthetic style in his lyric versions of both Tripos problems and various lectures by Tyndall and Tait. More directly instancing the 'dramatic vigour' that Boltzmann attributes to him, Maxwell's last poem '*A Paradoxical Ode*' (1878) parodies a song from Percy Bysshe Shelley's play *Prometheus Unbound*.

THE UNSEEN UNIVERSE

'*A Paradoxical Ode*' is an affable satire on the speculative system that Stewart and Tait developed in *The Unseen Universe, or Physical Speculations on a Future State* (1875) and its sequel, *Paradoxical Philosophy* (1878). Another of the poems that Maxwell wrote for Tait, it follows '(Cats) Cradle Song, *By a Babe in Knots*', which, addressed to 'Peter the Repeater', describes his childhood friend's current besotted state of beknottedness. The '*Ode*' opens with its protagonist announcing that 'My soul is an entangled knot / Upon a liquid vortex wrought.'[12]

Provoked by Tyndall's Belfast address, Stewart and Tait declare that they wrote *The Unseen Universe* 'to show that the presumed incompatibility of Science and Religion does not exist' and to liberate nature from the encumbrances of their factional foes: 'we have merely stripped off the hideous mask with which materialism has covered the face of nature to find underneath (what everyone with faith in anything at all must expect to find) something of surpassing beauty, but yet of inscrutable depth'.[13] 'A Paradoxical Ode' is addressed 'To Hermann Stoffkraft, Ph.D., the Hero of a recent work called "Paradoxical Philosophy"'. Stewart and Tait's book describes the jubilee meeting of the discussion club 'The Paradoxical Society', to which 'the well-known German Philosopher' has been invited by its president, Stephen Fairbank. 'In this country we should probably ... call him a materialist, but Dr Hermann Stoffkraft might no doubt be inclined to contest the propriety of the name', for, completing his resemblance to the 'Germaniser' Tyndall, he prefers to think of himself as a romantic philosopher, 'a votary of the goddess of Nature' and 'the Power which underlies [phenomenal] manifestations'.[14] As Daniel S. Silver observes, and Maxwell evidently appreciated, Stoffkraft's name derives from Ludwig Büchner's 1855 *Kraft und Stoff* (*Force and Matter*),[15] an ostensibly Newtonian formula describing the fundamental principles of mechanics that provide the engine for the materialist doctrines of Büchner and Tyndall alike. Whilst he sees the Hamiltonian to have recast and indeed largely superseded such terms, Maxwell is, as his 'Report on Tait's Lecture on Force' makes clear, careful to defend them from Tyndallic and other modern appropriations, 'Meanings most strange and various, fit to shock / Pupils of Newton.'

The *Paradoxical Philosophy* describes Stoffkraft's conversion to the doctrine of the Unseen Universe. As was noted in Chapter 6, both the Metropolitans and the North Britons use the principle of Continuity, or the uniformity of nature, to justify their respective schemes, and it is through this convergent point that Stoffkraft comes to accept the gospel of the Unseen Universe. Continuity grounds Tyndall's vast evolutionary cosmology in the unseen area 'between the microscope limit and the true molecular limit', as he puts it in his discourse on the imagination, whereas Stewart and Tait conversely require it to, as Maxwell's poem observes, reach heavenwards to the 'Unseen Universe': 'Still may thy causal chain, ascending, / Appear unbroken and unending.'[16]

While the first law of thermodynamics enables Continuity, the second law threatens to exhaust it. 'I fully agree with the modern theory of the dissipation of energy', Stoffkraft declares early in the *Paradoxical Philosophy*: 'It follows that a time must come when life, under any conditions conceivable to us, will be physically impossible. What then?'[17] Stewart and Tait devote an early chapter of *The Unseen Universe*, on 'The Present Physical Universe', to establishing this hypothesis, which anticipates a time when all energy will become unusable through entropy. Writing in the wake of the 1871 Paris Commune and subsequent working class unrest in England, they explain the second law by analogy with such threats to the status quo, as a cosmic contagion: 'heat is *par excellence* the communist of our universe, and it will no doubt ultimately bring the system to an end'. The chapter 'come[s] to the conclusion that [our universe] began in time and will in time come to an end. Immortality is therefore impossible in such a universe.'[18]

Having clarified 'the laws according to which the machine called the visible universe works', Stewart and Tait include a chapter on 'Development', in which they approvingly cite theories of Continuity that have been attributed to this limited physical system, including the nebular hypotheses of Kant and Laplace and the evolutionary ideas of Darwin, Wallace and Huxley.[19] 'A Paradoxical Ode' observes that this places them in an awkward alliance with Stoffkraft and the materialists, as in the second stanza the poem's protagonist endorses an apocalyptic cosmology that sanctions 'Evolution' and Spencer's notorious phrase 'the survival of the fittest':

> But when thy Science lifts her pinions
> In Speculation's wild dominions,
> I treasure every dictum thou emittest;
> While down the stream of Evolution
> We drift, and look for no solution
> But that of the survival of the fittest.
> Till in that twilight of the gods,
> When earth and sun are frozen clods,
> When, all its energy degraded,
> Matter in aether shall have faded;
> We, that is, all the work we've done,
> As waves in aether, shall for ever run
> In swift expanding spheres, through heavens beyond the sun.[20]

Vindicating Boltzmann's assessment of his style, the hypothesis of the universe's heat death is presented by Maxwell as a grand myth, indeed a bombastic aesthetic structure; the operatic cataclysm of the 'twilight of the gods', Wagner's *Götterdämmerung* having been first presented at Bayreuth two years earlier, in August 1876.

Maxwell's contemplation of science's promised end of days is all the more remarkable given the conditions in which he composed his poem. The prophecy of his entropic signature, with which he closes 'A Paradoxical Ode', was being realised as Maxwell was writing the poem, for he was dying of stomach cancer, as yet formally undiagnosed but probably not unsuspected, his mother having died of the same illness at the age he was then, forty-seven. 'Questions about the soul's immortality were', as Silver observes, 'no longer merely academic for Maxwell',[21] as they were for Stewart and Tait. Maxwell appears to have been occupied during his last days with thoughts of poetry, as he considered questions about some hexameters and a speech from Shakespeare's *The Merchant of Venice*. Garnett cites these examples in the obituary essay he wrote for *Nature*, as proof that 'the speculative character of his mind remained to the last'. The two letters published by *Nature* in response to the obituary both focus upon Maxwell's late thoughts on hexameters.[22]

The mechanistic physics and developmentalism of Stewart and Tait's scheme is carefully contained, and compensated for, by their supposition that such phenomena originally emerged from the Unseen Universe and will finally return to it. The epigraphs to the 'Development' chapter of *The Unseen Universe*, well-known passages from Tennyson's *In Memoriam* and Alexander Pope's *An Essay on Man*, prefigure this strategy, as romantic despair is met with neo-classical measure:

> 'Are God and Nature then at strife,
> That Nature lends such evil dreams?
> So careful of the type she seems,
> So careless of the single life;
> . . .
> '"So careful of the type"? but no,
> From scarped cliff and quarried stone
> She cries, "A thousand types are gone:
> I care for nothing, all shall go."' – TENNYSON.
>
> '*All nature is but art*, unknown to thee;
> *All chance, direction*, which thou canst not see,
> All discord, harmony not understood;
> All partial evil, universal good;
> And spite of pride, in erring reason's spite,
> One truth is clear, whatever is, is right.' – POPE.[23]

Stewart and Tait's scheme provides an optimistic reply to the rhetorical question that closes stanza LVI of *In Memoriam*, from which their first excerpt is taken; 'What hope of answer, or redress? / Behind the veil, behind the veil.'[24] The leaky engine of the Seen Universe is made whole by the Unseen Universe, supplemented and effectively restored to the pristine condition and perpetual motion of Carnot's cycle, a theoretical model that describes the reversible transformation and conservation of energy through the workings of an idealised steam engine.

In the first of their chapters on 'the machine called the visible universe', Stewart and Tait give a hypothetical case of two engines working together, an analogy for the complementary and completing relation of the Unseen to the Seen universe:

Suppose there could be an engine, M, more perfect than a reversible engine, N. Set the two to work together as a compound engine, M letting down heat from boiler to condenser, and doing work; N spending work in pumping back again the heat to the boiler. If N be made to restore to the boiler at every stroke exactly what M takes from it, the compound engine will do external work, for, by hypothesis, M is more perfect than N. Whence does the work come? Not from the boiler, for it remains as it was. Hence N must take more heat from the condenser than M gives it; *i.e.* you get work by cooling the condenser.

The authors see their preposterous case of the compound engine to be warranted by 'Maxwell's Demon'. They admit that their improvement of Carnot's reversible engine, in which a still 'more perfect' engine is able to 'take more heat from the condenser' than is returned to it, 'would seem to imply an ample *reductio ad absurdum*. But', they maintain, 'Clerk-Maxwell has shown it to be physically possible.' Transposing and multiplying Maxwell's 'being', Stewart and Tait hypothesise that the condenser is cooled through the discretionary efforts of an army of Demons, as they direct warmer molecules from the cooler condenser to the warmer chamber of the engine's boiler: 'by enlisting in our service conceivable finite beings (imagined by Clerk-Maxwell, and called demons by Thomson), it would be possible materially to alter this state of things, even though these beings should do absolutely no work'.[25] The review of *The Unseen Universe* that Clifford wrote for the *Fortnightly Review* lists 'molecular demons' amongst the book's 'machinery of Christian mythology', alongside 'angels, archangels, incarnation' and 'miracles'.[26] Maxwell also appears to have appreciated his friends' appropriation of his principle for he responded to it with a 'catechism', as Knott describes it, for Tait 'Concerning Demons', in which he good-naturedly clarifies the character and application of these 'Very small *but* lively beings'.[27]

In Stewart and Tait's grand thought experiment, the energy dissipated by mortals in the visible universe finds its perpetual afterlife in the Unseen. The title of *The Unseen Universe* is drawn from St Paul's Second Epistle to the Corinthians, the reference being cited on the book's title page: 'the things which are seen are temporal, but the things which are not seen are eternal'. The peculiar function described by the compound engine, of preserving and renewing energy between the Seen and the Unseen universes, is understood to occur through the propagating media of various ethers. While our physical world is subject to constant entropic degradation it is, Stewart and Tait argue, redeemed by being continuous with a series of ethers. Such media propagate the indestructible energy of an eternal and infinite 'Great Whole', an Unseen Universe that guarantees human immortality and also supports other, superior, forms of intelligence. Maxwell gives an account of Stewart and Tait's ethers in the review he wrote for *Nature* of the *Paradoxical Philosophy*:

> In the first place the luminiferous aether, the tremours of which are the dynamical equivalent of all the energy which has been lost by radiation from the various systems of grosser matter which it surrounds. In the second place a still more subtle medium, imagined by Sir William Thomson as possibly capable of furnishing an explanation of the properties of sensible bodies; on the hypothesis that they are built up of ring vortices set in motion by some supernatural power in a frictionless liquid: beyond which we are to suppose an indefinite succession of media, not hitherto imagined by any one, each manifoldly more subtle than any of those preceding it.[28]

Clifford glosses this regressive case with a couplet by the logician and mathematician Augustus de Morgan: 'Great fleas have little fleas, upon their backs to bite 'em; / Little fleas have lesser fleas, and so *ad infinitum*.'[29]

The subtle uses that the ethers are put to by Stewart and Tait can, as Graeme Gooday documents, be traced to journal articles that Stewart wrote in the late 1860s and early 1870s. The second of two papers on 'The Sun as a Type of the Material Universe', written with Lockyer for *Macmillan's Magazine* in August 1868, introduces Stewart's thesis that 'a great molecular delicacy of construction in the sun' gives it an 'unexpectedly "intimate" bond' with the planets.[30] Subtitled 'The Place of Life in a Universe of Energy', the paper extends this principle of 'delicacy' to the operations of the human mind, theorising thought and will as infinitesimal but eternally persistent expenditures of energy. Maxwell evidently read and remembered this essay, as he draws from it the reference he makes in his 1873 paper 'Science and Free Will', cited in Chapter 7, to Stewart's reflections on his topic.[31]

Most likely suggested or emboldened by recent applications of spectroscopy to astronomy, Stewart's idea of 'delicacy' is more fully developed in *The Unseen Universe*, where, through the propagation of the sun's light by the ether, 'continual photographs of all occurrences are thus produced and retained'. Correspondingly, 'every thought that we think is', he maintains, 'accompanied by a displacement and motion of the particles of the brain, and somehow – in all probability by means of the medium – we imagine that these motions are propagated throughout the universe'.[32] Stoffkraft's treasured dicta are accordingly figured as emissions in 'A Paradoxical Ode', while the soul is, Maxwell informs Tait in February 1878, faced with 'the prospect of an eternity of expansion in all directions at a rate of 3.004×10^{10} centimetres per second',[33] that is, the speed of light. The identification that Stewart and Tait make of the soul with energy is cast in Maxwell's poem as a simple equation, 'We, that is, all the work we've done'. Such traces escape their dissipation in the visible world as ever finer ethers transmit them deeper into the realms of the Unseen; 'As waves in aether, shall for ever run / In swift expanding spheres, through heavens beyond the sun.'

THE VORTEX ATOM SOUL

In his last letter to his friend, which he entitles '*Headstone* [i.e., Tête Peter] *in search of a new sensation*', Maxwell has Tait contemplate the personal practicalities of eternal existence in the Unseen. He is made to consider what to do in 'the aeonian ætherial phases of existence to which I am looking forward', only to discover that while time may be infinite, the possibilities for amusement are not; 'must the same chime be repeated with intolerable iteration through the dreary eternities of paradoxical existence?'[34] Tait is similarly cast in '*A Paradoxical Ode*' as the counterfactual subject of a thought experiment, the protagonist with the vortex atom soul.

'*A Paradoxical Ode*' parodies Asia's song from the close of Act II of *Prometheus Unbound*, which begins 'My soul is an enchanted boat, / Which, like a sleeping swan, doth float.'[35] Tait is figured in the '*Ode*' as Asia, the beloved of Prometheus, the titan who, having stolen fire from the gods to give to mankind, is conventionally identified with science. Asia's sisters Panthea and Ione (who is not present in this scene) suggest Tait's fraternal North Britons Stewart and Maxwell. The other of its *dramatis personae*, 'The Spirit of the Hour', who has been driving the sisters across the sky in a chariot, but brings it to rest for this scene, becomes in Maxwell's '*Ode*' the scientific *zeitgeist*, which can be identified with Stoffkraft and the popular prognosis described in the second stanza, in which 'We drift' 'down the stream of evolution' only to end with the whimpering apocalypse of entropic dissolution. This pessimistic trajectory, like the Spirit's chariot in *Prometheus Unbound*, has been arrested, stopped in its tracks by Tait's proxy for Shelleyan transcendentalism: 'Great Principle of all we see / Thou endless Continuity.'[36]

Devoted to describing the radiant transformation of Asia in appropriately fluid ethereal imagery, the final scene in Act II of *Prometheus Unbound* obligingly provides the grounding conceit for Maxwell's poem about his friend's apparent metamorphosis from professional scientist to speculative seer:

> SPIRIT.   . . . Apollo
> Is held in heaven by wonder; and the light
> Which fills this vapour, as the aëreal hue
> Of fountain-gazing roses fills the water,
> Flows from thy mighty sister.
> PANTHEA.   Yes, I feel –
> ASIA. What is with thee, sister? Thou art pale.
> PANTHEA: How thou art changed! I dare not look on thee.[37]

While Shelley's play looks forward to Asia's reunion with Prometheus, which occurs in the following act, Maxwell's poem correspondingly waits for Tait's return to science.

The title, 'To Hermann Stoffkraft, Ph.D. . . .: *A Paradoxical Ode*', echoes that of 'To the Chief Musician upon Nabla: *A Tyndallic Ode*'. The poem opens with Tait presenting himself, like his adversary in the earlier poem, as a riddle, *The Unseen Universe* and the *Paradoxical Philosophy* having each been published anonymously:

> My soul's an amphicheiral knot,
>   Upon a liquid vortex wrought
> By Intellect in the Unseen residing.[38]

Authorial identity is presented in the first three editions of *The Unseen Universe* and the first edition of the *Paradoxical Philosophy* as a graphic riddle, which the opening lines of Maxwell's poem translate into words. The earlier book has a figure of 'a trefoil knot, the symbol of the Vortex Atom imagined by Thomson',[39] the second, three interlocking rings, printed where the authors' names would conventionally appear on the spines and title pages. Tait writes to Robertson Smith in 1875 that in view of the various strong criticisms that the first edition of *The Unseen Universe* received, the authors need to adopt an alternative atomic persona for the forthcoming second edition: 'We must be *at first* a Lucretian Atom not a vortex ring, strong in solid singleness, not wriggling meanly away from the knife!'[40]

As if anticipating the cosmic importance that it later assumes for his friend, and indeed the jocular treatment it will receive in the '*Ode*', Maxwell writes to Tait in 1867 of the vortex atom theory that 'Thomson has set himself to spin the chains of destiny out of a fluid plenum'.[41] The theory is foundational for Stewart and Tait's speculations as it both reconciles what were previously assumed to be 'the opposing doctrines of atoms and of continuity' and, as

Maxwell further observes in his 1878 *Encyclopaedia Britannica* article on the 'Ether', disdains the principle of entropy: 'we conclude that the molecular vortices do not require a continual expenditure of work in order to maintain their motion, and that therefore this motion does not necessarily involve dissipation of energy'. Maxwell notes, however, that no ether adequate to this task has yet been hypothesised. His article closes by observing that an ether occupying 'the interplanetary and interstellar spaces' may 'as the authors of the *Unseen Universe* seem to suggest . . . constitute the material organism of beings exercising functions of life and mind as high or higher than ours are at present', but that it 'is a question far transcending the limits of physical speculation'.[42] While, as was noted in Chapter 7, the Liverpool address distinguishes the scientific credibility of Thomson's 'ring vortices' from the 'occult properties' of the Tyndallic particles, 'formed by investing the molecule with an arbitrary system of central force', in Maxwell's review of the *Paradoxical Philosophy* they are similarly seen to rest upon a mystical substrate, deemed to have been 'set in motion by some supernatural power in a frictionless liquid'. Tait's 'liquid vortex' soul is depicted in the poem as 'wrought / By Intellect in the Unseen residing', a phrase that echoes formulaic references to 'an Intelligence residing in the Unseen' that occur throughout *The Unseen Universe* and its sequel.[43]

In contrast to such chiral knots as the trefoil, 'An amphicheiral knot', as Tait's aptly named friend and biographer explains, 'is a knot which can be changed into its own mirror reflexion'.[44] A conceit to describe Tait's forsaking of Science for the Unseen Universe, the amphicheiral knot loses its scientific bearings in the first line of '*A Paradoxical Ode*' as, identified with the protagonist's soul, it becomes the mechanism by which he passes *Through the Looking-Glass*: into a preposterous world analogous to that of Carroll's book, which, as Chapter 2 noted, Maxwell asked his friend to send to him early in 1873.[45] Knots translate and embed curves from two-dimensional to three-dimensional space, while the most likely model for the protagonist's soul in the poem, 'the 4-fold amphicheiral knot'[46] that Tait introduces and explores in his 1877 paper 'On amphicheiral forms and their relations', suggests a more precise emblem for the fourth dimension, and indeed Riemannian n-fold aggregates the further dimensions of the Unseen Universe: the 'simplest form is that of 4-fold knottiness', Tait observes, 'All its forms have knottiness expressible as $4n$.'[47]

Amphicheiral knots demonstrate simple cases of homeomorphism, a term used to refer to the properties of objects that are preserved under continuous deformation. The study of such relations belongs to topology, a form of qualitative geometry, like projective geometry, that was devised by another of Gauss's students, Johann Benedict Listing, in 1847. Uninterested in metrical relations of distance, topology is accordingly able to appreciate a circle as identical to a closed line of any shape, a sphere to all other closed surfaces, and the equivalence of a doughnut to a coffee cup (as both the torus and the cup consist of a closed surface with a single hole).

Writing to Tait on 8 May 1871, Maxwell introduces a term from Listing that applies to the peculiar homeomorphic relations of amphicheiral knots, and hence to the vortex atom soul of the poem: 'Am I perverted? a mere man in a mirror, walking in a vain show? What saith the Master of Quaternions?' Having observed inconsistencies in the conventions used to describe motion along an axis and the rotation about it, Maxwell notes that in such cases Rowan Hamilton adopts the convention of a clockwise direction, whilst Thomson and Tait follow the rotation in the opposite direction.[48] 'If we confound the one with the other,' he explains to Tait a few days later, 'every figure will become *perverted* (a phrase of L[isting]) denoting an effect similar to that of reflexion in a mirror).'[49] The change that Maxwell makes in the drafts of his poem from describing the vortex-ring soul as an 'entangled' to an 'amphicheiral knot' suggests a poised punning judgement of the Unseen Universe hypothesis as a perverse application of Tait's scientific prowess and integrity. The amphicheiral soul of '*A Paradoxical Ode*' is the alter ego of the superlative scientific identity that Tait is similarly made to voice in Maxwell's final letter to him, cited in Chapter 5, through his spectral vision of the similarly mirroring identity of 'ALBAN' and 'NABLA'.

THE SPACE OF SPECULATION

Belonging to the Unseen Universe, Tait's soul is a knot that cannot be untied by Stoffkraft's workman-like efforts:

> While thou dost like a convict sit
> With marlinspike untwisting it
> Only to find my knottiness abiding,
> Since all the tools for my untying
> In four-dimensioned space are lying,
> Where playful fancy intersperses
> Whole avenues of universes,
> Where Klein and Clifford fill the void
> With one unbounded, finite homaloid,
> Whereby the Infinite is hopelessly destroyed.[50]

Stoffkraft is, as Silver explains, likened to convicts whose punishment required them to salvage usable hemp from old tarred and damaged rope,[51] an activity that furnishes an allegory for his demystifying but misguided efforts to reduce the vortex atom soul to matter and utility. The protagonist of the poem explains that the solution to his stubborn 'knottiness abiding', a phrase that suggests the grand speculative tangle of the Unseen Universe hypothesis itself, can be found in a realm that is currently inaccessible to Stoffkraft, beyond the veil of non-Euclidean space: 'Since all the tools for my untying / In four-dimensioned space are lying.' Knots translate and embed curves from two-dimensional to three-dimensional space, another way in which they furnish an emblem for the further dimensions of the Unseen Universe and the non-Euclidean geometry it is accordingly identified with. By invoking Klein's recent proof that any knot could be undone in four-dimensional space, these lines refer Tait's vortex atom soul to what he and Stewart describe as 'an Unseen whose matter has *four* dimensions'.[52] Like Sylvester, who sees it to extend the mathematical principle of local Continuity, Stewart and Tait see non-Euclidean space to enhance the physical principle of Continuity.

Frederick Fairbank, the eldest son of the president of the Paradoxical Society, announces in the final chapter of the *Paradoxical Philosophy* that 'Stoffkraft has given up for good his materialistic notions', while confirmation of his conversion to the Unseen Universe comes with the news that he has published 'three memoirs which are everywhere discussed in scientific circles'. The first of these, 'A Complete Theory of Canonizants', recalls Sylvester's preoccupations with the locally Continuous, the second Tait's current concerns with 'The Physical Determination of Beknottedness', while the third considers the Riemannian fourth dimension that engaged both men, 'Ueber Mannigfaltigkeiten in vier Dimensionen [On Manifoldness in four Dimensions]'.[53] Further developing and synthesising these ideas, and having, as Maxwell notes, 'quieted his spirit by a few evolutions in four dimensions', Stoffkraft settles down to write a work akin to the one he inhabits, an 'Exposition of the Relations between Religion and Science'.[54]

Riemann appreciates that the prevailing conception of dimension presupposes Continuity, an insight that Georg Cantor establishes some decades later.[55] The Continuity that Riemann theorises between Euclidean and non-Euclidean spaces aligns suggestively with Stewart and Tait's Seen and Unseen universes. The pair may also have been drawn to the place that his work gives to physics in explaining the variability of space. In trying to ascertain the nature of the geometry that describes physical space, Riemann observes 'that the metric relations of space in the infinitely small do not conform to the hypotheses of geometry'. Rather, 'in [such] a discrete manifoldness, the ground of its metric relations is given in the notion of it, while in a continuous manifoldness, this ground must come from outside ... in binding forces which act upon it ... This leads us into the domain of another science, of physic, into which the object of this work does not allow us to go today.'[56] Riemann argues that the spaces of constant curvature he hypothesises are determined by the gravitational field, an idea that Clifford appreciated and developed, and Einstein later confirmed with his general theory of relativity. Clifford concludes his paper 'On the Space-Theory of Matter' by explicating this idea with a topological analogy, which asserts 'That small portions of space *are* in fact of a nature analogous to little hills on a surface which is on the average flat', so 'that the ordinary laws of geometry are not valid for them'. He uses the analogy of the 'little hills' to build upon Riemann's principle in the passage from 'On the Space-Theory of Matter', cited in Chapter 9, where he writes that 'this property of being curved or distorted is continually being passed on from one portion of space to another after the manner of a wave'. Contemporary mathematics is, as Sylvester notes, a 'higher art [that] is again passing into science embodied in the sublimer theories of Physics and Elasticity'.

Clifford sees his wave-like 'variation of the curvature of space' to describe physical reality, to be 'what really happens in that phenomenon which we call the *motion of matter*, whether ponderable or etherial'.[57] This idea is suggestive for the elastic closed curves of the knotted vortex-tube atoms and their Continuity, and may have paralleled, or provided a catalyst for, Stewart and Tait's speculations. Clifford is in turn led to reflect upon these atoms through his reading of *The Unseen Universe*. Part II of his review essay of the book enforces a moratorium on mockery, as it considers the principle of the vortex atom seriously and substantively. While it begins with Helmholtz, mentions Tait's smoke-rings, and also considers Maxwell's molecular vortices, the account focuses appreciatively upon Thomson's 'brilliant conjecture'. Offering a rationale for Stewart and Tait's conflation of Thomson's ether with Riemannian space, Clifford concludes his discussion by suggesting that the vortex atom, integral to the ether it articulates, could be vindicated by Riemann's metrics, as 'distance or quantity may come to be expressed in terms of *position* in the wide sense of the *analysis sitûs*. And the theory of space-curvature hints at a possibility of describing matter and motion in terms of extension only.'[58]

Writing to Maxwell in November 1874, Tait defends the Riemannian metric against his friend's scepticism: 'Xplane why it is bosh to say the Riemannsche Idee may, if it be found true, give us *absolute* determination of position. Are not the max. & min. curvatures at any point of a given surface just as good coordinates as you could desire – though a little hard to work with,

Science on Parnassus 249

because (& only because) we are all, as yet, miserable duffers in mathematics? Why not, then, the three numbers corresponding in the Riemann affair?'[59] Maxwell's reply to his friend assumes that in making his inquiry Tait has in mind a method of specifying what Clifford refers to as the 'variation of the curvature of space'. Refuting this implicit premise, he identifies the non-Euclideans with a geometry of unyielding constant curvature:

> The Riemannsche Idee is not mine. But the aim of the space-crumplers is to make its curvature uniform everywhere, that is over the whole of space whether that whole is more or less than ∞. The *direction* of the curvature is not related to one of the $x\,y\,z$ more than another or to $-x\,-y\,-z$ so that as far as I understand we are once more on a pathless sea, starless, windless and poleless totus teres atque rotundus ['all smooth and round'].[60]

Maxwell finds the space of the Riemannian plane, an unbounded surface that is nonetheless finite, bleakly homogeneous, a sterile sphere. His poem attributes this idea to both the author of the proof about untying knots alluded to earlier in the poem and Riemann's translator: 'Klein and Clifford fill the void / With one finite, unbounded homaloid / And think the Infinite is now at last destroyed.'

Stoffkraft complains to Fairbank in the *Paradoxical Philosophy* that 'You are driving me back to Universe after Universe, to process after process, forming together an illimitable avenue, and you are quite determined we shall never get to the end.'[61] The place 'Where playful fancy intersperses / Long [or 'Whole'] avenues of universes',[62] 'four-dimensioned space' is presented in Maxwell's poem as involuted in the paradoxical manner of the Klein bottle, a recursive repository for the very ideas, such as those of Klein and Clifford, that bring it into being. The speculative congeries of the Unseen Universe is judged to be over-determined and over-crowded, like the scientific theories of the ether it builds upon, which, Maxwell notes in his 1878 *Britannica* article on this subject, have multiplied such hypothetical media 'till all space had been filled three or four times over with æthers'.[63]

NOVELS, HYMNS AND A PRAYER

The poem's 'Long avenues of universes'[64] suggest systems of Continuity, cosmologies as straight lines of cause and effect, of narrative, organised perhaps by taxon or genre, in a parallel grid. While Tait had earlier compared popular science lectures to sensation novels, Maxwell opens his review of the *Paradoxical Philosophy* by suggesting that his friends' book belongs generically with 'novels or theological works'.[65] As his Cambridge poems and essays make clear, Maxwell is wary of efforts to argue from religion to science and vice versa, a position that he reiterates in 1875: 'I think that the results which each man arrives at in his attempts to harmonize his science with his Christianity ought not to be regarded as having any significance except to the man himself and to him only for a time.'[66] A deeply religious man, Maxwell nevertheless follows Hume's modern example in not premising his philosophical and scientific thought upon the existence and nature of God. Having early distinguished the reverie of his 'Evening Hymn' from the metaphysical purpose of its companion poem 'Reflex Musings', Maxwell has no difficulty in recognising the genre of *The Unseen Universe*: 'It is said in Nature that U U is germinating into some higher form', he writes to Tait in September 1878; 'If you think of extending the collection of hymns given in the original work, do not forget to insert "How happy could I be with Ether".' This potent pun on MacHeath's song from John Gay's *The Beggar's Opera*[67] figures Tait's scientific work and his cosmic speculations respectively as wife and mistress. He is being unfaithful to science. *The Unseen Universe* and its sequel show that Tait is happy with either ether principle; the ideal medium in which Thomson's vortex atoms subsist,[68] or the spiritualist's appropriation of the concept, familiar from Maxwell's 1853 essay 'Idiotic Imps'.

'*A Paradoxical Ode*' closes with a prayer to Continuity:

> Great Principle of all we see,
> Thou endless Continuity!
> By thee are all our angles gently rounded,
> By thee are our misfits adjusted,
> And as I still in thee have trusted,
> So trusting, let me never be confounded!
> Oh never may direct Creation
> Break in upon my contemplation;
> Still may thy causal chain, ascending,
> Appear unbroken and unending,
> And where that chain is lost to sight
> Let viewless fancies guide my darkling flight
> Through atom-haunted worlds in series infinite.[69]

Far from being the fulfillment of Christian belief, Stewart and Tait's doctrine is recognised here as an alternative religion that rejects 'direct Creation', the story of Genesis, as a break in Continuity. As mock-heroic flights of fancy, in which 'Science lifts her pinions / In Speculation's wild dominions', both the materialist's terminal 'twilight of the gods' and Stewart and Tait's answer to it, the endless avenue of the Unseen Universe, recall the workings of the 'fancy scientific', 'the swift whirl with which we fly', in the closing stanzas of the '*Tyndallic Ode*', a further exercise of what 'Reflex Musings' describes as 'Fancy's power deceiving'.

The comparison that Maxwell makes in his review of the *Paradoxical Philosophy* to 'novels' can be read through his remarks on *Villette*, cited in the Chapter 1, where he glosses Brontë's novel by its characters as a set of premises for counterfactual explorations of social life and individual psychology. The review compares his friends' book to such works as George Berkeley's *Three Dialogues between Hylas and Philonous* and W. H. Mallock's *New Republic*. Maxwell identifies the *Paradoxical Philosophy* with a genre in which novelistic devices of character, dialogue and plot are subordinated, almost to the point of allegory, to counterfactual tasks of speculation, where various opinions and wrecklessly *a priori* physical theories are allowed to interact sociably: 'We shall therefore make the most of our opportunity when two eminent men of science, "driven", as they tell us, "by the exigencies of the subject", have laid down all the instruments of their art, shaken the very chalk from their hands, and, locking up their laboratories, have betaken themselves to those blissful country seats where Philonous long ago convinced Hylas that there can be no heat in the fire and no matter in the world; and where in more recent times, Peacock and Mallock have brought together in larger groups the more picturesque of contemporary opinions.' He compares the book favourably to Berkeley's *Alciphron, or the Minute Philosopher*, commending its authors for presenting characters that 'are no mere materialised spirits, or opinions labelled with names of the *Euphranor* or *Alciphron* type'.[70] Similarly, while in the '*Tyndallic Ode*' he accuses Tyndall of turning science into spiritualist spectacle for a popular audience, in his review Maxwell conversely congratulates Stewart and Tait for a Trojan horse strategy that introduces an unscientific readership to contemporary physics: 'they avail themselves of the general interest in theological dogmas to imbue their readers at unawares with the newest doctrines of science. There must be many who would never have heard of Carnot's reversible engine, if they had not been led through its cycle of operations while endeavouring to explore the Unseen Universe.'[71]

The example of Carnot's engine, which Stewart and Tait enhance with his Demons, would have reminded Maxwell that he, too, was no stranger to counterfactual speculations, that his friends' continuous principles of the Seen and the Unseen are an intemperate version of his own scientific efforts to understand the unseen by analogy with the seen. Like his apotheosis of 'Nonsense' in 'Molecular Evolution', '*A Paradoxical Ode*' appreciates the restless creativity, if not always the consequences, of audacious speculation.

GODS AND DEMONS

The scene from *Prometheus Unbound* that culminates with the verses Maxwell parodies in '*A Paradoxical Ode*' takes place as '*The [Spirit's] Car pauses within a Cloud on the top of a snowy Mountain.*'[72] While Tait's speculations may be comparable in their extravagance to those of the Metropolitans, '*A Paradoxical Ode*' situates them not in the lowly 'fields of fractured ice' of Tyndall's Alpine habitat, nor the constrained microscopic depths of his molecules, but in the airy Parnassian region that the Liverpool address identifies with Sylvester's Exeter address, 'those serene heights' described by Tennyson's 'Lucretius'. This is consistent with Maxwell's last letter to Tait, in which he renews his conceit of 'the Chief Musician upon Nabla'. Corresponding to Rowan Hamilton's characterisation of Newton ('as the mind of an artist calls up many forms, he meditated on many laws and caused many ideal worlds to pass before him'), Maxwell identifies his friends with the Parnassian prerogatives of imaginative reach and formal prowess in mathematics, qualities of 'bold phantasy and mathematical insight' that Planck, like Boltzmann, in turn appreciated in Maxwell's own work.[73]

Belonging to a latitude only slightly below that of the Parnassian North Britons, the Metropolitans are nonetheless relegated by Maxwell to the underworld. He summons the London-based agnostics in his 'Report on Tait's Lecture on Force' by citing the remark that Tait, directly attacking the rhetoric of the Belfast address, addressed to Tyndall in his original lecture, 'Are these thy gods, Oh Israel?':

> Are these the gods in whom ye put your trust,
>   Lordlings and ladies?
> The 'secret potency of cosmic dust'
>   Drives them to Hades.[74]

Tyndall's scientific materialism is figured as a hell of his own making. Unfurling the implications of the euphemistic 'empyrean fires', from which the lecturer hails in the '*Tyndallic Ode*', the reductionist doctrine of molecular forces is seen in the 'Report' to hurl its advocates southward to diabolically warmer climes. Tait's focus on force discloses the Laplacean nature of Tyndall's scientific materialism, which springing from the esoteric, indeed occult, '"secret potency of cosmic dust"' encompasses both the nebular origins he attributes to the universe and the deterministic engine of his molecular forces.

Laplace sees classical mechanics to entail the utter determinism of nature, which, as he observes in his *Philosophical Essay on Probabilities* (1814), a knowledgeable agent could fully appreciate and easily anticipate:

> An intelligence who at some given moment knew all the forces that animate nature, and the respective situation of the beings that compose it, if it were further sufficiently vast to submit these data to analysis, could embrace within a single formula, the movements of the largest bodies of the universe and those of the lightest atom: nothing would be uncertain for it, and the future, like the past, would be present to its eyes.[75]

This principle of 'intelligence', which became known as 'Laplace's Demon', finds its match half a century later in 'Maxwell's Demon'. Maxwell first described his hypothetical 'being' in a letter to Tait dated 11 December 1867, while the next letter he wrote to his friend, on 23 December, shows that he was also thinking at this time about the Laplacean principle it effectively pitches itself against: 'The sensationalist says "I am now going to grapple with the Forces of the Universe and if I succeed in this extremely delicate experiment you will see for yourselves exactly how the world is kept going".'[76] In contrast to the reversible and intricately determined forces that 'Laplace's Demon' presides over, 'Maxwell's Demon' upholds the irreversible second law of thermodynamics statistically, by affirming that particular molecules can act unpredictably in ways that contradict it. Entropy is understood to occur within a dynamic Hamiltonian system of energy, it is not the scrupulously even mechanistic process described by the materialists. Indeed, as Maxwell observes in his *Britannica* article on the 'Atom', irregularity is a fundamental condition of systems acting in an irreversible manner.[77]

As Tyndall explains in the Belfast address, his Laplacean science 'demands the radical extirpation of caprice and the absolute reliance upon law in nature'. He champions Democritus for 'his uncompromising antagonism to those who deduced the phenomena of nature from the caprices of the gods', whilst also acknowledging that the Greek atomist nonetheless believes in such supernatural beings:

> The gods were to him eternal and immortal beings, whose blessedness excluded every thought of care or occupation of any kind. Nature pursues her course in accordance with everlasting laws, the gods never interfering. They haunt
>
> > 'The lucid interspace of world and world,
> > Where never creeps a cloud or moves a wind,
> > Nor ever falls the least white star of snow,
> > Nor ever lowest roll of thunder moans,

> Nor sound of human sorrow mounts to mar
> Their sacred everlasting calm'.⁷⁸

Tyndall finds in the transcendent realm described in Tennyson's 'Lucretius' a good place to stow the ancient gods away from nature, the domain of the atoms that he thereby secures for modern science.

The contrast and contest between Maxwell's and Tyndall's science is registered nicely in the respective uses they make of the lines from 'Lucretius'. Crucial to Tyndall's purposes of banishing the gods from the domain of science, the first line of the extract, describing the Lucretian void, is omitted by Maxwell, who in his Liverpool address wishes to present 'those serene heights' to which Sylvester transported him at Exeter as an integral poetic vision of science, which he contrasts with the diabolical and rhetorically forceful region of his rival's Norwich address.

The following lines from Maxwell's poem 'British Association, 1874' describe the early part of Tyndall's argument that includes the lines from Tennyson:

> Yet they did not abolish the gods, but they sent them well out of the way,
> With the rarest of nectar to drink, and blue fields of nothing to sway.
> From nothing comes nothing, they told us, nought happens by chance, but by fate;
> There is nothing but atoms and void, all else is mere whims out of date!

The 'blue fields of nothing to sway' duly notes the calm cloudless region described in the extract from 'Lucretius'. It is, however, rendered summarily here as a poetic repository of not only the gods but also Tyndall's science, with an image that invokes his most celebrated scientific researches into the blueness of the sky, and indeed the lyrical phrase 'infinite azure' with which he famously closes his address, and Maxwell accordingly ends his poem.⁷⁹ The Sylvestrian Parnassus of Maxwell's original citation is reasserted gently and inclusively. Writing in the wake of the Belfast address, Maxwell opposes Tyndallic materialism by playfully insisting upon the re-enchantment of nature, with atoms and stars cast as gods in their transcendent realm, reserving the right to be capricious, to act on 'chance' and warrant the corresponding 'whims' of poetry and scientific speculation.

Tait's science is seen by Maxwell in his 'Report of Tait's Lecture on Force' to liberate us not only from Tyndall's 'occult' principle of molecular force, but also from the greater gothic principle it instances, the mind-forged manacle of Laplacean force itself: 'Impotent spectre! / Thy reign, O Force! is over.' The preceding line, which declares that Force's 'arrows now have lost their sting', establishes the stanza's echo of First Corinthians ('O death, where is they sting?'). 'Death is swallowed up in victory',⁸⁰ while, correspondingly, Force has been vanquished by the eternal life of energy.

Tait is congratulated for sublating the conflictive terms of Newton's third law of motion, 'Both Action and Reaction now are gone', in the state of tension: 'Stress joined their hands in peace, and made them one.'

The abolition of Force's tyranny, like the deposition of Jupiter in Shelley's *Prometheus Unbound*, inaugurates a new millennium:

> The Universe is free from pole to pole
> Free from all forces.
> Rejoice! ye stars – like blessed gods ye roll
> On in your courses.[81]

Maxwell mischievously commends Tait for a vision of the heavens that his friend explicitly rejects. Completing a brief catalogue of the 'fearfully absurd consequences' he sees as following from purely *a priori* approaches to physics, Tait observes early in his *Recent Advances in Physical Science* that 'Within the last fifty years we have had philosophers like Hegel saying that the motion of the heavenly bodies is not a being pulled this way and that: that they go along, as the antients said, like blessed gods.'[82] By opposing the tyranny of force with Hamiltonian dynamics, Tait is, however, seen by the poem to belong to Aristotle and Hegel's camp, to share Maxwell and Sylvester's Parnassian vision. No longer the Laplacean universal, force is found to be 'a mere / Space-Variation', a local principle, much as the parts of space itself are for Sylvester, as he similarly rejects what he considers to be the homogenising tyranny of Euclid in favour of Gauss and Riemann's ideas of Continuity.

PLEASURE AND POISE

The dynamic poise of 'Stress' or tension that the 'Report' commends Tait for restoring to the universe is for Maxwell also the essence of play, the principle that defines the workings of his toys and puns alike. In his essay on Plateau, Maxwell recognises the soap-bubble to be an especially lyrical instance of this tension, whilst also seeing it to demonstrate the continuity between childhood play and science. Indeed, as was noted in Chapter 1, he is quick to identify himself and his scientific peers with Plateau's figures of children blowing bubbles. Described in the essay as 'spheres of splendour', soap-bubbles bring into the concrete three-dimensional world of his experimental physics Maxwell's aesthetic and scientific appreciation of curved forms, which he expresses in his earliest scientific paper on tracing ellipses, 'bicentral sources of lasting joy'. He reproduces their three-dimensional forms with the stereograms of such figures as ellipsoids and spherical ellipses

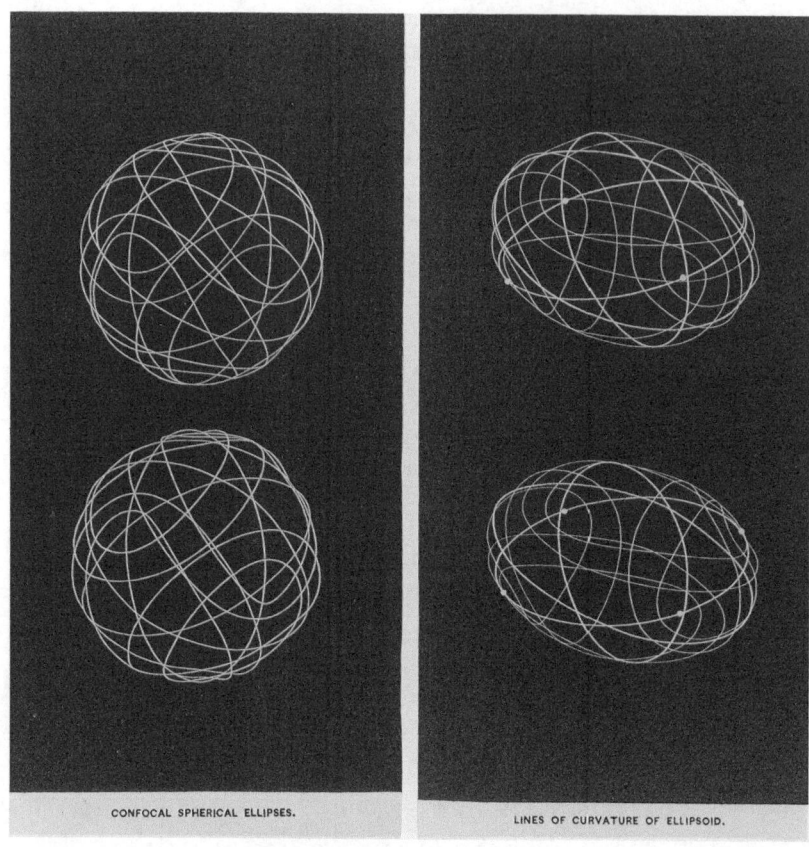

Figure 5 'The united images look like a real object in the air.' James Clerk Maxwell, Stereograms: Lines of Curvature of Ellipsoid, Concyclic Spherical Ellipse. Add.7655.V.i.II 1–2, Cambridge University Library. Reproduced by kind permission of the Syndics of Cambridge University Library.

he devised in 1867 for his improved stereoscope (Figure 5): 'The united images look like a real object in the air.'[83]

J. J. Thomson describes the devil-on-two-sticks as 'a home-made gyroscope with all the paradoxical properties of that instrument'. While Paris championed his thaumatrope as a punning machine, the devil, like the 'dynamical top' that, Thomson writes, 'no doubt it led [Maxwell] to', correspondingly conflates opposed principles of stasis and movement, of stability and instability, 'the properties of bodies in rotation'. Like Heraclitus, Maxwell finds in nature definitive tensions, unities composed of contradictory qualities, which have their formal parallel in the pun.[84]

Heraclitus' Fragment 115, 'The name of the bow is life, but its work is death', not only invokes the bow as an instance and emblem of physical tension, but correspondingly hinges upon the Greek word βίος, which with the accent on the first syllable is a word for life, while on the final syllable it becomes a word for bow.[85] Like Maxwell's entropic signature, Heraclitus' pun locates death paradoxically in the midst of life.

In his scientific memoir of Maxwell, Garnett observes that 'Of the problems suggested to him by the devil-on-two-sticks we have no account', whereas his work developing the 'dynamical top' is well documented. If the top could be perfected, its form evenly weighted and balanced, and set to spinning on an immaculately level and frictionless surface without resistance from the air, it could conceivably do so indefinitely. Indeed the spinning-top is one of the two examples that Newton gives in the *Principia* to illustrate his first law of motion. Bringing to his own teaching and research the playful principles of Paris's pedagogy, Maxwell improved his scientific toy to such an extent that his friends believed it could keep spinning from one day to the next:

The 'Dynamical Top', which was invented by Maxwell to illustrate dynamical propositions, technically so-called, was, in its final form, constructed of brass by Mr. Ramage of Aberdeen. It was this top which Maxwell brought with him to Cambridge when he came up for his M.A. degree in the summer of 1857, and exhibited to a tea-party in his room in the evening. His friends left it spinning, and next morning Maxwell, noticing one of them coming across the court, leapt out of bed, started the top, and retired between the sheets. It is needless to say that the spinning power of the top commanded as great respect as its power of illustrating Poinsot's *Theorie Nouvelle de la Rotation des Corps*.[86]

While the spinning-top demonstrates the Heraclitean unity of opposites, such poise eventually yields in the Lucretian manner to a new order. Like the laminar flow of atoms, the spinning-top is subject to slight deflections, which similarly disrupt its stable dynamic system, causing it to move in a large curve and eventually topple. Serres makes this parallel between the spinning-top and the Lucretian clinamen, noting that it also extends to the modern physics of thermodynamics,[87] as the stability of the Carnot cycle is undermined by the random deflections of heat-loss.

Appreciated as an emblem of irreversibility and chance, of statistical systems that defy the Laplacean prophecy of 'the twilight of the gods', the clinamen leaves one 'heavenly path' for another. It embodies the measured caprice of the 'stars – like blessed gods', which rejoice at having recently had their universe restored to them, freed 'from all forces', by Tait. Maxwell makes this correlation between the stars and the Lucretian atoms two years after writing the 'Report on Tait's Lecture', in an argument he mounts against reductionism in both his review of the *Paradoxical Philosophy* and his essay 'Psychophysik', which dates from February 1878.

The title 'Psychophysik' is drawn from G. T. Fechner's term for the study of relations between physical and mental states.[88] Tyndall is named by the essay as the most strident but not the most cogent advocate of such materialist conceptions of mind: 'The theory of Plastidule souls has been hinted at by several persons of whom Dr Tyndall has spoken loudest. A much clearer utterance', however, 'is that of Professor von Nägeli, of Munich, in an address delivered at the Munich meeting of the German Association in 1877.' The address was promptly translated and published in *Nature*, where Maxwell read it.[89] In his review of the *Paradoxical Philosophy*, he observes that Stewart and Tait endow Stoffkraft with von Nägeli's reductionist doctrine of sentient atoms: '"I feel myself compelled to believe," says the learned Doctor, "that all kinds of matter have their motions accompanied with certain simple sensations. In a word, all matter is, in some occult sense, alive".'[90] While Maxwell develops Tyndall's speculative doctrine of atomic forces as an anthropomorphic myth in his Liverpool address and the '*Tyndallic Ode*', he observes that von Nägeli's atoms require of him no such embellishment: 'To attribute life, sensation, and thought to objects in which these attributes are not established by sufficient evidence is nothing more than the good old figure of Personification.'[91]

In an account that is included word for word in both 'Psychophysik' and his review (and accordingly italicised in the following extract), Maxwell meets such solipsism, much as he does in 'Reflex Musings', by asserting the autonomy of the objects that physics studies:

Prof. von Nägeli must have forgotten his dynamics, or he would have remembered *that the molecules, like the planets, move along like blessed gods. They cannot be disturbed from the path of their choice by the action of any forces, for they have a constant and perpetual will to render to every force precisely* that *amount of deflexion which is due to it.*[92]

Maxwell's account of the molecules corresponds to Lucretius on atoms: 'those which are first-beginnings of things no force can quench'.[93] Such atoms are, like the gods, the other foundational principle of Lucretius' cosmology, invulnerable. Repudiating reductionism, and recalling his abiding and diverse practices of modelling reality figuratively, Maxwell's account of molecules asserts scientific truth through poetic analogy.

On the Nature of Things opens with an invocation that introduces the divine realm as not only the province, but also the consequence, of Venus:

Science on Parnassus 259

'Before thee, goddess, flee the winds, the clouds of heaven; before thee and thy advent.'[94] Nicely rendered by Gilliam van der Gouwen's engraving (Figure 6), Venus effortlessly deflects various other molar forces, just as Maxwell's planets do, while correspondingly, at the sub-microscopic level, his molecules 'cannot be disturbed from the path of their choice by the action

Figure 6 *Venus Physica*, engraving by Gilliam van der Gouwen of a print by Jan Goere, from Lucretius, *De Rerum Natura* (London: Tonson, 1712). Reproduced by kind permission of the Syndics of Cambridge University Library.

of any forces'. Free-willed and just, 'they have a constant and perpetual will to render to every force precisely that amount of deflexion which is due to it'. Lucretius' 'Hymn to Venus' is a prologue to the introductory accounts of atoms in Book I and of the clinamen in Book II of his poem. Maxwell reads the clinamen, Cicero's *concursus fortuitus*, through the 'Hymn', seeing the atoms to have 'by fortuitous embraces, / Engendered all that being hath'. Elsewhere, in his two poems on galvanometers, the 'Valentine by a Telegraph Clerk ♂ to Telegraph Clerk ♀' and the first of his 'Lectures to Women on Physical Science',[95] the electromagnetic field is represented as a plenum of Venusian energies, in which sexual and emotional tensions between their respective male and female protagonists are represented by the figure of electromagnetic induction.

Implicitly contrasted with the deterministic lust of the crude Tyndallic 'molecules with fierce desires' that 'Shiver in hot embraces', Maxwell's god-like molecules are impelled by a subtle desire, a discretionary will that (occasionally assisted by attendant 'Demons') expresses itself in inclinations and deflections. In his notes for Thomson 'On the History of the Kinetic Theory of Gases', Maxwell quotes the passage from *On the Nature of Things*, cited in Chapter 7, in which Lucretius opposes the clinamen, the 'minute swerving of first-beginnings', to the idea that 'all things be done by blows through as it were an outward force'. Maxwell evidently applies such criteria in distinguishing the clinamen from the Tyndallic atoms that 'clash together in fierce collision'. Like North British energy physics, the clinamen is seen to transcend the mechanistic tit-for-tat bickering of Newtonian forces. Evidently drawing upon the notes his friend wrote for him, Thomson explains to a Royal Institution audience in February 1879 that 'Clerk Maxwell's "demon" is a creature of imagination' that allows us to grasp the Lucretian character of atoms, it having been 'invented to help us to understand [that] the "Dissipation of Energy" . . . follows in nature from the fortuitous concourse of atoms'.[96]

The Lucretian clinamen opposes classical Laplacean mechanics with a baroque dynamics in which tiny events can have disproportionate repercussions, and stability is the resultant of elaborate instances of instability, systemic regularity the consequence of reverberant irregularity: 'The constancy and uniformity of the properties of the gaseous medium is', Maxwell observes, 'the direct result of the inconceivable irregularity of the motion of agitation of its molecules.'[97] The extravagantly dynamic but nonetheless poised composition of van der Gouwen's engraving of Venus, Pluto, Neptune, Jupiter, Cybele and other deities for the 'Hymn to Venus' provides an emblem for Maxwell's molecules, and highlights their kinship to Lucretius' atoms; 'though the first-beginnings of things are all in motion, yet the sum is seen to rest in supreme repose'.[98]

As a figure for 'Nonsense' in 'Molecular Evolution', with its metamorphosis of British Asses into wild Red Lions, Maxwell's principle of the clinamen is a baroque conceit akin to that which structures such rococo paintings as François Boucher's *The Rape of Europa* (1732–4), where Europa sits decorously on the bull, at the heart of an intricate, delicately balanced, composition, a harmonious cosmology that will be immediately devastated by Zeus's act of bestial violence. Nonsense asserts the right of the atoms of ideas and words to freedom of association through such forms as analogies and puns, to be *inflected* and so build new meanings; to jostle about with other such atoms and so escape semantic determinism, the fixed and preordained meanings of pedantry, solipsism and settled canonical theory. James Joyce also enacts this principle in the punningly over-determined mythic beginning to *Finnegans Wake*, in which the Genesis account and the Lucretian *even atoms* and their swerve each offer themselves alternately as Gestalt figure and ground: 'riverrun, past Eve and Adam's, from swerve of shore...'[99] Lucretius emphasises that without the clinamen there would be no principle of creation in the universe, but only the unrelieved homogeneity of atoms raining evenly downwards forever.[100] The clinamen offers Maxwell a proxy for the unruly play of the pun, the tense relations of analogy, and the variegated repetition of rhyme, which is for him, as for Sylvester, a model of lively knowledge; a transformative continuity between lines of verse to oppose simple reiterative (or indeed Euclidean) parallelism.

Maxwell appreciates 'those serene heights', '"Where never creeps a cloud, or moves a wind"', both scientifically and stylistically, for their freedom not only from Laplacean physical forces but also from the 'forcible expression of Dr. Tyndall'. The calculated metaphors of violence that Maxwell habitually applies to Tyndallic lectures, 'the forcible language and striking illustrations', identifies them with the brutal determinist ontology that Lucretius describes, in which 'all things be done by blows through as it were an outward force'. In contrast to the strident and coercive Tyndallic style, the Sylvestrian is presented by Maxwell as gentle and lyrical, identified with Lucretius' original poetic vision as it is further distilled by Tennyson. The domain of the gods, 'those serene heights', represents for Maxwell the Lucretian confluence of science and poetry, and hence Sylvester's refusal to separate the two. This Parnassian continuity yields a hermeneutic suppleness, an epistemological subtlety that is for both men predicated upon the infinitesimal *differential*: the slightest gestures of curvature, of inclination and deflection, of verbal inflection and play, the incipits of a variable and dynamic reality.

Supremely disdainful of the canons of causality, the deterministic enslavement that materialism wishes to subject them to, Maxwell's molecules are, 'like the planets', akin to Venus and the atoms that Lucretius models, and indeed the figure of the scientist as a child blowing bubbles: 'They must therefore,' he concludes, 'enjoy a perpetuity of the highest and most unmixed pleasure.'[101] With his entropic signature and metaphysical wit, Maxwell both celebrates and satirises the free play enshrined in nonsense, by which Victorian science discovers and amuses itself.

Notes

CHAPTER 1

1. Graves, *Life of Hamilton*, I, 262.
2. Hankins, *Sir William Rowan Hamilton*, 50.
3. Graves, *Life of Hamilton*, I, 269.
4. *Ibid.*, II, 591; I, 267, 491–2.
5. *Ibid.*, I, 313.
6. Bk IV, ll. 1251–63. Wordsworth, *The Poems*, II, 155.
7. Graves, *Life of Hamilton*, I, 316–17.
8. See Stimson, *Scientists and Amateurs*, 190–3; Howarth, *The British Association*, 3–6; Morrell and Thackray, *Gentlemen of Science*, 47–52.
9. Whewell, 'On the Connexion of the Physical Sciences', 59; Levere, 'Coleridge and the Sciences', 296; Whewell, 'On the Connexion of the Physical Sciences', 60, 59.
10. Morrell and Thackray, *Gentlemen of Science*, 96.
11. Huxley, *Life and Letters of Thomas Henry Huxley*, I, 96.
12. Whewell, *Philosophy of the Inductive Sciences* (1840), I, lxxxi.
13. Cited in Morrell and Thackray, *Gentlemen of Science*, 276.
14. Todhunter, *William Whewell*, II, 127.
15. *Report of the Third Meeting of the BAAS at Cambridge 1833*, xi.
16. *Ibid.*, xii, xiii, xiv, xix.
17. Cited in Howarth, *The British Association*, 5, 6.
18. Morrell and Thackray, *Gentlemen of Science*, 276.
19. See Gill, *Wordsworth and the Victorians*, and Reid, *Wordsworth and the Formation of English Studies*.
20. Cited in Morrell and Thackray, *Gentlemen of Science*, 277.
21. For some recent studies that depart from this traditional model, see Beer, *Open Fields*; Brown, *Hopkins' Idealism*; Gold, *ThermoPoetics*; Gooday, 'Spotwatching, Bodily Postures and the "Practised Eye"'; Jenkins, *Space and the 'March of Mind'*; O'Gorman, 'John Tyndall as Poet'; Silver, 'The Last Poem of James Clerk Maxwell'.
22. Whewell, 'On the Connexion of the Physical Sciences', 59; Whewell, *Philosophy of the Inductive Sciences* (1840), I, cxiii.
23. Graves, *Life of Hamilton*, I, 317.
24. *Ibid.*, 502–3; Wordsworth, *Prelude* (1850), Bk III, l. 63 (103).
25. Graves, *Life of Hamilton*, I, 145.

26. *Ibid.*, I, 145–6, 146, 127.
27. Lecercle, *Philosophy of Nonsense*, 202.
28. Graves, *Life of Hamilton*, I, 193.
29. Smith, *Darwin and Victorian Visual Culture*, 94.
30. *Ibid.*, 33–8.
31. Hyman, *Lear's Birds*, 20. See also Lear, *Selected Letters*, 14.
32. Noakes, *Edward Lear*, 272n.
33. Lear, *Selected Letters*, 20.
34. *Ibid.*, 19.
35. Lear, *Complete Nonsense*, 129.
36. Darwin, *The Botanic Garden: A Poem in Two Parts*, II, 17, 37–9.
37. Lear, *Complete Nonsense*, 123.
38. *Ibid.*, 149.
39. Max Müller, *Lectures on the Science of Language*, 361–8.
40. Lear, *Complete Nonsense*, 149.
41. *Ibid.*, 141, 148, 141.
42. *Ibid.*, 35.
43. *Ibid.*, vii–viii. See, e.g., Prickett, *Victorian Fantasy*, 114–19.
44. Lecercle, *Philosophy through the Looking-Glass*, 140.
45. Bennett, 'Macropus Parryi', 300. It is reproduced in Hyman, *Lear's Birds*, 34.
46. Hyman, *Lear's Birds*, 35.
47. *Ibid.*, 82.
48. Shaw and Nodder, *Naturalist's Miscellany*, VI.
49. Bennett, 'Macropus Parryi', 295.
50. Main, 'History of Discovery of Anura', 3.
51. Reproduced in Hyman, *Lear's Birds*, 48.
52. Colley, 'Lear's Anti-Colonial Bestiary', 114.
53. Lear, *Complete Nonsense*, 51, 195, 180, 198, 165, 189, 42.
54. See Deleuze and Guattari, *Kafka*, also *1000 Plateaus*, 232–309.
55. Lear, *Selected Letters*, 16.
56. See Paradis, 'Satire and Science', and Browne, 'Darwin in Caricature'.
57. Hopkins, *Journals and Papers*, 204.
58. Lear, *Complete Nonsense*, 128, 221.
59. *Ibid.*, 37, 49, 201, 23.
60. *Ibid.*, 27, 45, 16, 55, 3, 34.
61. *Ibid.*, 47, 11, 159.
62. *Ibid.*, e.g., 4, 8, 3, 36.
63. Reproduced in Hyman, *Lear's Birds*, 47.
64. Lear, *Complete Nonsense*, 58, 5, 39.
65. Lear, *Selected Letters*, 16.
66. Lear, *Complete Nonsense*, 19.
67. Ede, 'Lear's Limericks and their Illustrations', 112.
68. Lear, *Complete Nonsense*, 12.
69. Lecercle, *Philosophy through the Looking-Glass*, 140–1.
70. Lear, *Complete Nonsense*, 13.
71. *Ibid.*, 140.

72. *Ibid.*, 155.
73. *Ibid.*, 218. For other uses of 'scroobious', see 177 and 198.
74. Lewis, 'Causation', 556.
75. Kim and Maslen, 'Counterfactuals', 82, 84.
76. Maxwell, *Theory of Heat*, 308–9.
77. See Lewis, *Counterfactuals*, and *On the Plurality of Worlds*.
78. Benjamin, *Charles Baudelaire*, 48.
79. Campbell and Garnett, *Life of Maxwell*, 190.
80. Graves, *Life of Hamilton*, I, 183.
81. *Ibid.*, I, 502–3, 503.
82. *Ibid.*, I, 141.
83. Edgeworth, *Harry and Lucy Concluded*, I, viii.
84. Paris, *Philosophy in Sport*, x, II, 7, 13, 8.
85. Campbell and Garnett, *Life of Maxwell*, 189.
86. Maxwell, *Letters and Papers*, III, 516. See also *ibid.*, III, 747. Cf. Paris, *Philosophy in Sport*, 211–12.
87. Campbell and Garnett, *Life of Maxwell*, 429.
88. *Ibid.*, 37, 37, 37–8. For this volume, Campbell wrote the biography (Part I), while Garnett wrote 'Contributions to Science' (Part II). See *ibid.*, vii.
89. *Ibid.*, 36–7, 484–5, 210, 499–500; Maxwell, *Letters and Papers*, I, 499–501.
90. Kim and Maslen, 'Counterfactuals', 82.
91. Campbell and Garnett, *Life of Maxwell*, 135, 251.
92. Maxwell, *Scientific Papers*, II, 243.
93. J. J. Thomson, 'James Clerk Maxwell', 6.
94. Maxwell, *Scientific Papers*, I, 248.
95. Campbell and Garnett, *Life of Maxwell*, 37, 485. See also Maxwell, *Letters and Papers*, II, 444–7, plates XIII and XIV.
96. Campbell and Garnett, *Life of Maxwell*, 484–5.
97. Lamb, 'Maxwell as Lecturer', 145–6.
98. Campbell and Garnett, *Life of Maxwell*, 429.
99. Paris, *Philosophy in Sport*, 18, xi.
100. Maxwell, *Scientific Papers*, II, 393.
101. Paris, *Philosophy in Sport*, 209.
102. Plateau, *Statique Expérimentale*, II, 120.
103. Nietzsche, *Beyond Good and Evil*, 62.
104. Maxwell, *Scientific Papers*, II, 393.
105. Campbell and Garnett, *Life of Maxwell*, 648.
106. *Ibid.*, 174.
107. Rudy, *Electric Meters*, 3.
108. Hayley, *Life and Writings of Cowper*, III, 196.

CHAPTER 2

1. Knott, *Life of Tait*, 101.
2. Snow, *Two Cultures*, 14.
3. Campbell and Garnett, *Life of Maxwell*, 223.

4. Maxwell, *Letters and Papers*, I, 376.
5. *Romeo and Juliet*, II.ii.3.
6. *Richard III*, I.i.1–2.
7. Maxwell, *Letters and Papers*, II, 832.
8. Carroll, *Alice's Adventures*, 137, 229–30.
9. Paris, *Philosophy in Sport*, 8, 8–9.
10. Garnett, 'Maxwell, F.R.S.', 44.
11. Campbell and Garnett, *Life of Maxwell*, 629.
12. Dallas, 'Student Life in Scotland', 375.
13. Campbell and Garnett, *Life of Maxwell*, 629. See also *ibid.*, 167.
14. Maxwell, *Letters and Papers*, I, 376.
15. Lecercle, *Philosophy of Nonsense*, 4.
16. Campbell and Garnett, *Life of Maxwell*, 625.
17. *Ibid.*, 591.
18. *Ibid.*, 75.
19. 'On the Description of Oval Curves and those having a plurality of Foci', *Proceedings of the Royal Society of Edinburgh*, II (1846), 89–93. Rpt in Campbell and Garnett, *Life of Maxwell*, 76–9; Maxwell, *Scientific Papers*, I, 1–3.
20. Tait, 'Clerk-Maxwell's Scientific Work', 317.
21. Maxwell, *Letters and Papers*, I, 69.
22. Dallas, 'Student Life in Scotland', 377–8.
23. Davie, *The Democratic Intellect*, 169–72.
24. Maxwell, *Letters and Papers*, I, 72; W. Hamilton, *Works of Reid*, 710.
25. Davie, *The Democratic Intellect*, 111.
26. Hamilton, *Works of Reid*, 710.
27. Tait, 'Spencer versus Thomson,' 402.
28. Davie, *The Democratic Intellect*, 197, 196.
29. Campbell and Garnett, *Life of Maxwell*, 422.
30. Rouse Ball, *Mathematics at Cambridge*, 214–15; Warwick, *Masters of Theory*, 139.
31. Wilson, 'The Educational Matrix', 22.
32. Goodwin, *Elementary Course of Mathematics* (4th edn), vii.
33. Warwick, *Masters of Theory*, ch. 3.
34. Ball, *Mathematics at Cambridge*, 213; Whewell, *Of a Liberal Education* (1845), 154.
35. See Whewell, *Of a Liberal Education* (1845), 149–51, also the account he cites from Jebb (*ibid.*, 169–76).
36. Smith, 'Trinity College Annual Examinations', 129. Galton records that honours students in the 1860s would be graded out of a total of 17,000 points (*Hereditary Genius*, 19).
37. Ireland, *Hogarth*, 152–3; Harrison, *The Humourist*, 2; Grandville, *Un Autre Monde*; e.g., Brough (ed.), *Comic Almanack*, 21.
38. Cited in Warwick, *Masters of Theory*, 114.
39. Campbell and Garnett, *Life of Maxwell*, 615.
40. *Ibid.*, 594.

41. Hamilton, *Works of Reid*, 709.
42. Campbell and Garnett, *Life of Maxwell*, 648.
43. Paris, *Philosophy in Sport*, 8, 17.
44. *Ibid.*, 377, 382, 381.
45. Ceram, *Archaeology of the Cinema*, 64–5.
46. Paris, *Philosophy in Sport*, 389.
47. Crary, *Techniques of the Observer*, 69.
48. Campbell and Garnett, *Life of Maxwell*, 625, 614.
49. *Ibid.*, 174–5.
50. Burns, *Poems and Songs*, III, 1503. The first setting is given at II, 843–4.
51. Johnson, *Scots Musical Museum*, V, no. 418.
52. Campbell and Garnett, *Life of Maxwell*, 174–5; 630, 626.
53. Cited in Warwick, *Masters of Theory*, 214. See also Craik, *Mr Hopkins' Men*, 21–3.
54. Campbell and Garnett, *Life of Maxwell*, 623.
55. Cited in Harman, *Natural Philosophy of Maxwell*, 20, 19–20.
56. Campbell and Garnett, *Life of Maxwell*, 150, 195.
57. *Ibid.*, 616, 173.
58. Maxwell, *Letters and Papers*, I, 356. See 'On Faraday's Lines of Force', Maxwell, *Scientific Papers*, I, 155–229.
59. Tait, 'Clerk-Maxwell's Scientific Work', 317.
60. Maxwell, *Letters and Papers*, I, 355.
61. Cited in Harman, *Natural Philosophy of Maxwell*, 88.
62. Cited in *ibid.*, 71.
63. Locke, *Essay Concerning Human Understanding*, II.1.2. See also II.1.3–5.
64. See Hamilton, *Lectures*, I, 234–8; II, 200–1, and Whewell, *Philosophy* (1847), I, 27–9, 171–4; II, 308–19.

CHAPTER 3

1. Maxwell, *Letters and Papers*, III, 598.
2. Maxwell Papers, 'Logic No. 2'.
3. Hendry, *Maxwell and the Theory of the Electromagnetic Field*, 28.
4. Dallas, 'Student Life in Scotland', 378.
5. Maxwell, *Letters and Papers*, I, 69.
6. Campbell and Garnett, *Life of Maxwell*, 108.
7. Hamilton, *Lectures*, I, 288. This edition offers an accurate version of the lectures that Maxwell attended, for as Veitch notes, after their first composition in the 1830s they 'were never substantially changed; they received only occasional verbal alterations' (*Hamilton*, 16). See also Harman, *Natural Philosophy of Maxwell*, 29.
8. Maxwell, *Letters and Papers*, I, 228.
9. Campbell and Garnett, *Life of Maxwell*, 227.
10. Maxwell, *Letters and Papers*, I, 228.
11. Quoted in Harman, *Natural Philosophy of Maxwell*, 31.
12. Campbell and Garnett, *Life of Maxwell*, 593–4.

13. Scott, *Marmion*, 349.
14. Coleridge, *Poetical Works*, 456.
15. Campbell and Garnett, *Life of Maxwell*, 595; 594; 594.
16. Maxwell, *Letters and Papers*, I, 228.
17. Herschel Papers, Portfolio F. No. 24. Herschel, *Essays*, 737.
18. Campbell and Garnett, *Life of Maxwell*, 594; 596.
19. Maxwell, *Letters and Papers*, I, 227.
20. Hamilton, *Lectures*, III, 1; III, 60. See also I, 62.
21. *Ibid.*, I, 58. See Newton, *Principia*, Bk III, 'Rules of Reasoning in Philosophy', 1 and 2.
22. Maxwell, *Letters and Papers*, I, 227.
23. Hamilton, *Lectures*, II, 401–13; 404; 405–6; 375.
24. Paley, *Natural Theology*, 1–18.
25. Maxwell, *Letters and Papers*, I, 382.
26. Kant, *Critique of Practical Reason*, 312, in *Kant's Critique of Practical Reason*, 260. Hamilton cites the passage at *Lectures*, II, 515.
27. Campbell and Garnett, *Life of Maxwell*, 596.
28. Maxwell, *Scientific Papers*, I, 155–6.
29. Maxwell, *Letters and Papers*, I, 382.
30. *Ibid.*, I, 376–7.
31. Hume, *Enquiry*, 211.
32. Nietzsche, *Beyond Good and Evil*, 7.
33. Maxwell, *Letters and Papers*, I, 209.
34. *Ibid.*, I, 673.
35. *Ibid.*, I, 227, 379.
36. Hamilton, *Lectures*, II, 390.
37. Hamilton, *Lectures*, II, 376–413, esp. 390–2.
38. Maxwell, *Letters and Papers*, I, 380.
39. Hamilton, *Lectures*, II, 390, 154.
40. Harman, *Natural Philosophy of Maxwell*, 30.
41. Maxwell, *Letters and Papers*, I, 112.
42. Hamilton, *Lectures*, II, 114. See also note 'D' to Hamilton's Dissertations, in Hamilton (ed.), *Works of Reid*, II, 861.
43. Maxwell, *Letters and Papers*, I, 111.
44. Campbell and Garnett, *Life of Maxwell*, 108.
45. Maxwell, *Letters and Papers*, I, 378. See also I, 377.
46. Maxwell, *Scientific Papers*, I, 489–502.
47. Hamilton, *Lectures*, II, 115; Maxwell, *Letters and Papers*, I, 72.
48. Hamilton (ed.), *Works of Reid*, II, 858–9.
49. *Ibid.*, II, 848.
50. Maxwell, *Letters and Papers*, I, 249.
51. *Ibid.*, I, 227; II, 957.
52. Cited in Davie, *The Democratic Intellect*, 127.
53. Exodus 2.22.
54. Isaiah 45.15.

55. Maxwell, *Scientific Papers*, II, 216.
56. Campbell and Garnett, *Life of Maxwell*, 595.
57. Maxwell, *Letters and Papers*, I, 378.
58. Campbell and Garnett, *Life of Maxwell*, 601.
59. *Ibid.*, 114.
60. Descartes, *Meditations on First Philosophy*, 19, in *Philosophical Writings*, II, 13.
61. Campbell and Garnett, *Life of Maxwell*, 599.
62. Carroll, *Alice's Adventures*, 127.
63. E.g., Hamilton, *Lectures*, I, 329.
64. *Ibid.*, I, 323, 324–34, 324.
65. Maxwell, *Letters and Papers*, I, 72.
66. Freud, *Interpretation of Dreams*, 138–259.
67. See Maxwell, *Letters and Papers*, I, 208, 315, 378n., 411, 421n., 519n., 668.
68. Campbell and Garnett, *Life of Maxwell*, 268.
69. Hamilton, *Lectures*, I, 349, 339.
70. See Plato, *Meno*, 81c–86c; *Phaedo*, 72e–77a.
71. Hamilton, *Lectures*, I, 345.
72. *Ibid.*, I, 349, 348.
73. Campbell and Garnett, *Life of Maxwell*, 113.
74. *Ibid.*, 229, 300.
75. Cf. Maxwell's review of Plateau: 'it is *our* breath which is turning dirty soap-suds into spheres of splendour' (*Scientific Papers*, II, 393).
76. Schaffer, 'A Science Whose Business Is Bursting', 163.
77. Cited in *ibid.*, 168. Maxwell refers to Thomson's work on soap-bubbles in his Liverpool address (*Scientific Papers*, II, 222).
78. Maxwell, *Letters and Papers*, I, 443.
79. See Freud, *Interpretation of Dreams*, and *Jokes and their Relation to the Unconscious*.
80. Carroll, *Alice's Adventures*, 187.
81. Hilts, 'A Guide to Francis Galton's English Men of Science', 59.
82. Campbell and Garnett, *Life of Maxwell*, 73–4, 74.
83. Hay, 'Description of a Machine for Drawing the Perfect Egg-oval'.
84. J. J. Thomson, 'James Clerk Maxwell', 4.
85. Maxwell, *Letters and Papers*, I, 248–9, 248.
86. Kant, *Critique of Judgement*, 219, 218, 191 (60, 59, 32). See also 242–4 (87–9).
87. Hamilton, *Lectures*, II, 509. See II, 498–513.
88. Maxwell, *Letters and Papers*, I, 381–2.
89. Cited in Hendry, *Maxwell and the Theory of the Electromagnetic Field*, 28.
90. See Chapters 8 and 9.
91. Campbell and Garnett, *Life of Maxwell*, 215.
92. Maxwell, *Letters and Papers*, I, 377.
93. *Ibid.*, I, 383.
94. Cited in Fisch, *Whewell*, 71.
95. Campbell and Garnett, *Life of Maxwell*, 618.

96. Indeed, clarifying this principle of reciprocation in his essay 'Has Everything Beautiful in Art its Original in Nature?', Maxwell suggests that science is not only the paradigm of truth but also the radical precondition for creating beauty, indeed the type for art: 'Nothing beautiful can be produced by Man except by the laws of mind acting in him as those of Nature do without him; and therefore the kind of beauty he can thus evolve must be limited by the very small number of correlative sciences which he has mastered' (Maxwell, *Letters and Papers*, I, 248).
97. Maxwell, *Scientific Papers*, II, 227; I, 452, 492, 486.
98. Maxwell, *Letters and Papers*, II, 337.
99. Maxwell, *Scientific Papers*, II, 227.
100. Harman, *Natural Philosophy of Maxwell*, 5.
101. Maxwell, *Scientific Papers*, I, 500. See also 'A Dynamical Theory of the Electromagnetic Field' (1864), *ibid.*, I, 526–97.
102. Maxwell, *Letters and Papers*, I, 320, 322, 376.
103. See Chapters 7 and 10.

CHAPTER 4

1. Graves, *Life of Hamilton*, I, 317, 316.
2. *Ibid.*, I, 194; II, 151.
3. Geikie, *Memoir of Ramsay*, 62; Wilson and Geikie, *Memoir of Edward Forbes*, 247. See Morrell and Thackray, *Gentlemen of Science*, 138.
4. Geikie, *Memoir of Ramsay*, 58.
5. Wilson and Geikie, *Memoir of Edward Forbes*, 379–81, 388. Gay and Gay, 'Brothers in Science', 434.
6. T. H. Huxley, 'Professor Tyndall', 5–6.
7. Wilson and Geikie, *Memoir of Edward Forbes*, 248.
8. Huxley, *Life and Letters of Thomas Henry Huxley*, I, 91.
9. Wilson and Geikie, *Memoir of Edward Forbes*, 248, 456.
10. Rpt in Gardiner, 'Edward Forbes, Richard Owen and the Red Lions', 354.
11. Cited in Cowen, *Relish*, 224. See 223–4.
12. Geikie, *Memoir of Ramsay*, 179–80.
13. Morrell and Thackray, *Gentlemen of Science*, 112–13, 76–7; Todhunter, *William Whewell*, II, 129.
14. Morrell and Thackray, *Gentlemen of Science*, 110.
15. Geikie, *Memoir of Ramsay*.
16. Cited in Gay and Gay, 'Brothers in Science', 432.
17. Huxley, 'Professor Tyndall', 6.
18. Morrell and Thackray, *Gentlemen of Science*, 273. See also 276–96.
19. Huxley, *Life and Letters of Thomas Henry Huxley*, I, 94.
20. Gardiner, 'Edward Forbes, Richard Owen and the Red Lions', 364–5.
21. Wilson and Geikie, *Memoir of Edward Forbes*, 421.
22. Cited in Gardiner, 'Edward Forbes, Richard Owen and the Red Lions', 364.
23. Wilson and Geikie, *Memoir of Edward Forbes*, 421.

24. Campbell and Garnett, *Life of Maxwell*, 489.
25. Lodge, *Advancing Science*, 154–5.
26. *Ibid.*, 154.
27. Tait, 'Clerk-Maxwell's Scientific Work', 321.
28. Tennyson, *Poetical Works*, 130–3.
29. Knott, *Life of Tait*, 171.
30. Maxwell, 'To the Chief Musician upon Nubla', 291. The typographical error follows from Maxwell's handwriting, see Maxwell Papers, Add. 7655/VL/3 (v).
31. Geikie, *A Long Life's Work*, 147.
32. Knott, *Life of Tait*, 171, 143. The *OED* cites William Thomson as attributing the application to 'a humorous suggestion of Maxwell's' ('Nabla', *OED*).
33. *Ibid.*, 172.
34. *Report of the BAAS at Edinburgh 1871*, Notices and Abstracts, 2.
35. Knott, *Life of Tait*, 145. Maxwell, *Letters and Papers*, II, 590.
36. Huxley, *Life and Letters of Thomas Henry Huxley*, I, 94–5.
37. Desmond, *Huxley: The Devil's Disciple*, 157.
38. Huxley Papers, MSS 79, fos. 1–3.
39. Smith, 'North British network'.
40. Cited in Knott, *Life of Tait*, 11; Campbell and Garnett, *Life of Maxwell*, 642–4.
41. Tyndall Papers, JT/1/TYP/9, T. H. Huxley to Tyndall, 25 February 1853 (2859).
42. Cited in Barton, '"Huxley, Lubbock, and Half a Dozen Others"', 411.
43. MacLeod, 'The X-Club', 319.
44. Turner, 'The Victorian Conflict between Science and Religion', 356–76.
45. Smith, *The Science of Energy*, 171.
46. Huxley, 'Professor Tyndall', 2.
47. Tennyson, *Poetical Works*, 130.
48. Campbell and Garnett, *Life of Maxwell*, 133.
49. *Ibid.*, 79, 80, 181.
50. Geikie, *Memoir of Ramsay*, 251.
51. Hevly, 'The Heroic Science', 81.
52. Rowlinson, 'The Theory of Glaciers', 194–7; Hevly, 'The Heroic Science', 80.
53. Rowlinson, 'The Theory of Glaciers', 196–7.
54. Cited in Smith, *The Science of Energy*, 180.
55. Knott, *Life of Tait*, 209.
56. See Smith, *The Science of Energy*, 180–91.
57. *Ibid.*, 172.
58. Tyndall, 'Remarks on an article entitled "Energy"', 220–4; Tait, 'Reply to Prof. Tyndall's remarks', 263–6.
59. Tyndall, 'Remarks on the Dynamical Theory of Heat', 268–87; Tait, 'The Dynamical Theory of Heat', 40–69; Tait, 'Energy', 337–68.
60. Knott, *Life of Tait*, 210.
61. Campbell and Garnett, *Life of Maxwell*, 343.
62. Barton, 'Scientific Authority and Scientific Controversy', 232.
63. J. Ruskin, *Fors Clavigera*, letter no. 34, October 1873; rpt Rendu (ed.), *Theory of the Glaciers*, 163.

64. Maxwell, *Letters and Papers*, II, 915. Maxwell contributed a note on Forbes's work on colours to Shairp, Tait and Adams-Reilly, *Life of Forbes*, 464–5 (rpt in *Letters and Papers*, II, 774–5).
65. Tait, 'Tyndall and Forbes', 381–2, and Tyndall, 'Tyndall and Tait', 399; [Lockyer], 'Tait and Tyndall', 431.
66. Knott, *Life of Tait*, 18.
67. Tait, Scrapbook.
68. Maxwell Papers, IIIb, no. 12, fos. 2–2v.
69. Mahon, *The Man Who Changed Everything*, 131–2.
70. Smith, *The Science of Energy*, 181.

CHAPTER 5

1. Tait, Scrapbook.
2. Maxwell, *Letters and Papers*, II, 617.
3. [Anon.], 'Professor Tyndall's Lectures at the Royal Institution', 243.
4. Maxwell, *Scientific Papers*, II, 243.
5. *Ibid.*, 267–79.
6. Maxwell, *Letters and Papers*, II, 617.
7. *Ibid.*, II, 578.
8. Maxwell, 'Tait's *Recent Advances*', 462.
9. Maxwell, *Scientific Papers*, II, 268; 279.
10. BAAS Papers, 94, It. 159, It. 186. See also 'Report of the Committee on Science-Lectures and Organization' and 'Second Report' of this committee, in *Report of BAAS at Bradford 1873*, 495–507, 507–8. Tait is listed as one of the committee members in both reports, while Maxwell's name is included only in the second.
11. Lockyer, 'Science Lectures for the People', 81.
12. See: Lockyer, 'Lectures to Ladies'; Lockyer, 'The Scientific Education of Women'; Stuart, 'The Scientific Education of Women'; Dawson, 'Thoughts on the Higher Education of Women'; [Lockyer], 'Science for Women'; Kitchener, 'Instruction in Science for Women'.
13. Maxwell, 'A Lecture on Thomson's Galvanometer'.
14. David Mather Masson, *Edinburgh Ladies' Educational Association: Introductory lectures of the second session* [by] *Prof. Masson, Prof. Tait, Prof. Fraser* (Edinburgh: 1868); Tait, Scrapbook.
15. *Report of BAAS at Bradford 1873*, Notices and Abstracts, 495.
16. Maxwell's conception of the first five stanzas as each representing a distinct lecture is indicated also by his drafts, in which he alters their sequence several times (Maxwell Papers, VL/3(v) and (vi)); Beer, *Open Fields*, 308.
17. Maxwell, *Letters and Papers*, III, 99.
18. Cambridge U. L. Add. 7655/VL/3, Royal Institution MS JT/B/2; Maxwell, *Letters and Papers*, II, 652.
19. He also congratulates Sylvester on his 1869 Address at Exeter, 'the only successful attempt to invest mathematical reasoning with a halo of glory'. Tait, *Scientific Papers*, I, 168.

20. Maxwell, *Letters and Papers*, II, 760.
21. Cited in Hankins, 'Triplets and Triads', 176.
22. Maxwell, *Letters and Papers*, II, 715.
23. *Ibid.*, II, 755.
24. Campbell and Garnett, *Life of Maxwell*, 625–7.
25. Sylvester, *Laws of Verse*, 105.
26. *Ibid.*, 583–5; Tait, Scrapbook. See also Maxwell, *Letters and Papers*, II, 710.
27. Knott, *Life of Tait*, 244–5; rpt in Maxwell, *Letters and Papers*, III, 840–1.
28. McMillan and Meehan, *Tyndall*, 50; Tyndall, *Fragments of Science*, I, 327.
29. Tait, *Scientific Papers*, I, 168.
30. Maxwell, *Scientific Papers*, II, 220.
31. Tyndall, *Glaciers of the Alps*, v.
32. Tait, *Lectures on Recent Advances*, xv.
33. Knott, *Life of Tait*, 349, 347, 348.
34. Maxwell, 'Tait's *Recent Advances*', 462.
35. Campbell and Garnett, *Life of Maxwell*, 260, 314.
36. Maxwell, *Letters and Papers*, II, 245–8, 253.
37. *Ibid.*, II, 921.
38. Tyndall, 'Tyndall and Tait', 399. See also Tait, 'Tyndall and Forbes', 381–2.
39. [Anon.], 'Professor Tait on Popular Science', *The Edinburgh Courant*, 1 November 1877, in Tait, Scrapbook.
40. Campbell and Garnett, *Life of Maxwell*, 646; Knott, *Life of Tait*, 253.
41. Knott, *Life of Tait*, 253.
42. Lodge, *Advancing Science*, 42.
43. Knott, *Life of Tait*, 253.
44. *The Edinburgh Courant*, 9 September 1876, in Tait, Scrapbook.
45. Tait, *Lectures on Recent Advances*, 338.
46. See Knott, *Life of Tait*, 283.
47. 'Professor Tait on "Force"', *The Edinburgh Courant*, 9 September 1876. *The Northern Whig*, 16 September 1876, in Tait, Scrapbook.
48. Knott, *Life of Tait*, 253.
49. Tait, Scrapbook.
50. *Report of the BAAS at Belfast 1874*, lxvii.
51. Tait, *Lectures on Recent Advances*, 340.
52. Gold, *ThermoPoetics*, 115–16, 122–5.
53. Campbell and Garnett, *Life of Maxwell*, 647; Knott, *Life of Tait*, 254–5.
54. Tait, *Lectures on Recent Advances*, 349.
55. Campbell and Garnett, *Life of Maxwell*, 647–8; Knott, *Life of Tait*, 254.
56. Serres, *Hermes*, 141.
57. Maxwell, *Letters and Papers*, II, 945.
58. Tyndall, 'On the Bending of Glacier Ice', 447.
59. Carlyle, *Sartor Resartus*, 271.
60. [Anon.], 'Man of the Day, No. 43: Professor John Tyndall, F.R.S.', *Vanity Fair* 7 (6 April 1872), 111.
61. Hevly, 'The Heroic Science', 66.

62. See Tyndall, *Glaciers of the Alps*, 346–52; *Heat considered as a Mode of Motion* (2nd edn), 202–22; *Hours of Exercise*, 339–404.
63. Tyndall, *Hours of Exercise*, 383. See also Tyndall, *Forms of Water*, 169, and Rowlinson, 'The Theory of Glaciers', 192–4.
64. Tyndall, *Hours of Exercise*, 378.
65. E.g., *ibid.*, 359.
66. *Ibid.*, 355, 382, 384. Tyndall also refers to it as 'the fracture theory'.
67. Tyndall, *Hours of Exercise*, 359; see Mathews, 'Mechanical Properties of Ice'.
68. Bonney, 'Tyndall's Hours of Exercise', 199.
69. Tyndall, 'On the Bending of Glacier Ice', 447.
70. Tyndall, *Heat considered as a Mode of Motion*, 109–10.
71. Tait, Scrapbook.
72. Pollock, *Personal Remembrances*, II, 128–33. On the *Reader*, see also Barton, 'Scientific Authority and Scientific Controversy', 224–5.
73. Pollock, *Personal Remembrances*, II, 134; Pollock, 'The Ice Flower', 168; Tyndall Papers, JT/8/3.
74. Pollock, *Personal Remembrances*, II, 135.
75. Tyndall, *Heat considered as a Mode of Motion*, 110.
76. Cited in James, 'Reporting Royal Institution Lectures', 70.
77. See Morus, *When Physics became King*, 87–122.
78. James, 'Reporting Royal Institution Lectures', 67–79.
79. Tyndall Papers, J. D. Hooker to Tyndall, 27 Dec. 1871.
80. McMillan and Meehan, *Tyndall*, 49.
81. Lightman, *Victorian Popularizers of Science*, 213–14.
82. Cited in Howard, '"Physics and fashion"', 746.
83. Maxwell, *Letters and Papers*, II, 617–18.
84. Campbell and Garnett, *Life of Maxwell*, 381.
85. Maxwell, *Scientific Papers*, II, 216; Maxwell, *Theory of Heat*, vi.
86. Tait, 'The Dynamical Theory of Heat', 67–8.
87. Tait, 'Sensation and Science', 177. See Knott, *Life of Tait*, 354.
88. Tait, 'Sensation and Science', 178.
89. Tait, 'Sensation and Science [II]', 177.
90. [Anon.], 'Tyndall's Lectures on Electrical Phenomena', 244.
91. Lockyer, *The Spectroscope*, 108, 96.
92. *Report of the BAAS at Edinburgh 1871*, xcv–xcviii.
93. Maxwell, *Letters and Papers*, II, 817.
94. Pollock, *Personal Remembrances*, II, 192.
95. Tyndall, 'On a New Series of Chemical Reactions', 93, 92.
96. *Ibid.*, 97.
97. Pesic, *Sky in a Bottle*, 105.
98. Tyndall, 'On a New Series of Chemical Reactions', 99.
99. Maxwell, *Scientific Papers*, II, 221.
100. Tyndall, 'On a New Series of Chemical Reactions,' 101–2, 102.
101. *Ibid.*, 93.
102. Tyndall, *Imagination in Science*, 33.

103. Tyndall, 'On a New Series of Chemical Reactions', 93.
104. Tait, *Lectures on Recent Advances*, 17. See 'Notes', *Nature*, 1 (23 Dec. 1869), 219.
105. Knott, *Life of Tait*, 173; Campbell and Garnett, *Life of Maxwell*, 635.
106. Maxwell, *Letters and Papers*, I, 225.
107. Campbell and Garnett, *Life of Maxwell*, 156n.
108. Maxwell, *Letters and Papers*, I, 224, 225, 225, 222.
109. *Ibid.*, I, 224.
110. Campbell and Garnett, *Life of Maxwell*, 189.
111. Lightman, *Victorian Popularizers of Science*, 202–8.
112. Howard, '"Physics and fashion"', 746.
113. Tyndall, *Imagination in Science*, 15. See also the extract from the *Record* at p. 5, which refers to 'His imaginary picture of the occult operations of light'.
114. Tyndall, *Fragments of Science*, I, 322–8.
115. Zerffi, *Spiritualism and Animal Magnetism*, 103.
116. Quoted in Lamont, 'Spiritualism and a Mid-Victorian Crisis of Evidence', 911.
117. Crookes, 'Experimental Investigation of a New Force'; 'Some Further Experiments on Psychic Force'.
118. Podmore, *Modern Spiritualism*, II, 145–6.
119. Lamont, 'Spiritualism and a Mid-Victorian Crisis of Evidence', 914.
120. See Crookes, 'Spiritualism Viewed by the Light of Modern Science', *Quarterly Journal of Science*, 7 (July 1870), rpt in Medhurst and Barrington (eds.), *Crookes and the Spirit World*, 15–21.
121. Maxwell, *Scientific Papers*, II, 759.
122. Rpt in Medhurst and Barrington (eds.), *Crookes and the Spirit World*, 22–31. See also Crookes's 'Memoranda' of this sitting, *ibid.*, 172–4.
123. Earwaker, and Fraser, each under the title 'The New Psychic Force'.
124. *Report of the BAAS at Edinburgh 1871*, xcviii–xcix.
125. Cited in Christiansen, *The Victorian Visitors*, 154.
126. Earwaker, 'The New Psychic Force', 279.
127. *Report of the BAAS at Edinburgh 1871*, Transactions, 7. Cf. Tait, *Lectures on Recent Advances*, 25.
128. *Ibid.*, 121. The report in *Nature* substitutes 'foolish' for 'delusive', see Thomson, 'Section D; Opening Address by the President', 296.
129. Cited in Medhurst and Barrington (eds.), *Crookes and the Spirit World*, 48.
130. William Benjamin Carpenter, 'Spiritualism and its Recent Converts', *Quarterly Review*, 131 (October 1871), 301–53; Crookes, 'Psychic Force and Modern Spiritualism: A Reply to the "Quarterly Review"', rpt in Medhurst and Barrington (eds.), *Crookes and the Spirit World*, 62–4.
131. Medhurst and Barrington (eds.), *Crookes and the Spirit World*, 47, 48.
132. Maxwell, *Letters and Papers*, III, 126n.
133. Maxwell, *Scientific Papers*, II, 681–712.
134. Maxwell, *Letters and Papers*, III, 66. See *ibid.*, III, 39–41.
135. Tyndall, *Sound*, 217–54, 232, 241.
136. Tait, Scrapbook.

137. Pollock, 'Callous, cruel, clever Tyndall!!!', Tyndall Papers, JT/8/3.
138. Tyndall, *Sound*, 242, 232.

CHAPTER 6

1. Reichenbach, *Philosophy of Space and Time*, 10.
2. Sylvester, *Laws of Verse*, 114, 113.
3. Parshall, *Sylvester: Life and Work in Letters*, 255.
4. Abbott, *Flatland*, v.
5. Hinton, *Selected Writings*, 1–22.
6. Beer, 'Alice in Space'.
7. See p. 31.
8. Sylvester, *Laws of Verse*, 114.
9. Hinton, *Selected Writings*, 22.
10. Ingleby, 'Transcendent Space' (13 January 1870), 289.
11. Ingleby, 'Transcendent Space' (17 February 1870), 407.
12. Wilde, *Lord Arthur Savile's Crime*, 75.
13. Tait, Scrapbook.
14. Maxwell, *Letters and Papers*, II, 578.
15. Tait, 'Imagination in Science', 395.
16. *Report of the BAAS at Norwich 1868*, Notices, 3.
17. *Report of the BAAS at Belfast 1874*, xcvi; Maxwell, *Letters and Papers*, III, 118; 119. See McMillan and Meehan, *Tyndall*, 44.
18. *Report of the BAAS at Norwich 1868*, Notices, 3, 4.
19. *Report of the BAAS at Liverpool 1870*, lxxxiii–lxxxiv, lxxxiii.
20. Huxley, *Life and Letters of Thomas Henry Huxley*, I, 409.
21. *Report of the BAAS at Liverpool 1870*, lxxxiii.
22. Tyndall, *Imagination in Science*, 47.
23. Maxwell, *Scientific Papers*, II, 216.
24. Dawson, *Darwin, Literature and Victorian Respectability*, 106.
25. Maxwell, *Scientific Papers*, II, 254.
26. Tyndall, *Imagination in Science*, 15; Tyndall, *Fragments of Science*, II, 397.
27. Tyndall, *Imagination in Science*, 40–1.
28. Maxwell, *Scientific Papers*, II, 376, 377.
29. Paley, *Natural Theology*, 1–18.
30. *Report of the BAAS at Norwich 1868*, Notices, 2.
31. Maxwell, *Letters and Papers*, II, 840, also II, 617n. See Gillian Beer, *Open Fields*, 233–5. Maxwell may also have known Max Müller's lecture to the Royal Institution, 'On the Philosophy of Mythology', which was delivered soon after his own 'On Colour Vision', early in 1871.
32. Tyndall, *Imagination in Science*, 13.
33. Maxwell, *Letters and Papers*, II, 578.
34. Tyndall, *Imagination in Science*, 13, 14, 15.
35. Bain, *Logic*, 221–42, esp. 239–42.
36. See also Bain, 'Professor Tait on Bain's Logic'.
37. Tyndall, *Imagination in Science*, 16.

38. 'Dejection; an Ode', ll. 85–6. Coleridge, *Poetical Works*, 366.
39. Tyndall, *Imagination in Science*, 21–2, 50.
40. Knott, *Life of Tait*, 173. Cf. Campbell and Garnett, *Life of Maxwell*, 635–6.
41. *Report of the BAAS at Norwich 1868*, Notices, 3; *Fragments*, II, 397.
42. Smith and Smith, *Rejected Addresses*, 119, 114.
43. Genesis 11.4.
44. Tyndall, *Imagination in Science*, 19–20.
45. Descartes, *Philosophical Writings*, I, 114.
46. Maxwell, *Letters and Papers*, I, 673.
47. Tait, 'Imagination in Science', 395.
48. Colston, *The Edinburgh and District Water Supply*, 135.
49. Maxwell Papers, IIIb, no. 13a.
50. 'Yarrow Unvisited', ll. 43–5, 50–1, in Wordsworth, *The Poems*, I, 601–2.
51. The MS notebook containing the drafts are bound in with another that makes reference to an excerpt from the 1874 'Apology for the Belfast Address' as having been written four years earlier (Tyndall Papers, JT/8/3). There is a chance that Maxwell could have seen a draft version of the poem, as Tyndall is known to have sent at least one of his poems to him, a copy of some early verses on Luther, in June 1871 (Maxwell Papers, Add. 7655/VL/3 (vii); Tyndall Papers, JT/B/2).
52. Tyndall, *New Fragments*, 498.
53. Line 28. Coleridge, *Poetical Works*, 101.
54. Carlyle, *Sartor Resartus*, 269. See Tyndall, *New Fragments*, 355–6.
55. Tyndall, *New Fragments*, 499.
56. *Ibid.*, 348.
57. See Haugrud, 'Tyndall's Interest in Emerson', 507; Turner, 'Victorian Scientific Naturalism', esp. 328–30; Barton, 'John Tyndall, Pantheist', 126.
58. Tyndall, *Fragments*, II, 592, 593.
59. Huxley, *Life and Letters of Thomas Henry Huxley*, I, 220.
60. Tyndall, *New Fragments*, 355.
61. Tyndall Papers, JT/3/44. See Tyndall, *Fragments*, II, 474.
62. Shelley, *Poetical Works*, 532.
63. Tyndall, *New Fragments*, 498–9.
64. Tyndall, *Imagination in Science*, 72.
65. Tyndall Papers, JT/3/44. Some material from the early drafts is reproduced in O'Gorman, 'John Tyndall as Poet', 353.
66. *Report of the BAAS at Belfast 1874*, xcvii.
67. Joyce, *Ulysses*, 415.
68. Huxley, *Poems*, 22.

CHAPTER 7

1. Maxwell, *Letters and Papers*, I, 223.
2. Maxwell, *Letters and Papers*, III, 100; Duncan, *Life of Spencer*, 426.
3. Spencer, *Autobiography*, II, 21.
4. Duncan, *Life of Spencer*, 377. See Spencer, *First Principles* (London 1862), 362–7.
5. Maxwell, *Letter and Papers*, I, 376–7.

6. Campbell and Garnett, *Life of Maxwell*, 637–8; Maxwell, *Scientific Papers*, II, 376.
7. Lucretius, *De Rerum Natura*, Bk II, ll. 218–20. Munro (trans.), II, 33.
8. Tyndall, *Fragments of Science*, I, 495.
9. Maxwell, *Letters and Papers*, III, 135.
10. *Ibid.*, III, 119.
11. *Report of the BAAS at Belfast 1874*, lxx.
12. *Ibid.*, lxxix; lxxxi.
13. Cited in Dawson, *Darwin, Literature and Victorian Respectability*, 88.
14. 'Ancient materialism and modern science: Lucretius among the Victorians', in Turner, *Contesting Cultural Authority*, 262–83.
15. Jenkin, 'Atomic Theory of Lucretius', 211.
16. Maxwell, *Letters and Papers*, I, 113.
17. Jenkin, 'Atomic Theory of Lucretius', 220, 222. See Lucretius, *De Rerum Natura*, Munro (trans.), II, 33.
18. Maxwell Papers, Add. 7655/II/51, Jenkin to Maxwell, 28 October 1871.
19. Maxwell, *Letters and Papers*, II, 250.
20. Maxwell, *Scientific Papers*, II, 27–8.
21. Lucretius, *De Rerum Natura*, Bk II, ll. 217–18. Munro (trans.), II, 33.
22. Maxwell, *Letters and Papers*, II, 654–5. *De Rerum Natura*, Bk II, ll. 284–93, trans. cited in 655n.
23. Maxwell, *Scientific Papers*, II, 373.
24. Maxwell, *Letters and Papers*, II, 121.
25. Cicero, *Of the Nature of the Gods*, xxiv.
26. Lucretius, *De Rerum Natura*, Bk II, 95–8; 105–7, 101. Munro (trans.), II, 30.
27. Fowler, *Lucretius on Atomic Motion*, 181.
28. Maxwell, *Scientific Papers*, I, 506.
29. Thompson, *Life of Thomson*, I, 514–15.
30. Maxwell, *Letters and Papers*, II, 446. See Plate XIII opposite, a photograph of the zoetrope and the vortex rings strip.
31. Maxwell, *Scientific Papers*, II, 470, 471.
32. Jenkin, 'Atomic Theory of Lucretius', 220.
33. Maxwell, *Scientific Papers*, II, 223.
34. Tait, *Lectures on Recent Advances*, 349. See *Report of the BAAS at Belfast 1874*, xcii.
35. Maxwell, *Scientific Papers*, II, 461.
36. Maxwell, *Letters and Papers*, I, 555.
37. *Ibid.*, II, 820.
38. Campbell and Garnett, *Life of Maxwell*, 633.
39. Maxwell, *Letters and Papers*, II, 492–5.
40. Lucretius, *De Rerum Natura*, Bk IV, ll. 13–17. Leonard (trans.), 220. Cf. Munro (trans.), I, 110. See also the proem to Bk I, the Hymn to Venus, esp. ll. 1–20.
41. Locke, *Essay concerning Human Understanding*, Bk II, ch. XXXIII (394–401).
42. Hume, *Treatise*, Bk I, pt I, sect. iv (12–13).

43. Maxwell, *Scientific Papers*, II, 221.
44. *Report of the BAAS at Norwich 1868*, Notices and Abstracts, 6.
45. Hume, *Treatise*, Bk I, pt I, sect. iv (10).
46. Deleuze, *Pure Immanence*, 38, 37. See also Deleuze, *Logic of Sense*, 266–79.
47. Maxwell, *Letters and Papers*, 119.
48. Cf. Kuhn's model of paradigm shifts, in his *The Structure of Scientific Revolutions*.
49. Maxwell, *Scientific Papers*, II, 226.
50. Line 15, Horace, *Odes I*, 104–5.
51. Cited in Menninghaus, *In Praise of Nonsense*, 1.
52. Deleuze, *Pure Immanence*, 43, 42.
53. Line 2, Horace, *Odes I*, 160–1.
54. Whewell, *Astronomy and General Physics*, 305.
55. Valdés and Guyon, 'Serendipity in Poetry and Physics', 30.
56. Line 43, in Russell and Winterbottom, *Ancient Literary Criticism*, 283.
57. Such iconography for the Red Lions can be traced to the triangular insignia of the Metropolitan Red Lions, which includes the figure of a rampant red lion holding a mug of wine and a long pipe (see Wilson and Geikie, *Memoir of Edward Forbes*, 458). The president or Lion King at the annual meetings of the original group was also crowned with a 'velvet cap, embroidered with the effigies of rampant red lions' (*ibid.*, 443).

CHAPTER 8

1. Sylvester, *Laws of Verse*, 124–5, 14–15.
2. Kant, *Critique of Pure Reason*, A163/B204 (199).
3. Friedman, *Kant and the Exact Sciences*, 55.
4. *Address to the Mathematical and Physical Section of the British Association. Exeter, August 19th, 1869*, 'Proof' copy, 3. Papers of the BAAS collection, Bodleian Library, Oxford.
5. Sylvester, 'A Plea for the Mathematician', 238; *Laws of Verse*, 111.
6. *Nature* 1 (1870), 387.
7. Sylvester, *Mathematical Papers*, II, 5.
8. Kant, *Critique of Pure Reason*, A25/B39 (69).
9. Sylvester, *Laws of Verse*, 108.
10. Plato, *The Republic*, Bk VII, 514a–520a.
11. Sylvester, *Mathematical Papers*, II, 7.
12. *Ibid.*, II, 5.
13. Kant, *Critique of Pure Reason*, B41 (70).
14. Euclid, Bk I, Prop. 32; Kant, *Critique of Pure Reason*, A25/B39 (69), A715–17/B743–5 (578–9). See Friedman, *Kant and the Exact Sciences*, 56–7.
15. Sylvester, *Laws of Verse*, 122, 120.
16. *The Tempest*, IV.i.61–2 (116).
17. Kant, *Critique of Pure Reason*, A162–3/B203–4, cited in Friedman, *Kant and the Exact Sciences*, 58.

18. Sylvester, *Mathematical Papers*, II, 6.
19. Serres, *Hermes*, 44.
20. Michel Chasles, 'On descriptive geometry', in Fauvel and Gray (eds.), *History of Mathematics*, 544.
21. Fauvel and Gray (eds.), *History of Mathematics*, 541.
22. Cited in Struik, *A Concise History of Mathematics*, 164.
23. See Richards, *Mathematical Visions*, 120–1, and Bell, *Men of Mathematics*, 210–11.
24. Richards, *Mathematical Visions*, 121–2.
25. Temple, *100 Years of Mathematics*, 46.
26. See Moktefi, 'Geometry', 326.
27. Sylvester, *Laws of Verse*, 120.
28. Bell, *Men of Mathematics*, 390–1.
29. Quoted in Parshall, *Sylvester: Jewish Mathematician*, 108.
30. Cited in Crilly, *Arthur Cayley*, 236.
31. Richards, 'The Geometrical Tradition', 466–7.
32. Sylvester, *Laws of Verse*, 112–15, 113.
33. *Ibid.*, 111.
34. Riemann, 'On the Hypotheses', 14.
35. *Ibid.*, 15.
36. *Ibid.*
37. Helmholtz, 'Axioms of Geometry', 129.
38. Riemann, 'On the Hypotheses', 15.
39. Sylvester, *Laws of Verse*, 111.
40. Riemann, 'On the Hypotheses', 17.
41. *Ibid.*, 36.
42. Helmholtz, 'Axioms of Geometry', 128. See also Reichenbach, *Philosophy of Space and Time*, 7–10.
43. Riemann, 'On the Hypotheses', 14.
44. Sylvester, *Laws of Verse*, 113.
45. *Ibid.*, 114.
46. Ingleby, 'Transcendent Space', 407.
47. Others adopted Sylvester's term, e.g., Rodwell, 'On Space of Four Dimensions', 9.
48. Clifford, 'The Unreasonable', 282; Ingleby, 'Prof. Clifford on Curved Space', 282–3.
49. Ingleby, 'The Antimonies of Kant', 262; Ingleby, 'The Unreasonable', 302–3.
50. Sylvester, *Mathematical Papers*, II, 9.
51. Graves, *Life of Hamilton*, I, 194, 314.
52. Knott, *Life of Tait*, 63.
53. Stewart and Tait, *The Unseen Universe* (1875), 169.
54. Sylvester, *Laws of Verse*, 116, 117.
55. Tyndall Papers, JT/1/TYP/9, Huxley to Tyndall, 14 June 1870 (2934).
56. Maxwell, *Scientific Papers*, II, 216.
57. Tennyson, *Poetical Works*, 151.

58. Parshall, *Sylvester: Jewish Mathematician*, 213.
59. Sylvester, *Laws of Verse*, 81, 92, 91, 83.
60. 'The Lily Fair of Jasmin Dean', 251, 252, 253.
61. Sylvester, *Laws of Verse*, 19, 89, 90.
62. Cf. Milton, *Paradise Lost*, Bk 11, l. 1013.
63. Sylvester, *Laws of Verse*, 86, 88.
64. Sylvester, *Mathematical Papers*, II, 377–479.
65. Parshall, *Sylvester: Jewish Mathematician*, 184.
66. Archibald, 'Unpublished Letters of Sylvester', 129.
67. Cited in *ibid.*, 112, 111, 112.
68. Sylvester, 'Sonnet to the Savilian Professor of Astronomy', 516. See also 'Note on Sonnet to Pritchard', 558.
69. Archibald, 'Unpublished Letters of Sylvester', 109.
70. Sylvester Papers, w.1, Box 1, 'Ten Sonnets by X', 'To Professor Pritchard on his investigations of the relative brightness of the fixed stars and the proper motions of 40 stars in the Pleiades'.
71. Sylvester, *Mathematical Papers*, IV, 1–83, 1, 26, 59.
72. *Ibid.*, IV, 304.
73. *Ibid.*, IV, 285. Quoted in Parshall, *Sylvester: Life and Work in Letters*, 259.
74. Sylvester, *Mathematical Papers*, IV, 302, 292, 293.
75. *Ibid.*, III, 83. See also Parshall, *Sylvester: Jewish Mathematician*, 68–9, 79, 143, 217–18.
76. Sylvester, *Laws of Verse*, 120, 43.
77. Sylvester, *Mathematical Papers*, III, 81n.; Cantor, *Quakers, Jews, and Science*, 85; cited 84–5.
78. Sylvester, *Laws of Verse*, 102.
79. *Ibid.*, 118.
80. *Ibid.*, 89, 91, 90.

CHAPTER 9

1. Sylvester, *Laws of Verse*, 10.
2. *Ibid.*, 11, 64–5.
3. *OED*, 'Synectic', b. 1888 B. Williamson.
4. Peirce, *The Essential Peirce*, 1.
5. Sylvester, *Laws of Verse*, 64, 11.
6. *Ibid.*, 13.
7. *OED*, 'Syzygy', 4.
8. Sylvester, *Laws of Verse*, 44.
9. Lanier, *The Science of English Verse*, 307.
10. Sylvester, *Laws of Verse*, 10.
11. Poe, *Works*, II, 215, 221, 236.
12. *Ibid.*, II, 246–7, 239, 246.
13. Lanier, *Science of English Verse*, iii, 41–6.
14. Sylvester, *Laws of Verse*, 44.

15. Sylvester Papers, Box 2, Folder 5, It. 114. Lanier to Sylvester, 31 October 1878.
16. Lanier, *Science of English Verse*, vi, xv, 305–8.
17. Sylvester, *Laws of Verse*, 45.
18. Ingleby, 'Neologisms', 385.
19. Sylvester, *Laws of Verse*, 11.
20. *Ibid.*, 30, 31.
21. Lanier, *Science of English Verse*, xxii.
22. Sylvester, *Laws of Verse*, 45.
23. *Ibid.*, 45–6, 46, 11–12, 13.
24. There are, however, a few lines that do not sustain this rhyme, e.g. ll. 51–2 (Sylvester, *Fliegende Blätter*, 9).
25. Gilman, *Launching of a University*, 67.
26. Sylvester, *Mathematical Papers*, 1, 132.
27. Carroll, *Picture Book*, 279; 290.
28. Sylvester, *Fliegende Blätter*, 37n., 9.
29. Carroll, *Picture Book*, 298.
30. *Ibid.*, 299–300. See Tennyson, *Poetical Works*, 3–5.
31. Locker, *London Lyrics*, 125–7.
32. Sylvester, *Fliegende Blätter*, 5.
33. Locker, *London Lyrics*, 126.
34. Sylvester, *Fliegende Blätter*, 6, 26.
35. See Crilly, *Cayley*, 172.
36. Carroll, *Picture Book*, 294.
37. *Ibid.*, 295, 293.
38. Shakespeare, *As You Like It*, III.ii.86–92, 94–5 (pp. 64–5, 65).
39. Lear, *Selected Letters*, 1–3.
40. Sylvester, *Fliegende Blätter*, 5, 42.
41. Cited in Marshall, *Shakespeare and Victorian Women*, 87. See *ibid.*, 87–91.
42. See Sylvester, *Fliegende Blätter*, 6–7.
43. Locker, *London Lyrics*, 127.
44. Sylvester Papers, Box 2, Folder 5, It. 90. S. H. Hodgson to Sylvester, 29 November 1876.
45. Cf. Touchstone's 'Sweetest nut hath sourest rind', Shakespeare, *As You Like It*, III.ii.107 (p. 66).
46. Gilman, *Launching of a University*, 68; Sylvester, *Fliegende Blätter*, 28.
47. Shakespeare, *As You Like It*, II.vii.147–8 (56). See also II.ii.23–4 (38); III.ii.297–8 (75); v.ii.33 (116); v.ii.83 (117).
48. Sylvester, *Fliegende Blätter*, 7, 8.
49. Wilde, *Oscar Wilde*, 85–7.
50. Shakespeare, *A Midsummer Night's Dream*, v.i.7–8 (103).
51. *Ibid.*, 104; Sylvester, *Mathematical Papers*, II, 380, 381.
52. Sylvester, *Fliegende Blätter*, 11.
53. Maxwell Papers, Add. 7655/VI/3viii.
54. Campbell and Garnett, *Life of Maxwell*, 637.

55. Maxwell, *Letters and Papers*, II, 294.
56. Sylvester, *Laws of Verse*, 125.
57. Parshall, *Sylvester: Life and Work in Letters*, 113.
58. Sylvester, *Fliegende Blätter*, 5.
59. Lanier, *Science of English Verse*, 308.
60. Sylvester, *Fliegende Blätter*, 16.
61. *Ibid.*, 6.
62. Sylvester, *Laws of Verse*, 45.
63. Sylvester, *Fliegende Blätter*, 6.
64. Sylvester, *Mathematical Papers*, IV, 588.
65. Crilly, *Cayley*, 469.
66. Cayley, *Mathematical Papers*, II, 381.
67. Sylvester, *Fliegende Blätter*, 26.
68. Sylvester, *Mathematical Papers*, I, 293.
69. Parshall, *Sylvester: Life and Work in Letters*, 111.
70. *Ibid.*, 113.
71. Parshall, *Sylvester: Jewish Mathematician*, 218–19.
72. Sylvester, *Mathematical Papers*, II, 419; I, 190.
73. *Ibid.*, I, 284.
74. Maxwell Papers, Add. 7655/VI/3viii. Cf. Campbell and Garnett, *Life of Maxwell*, 636.
75. Sylvester, *Laws of Verse*, 46–8.
76. Cited in Parshall, *Sylvester: Jewish Mathematician*, 102.
77. Sylvester, *Laws of Verse*, 48.
78. Sylvester, *Mathematical Papers*, I, 284.
79. *Ibid.*, I, 284, 247, 211.
80. Sylvester, *Spring's Début*, 27.
81. Sylvester, *Mathematical Papers*, I, 582.
82. Sylvester, *Spring's Début*, 27.
83. Sylvester, *Laws of Verse*, 44.
84. Sylvester, *Fliegende Blätter*, 26.
85. Clifford, 'On the space-theory of matter', 158.
86. Sylvester, *Fliegende Blätter*, 27.
87. *Ibid.*, 6, 7.
88. *Ibid.*, 9, 7, 9, 33, 29.
89. Stanzas 24, 25, Hopkins, *Poems*, 63.
90. Sylvester, *Fliegende Blätter*, 38, 15, 14, 15.
91. *Ibid.*, 38. See also, Sylvester, *Spring's Début*, 27.
92. Hopkins, *Letters of Hopkins to Bridges*, 51–2.
93. Hopkins similarly likens words to 'heavy bodies' (Hopkins, *Journals and Papers*, 269).
94. See Brown, *Hopkins' Idealism, Gerard Manley Hopkins*, ch. 7.
95. Sylvester, *Laws of Verse*, 45, 12–13.
96. *Ibid.*, 44–5.

CHAPTER 10

1. Sylvester, *Laws of Verse*, 116.
2. Crilly, *Cayley*, 196–8.
3. Sylvester, *Fliegende Blätter*, 5.
4. Sylvester, *Laws of Verse*, 105.
5. Parshall, *Sylvester: Jewish Mathematician*, 86.
6. [Owen], 'Oken', 498, 498–9.
7. Sylvester, *Mathematical Papers*, III, 73.
8. Pater, *The Renaissance*, 140.
9. Sylvester, *Mathematical Papers*, III, 78; II, 419.
10. Weber, *A Random Walk in Science*, 43. Cf. J. J. Thomson, 'James Clerk Maxwell', 31; Planck, 'Maxwell's Influence', 55–6.
11. Cercignani, *Boltzmann*, 17.
12. Campbell and Garnett, *Life of Maxwell*, 649.
13. Stewart and Tait, *Unseen Universe* (1875), vii; *ibid.* (1876), xiii. See Knott, *Life of Tait*, 236.
14. Stewart and Tait, *Paradoxical Philosophy*, 12.
15. Silver, 'Last Poem of Maxwell', 1269. See Maxwell, *Scientific Papers*, II, 757.
16. Campbell and Garnett, *Life of Maxwell*, 651.
17. Stewart and Tait, *Paradoxical Philosophy*, 73.
18. Stewart and Tait, *Unseen Universe* (1875), 91, 93.
19. *Ibid.*, 120; see 120–39, also 176–8; Stewart and Tait, *Paradoxical Philosophy*, 79, 106–7.
20. Knott, *Life of Tait*, 243. Cf. Campbell and Garnett, *Life of Maxwell*, 650.
21. Silver, 'Last Poem of Maxwell', 1269. For Maxwell's stoicism, see his 1876 essay 'On Modified Aspects of Pain' (Campbell and Garnett, *Life of Maxwell*, 444–52, esp. 447).
22. Garnett, 'Maxwell, F.R.S.', 45; Walker, 'The "Hexameter"', 57, Ingleby, 'The "Hexameter"', 81.
23. Stewart and Tait, *Unseen Universe* (1875), 120.
24. Tennyson, *Poetical Works*, 243.
25. Stewart and Tait, *Unseen Universe* (1875), 87, 89.
26. Cited in *ibid.*, 133.
27. Maxwell, *Letters and Papers*, III, 185. See also Smith, *Science of Energy*, 251.
28. Maxwell, *Scientific Papers*, II, 758. See also Stewart and Tait, *Paradoxical Philosophy*, 94–5.
29. Clifford, 'The Unseen Universe', 776.
30. Cited in Gooday, 'Sunspots, Weather, and the Unseen Universe', 125.
31. Maxwell, *Letters and Papers*, II, 820.
32. Stewart and Tait, *Unseen Universe* (1875), 156.
33. Maxwell, *Letters and Papers*, III, 609.
34. *Ibid.*, III, 840.
35. Lines 72–3, Shelley, *Poetical Works*, 241.
36. Campbell and Garnett, *Life of Maxwell*, 650.

37. *Prometheus Unbound* II.v.10–16, Shelley, *Poetical Works*, 240.
38. Knott, *Life of Tait*, 242–3. Cf. Campbell and Garnett, *Life of Maxwell*, 649–50.
39. Knott, *Life of Tait*, 236.
40. Stewart and Tait, *Unseen Universe* (1875), 103; Knott, *Life of Tait*, 239. The description of the atom as 'strong in solid singleness' occurs repeatedly in Bk I and again in Bk II of Lucretius' poem. See Munro (trans.), 1, 13, 14, 15, 32.
41. Maxwell, *Letters and Papers*, II, 321.
42. Maxwell, *Scientific Papers*, II, 445, 774, 775.
43. Stewart and Tait, *Paradoxical Philosophy*, 104; see also 94, 101, 192, 197, 226, *Unseen Universe* (1875), 167, 169, 172, 199.
44. Knott, *Life of Tait*, 242.
45. Maxwell, *Letters and Papers*, II, 832.
46. Tait, 'On amphicheiral forms', 392.
47. *Ibid.*, 391.
48. Maxwell, *Letters and Papers*, II, 637.
49. Cited in Harman, *Natural Philosophy of Maxwell*, 157. See also Maxwell, *Letters and Papers*, II, 642.
50. Knott, *Life of Tait*, 242–3. Cf. Campbell and Garnett, *Life of Maxwell*, 649–50.
51. Silver, 'Last Poem of Maxwell', 1270.
52. *Ibid.*, 1268; Stewart and Tait, *Unseen Universe* (1876), 220.
53. Stewart and Tait, *Paradoxical Philosophy*, 205, 229.
54. Maxwell, *Scientific Papers*, II, 757, see Stewart and Tait, *Paradoxical Philosophy*, 233.
55. Monastyrsky, *Riemann, Topology, and Physics*, 33.
56. Riemann, 'On the Hypotheses', 37.
57. Clifford, 'On the space-theory of matter', 158.
58. Clifford, 'Unseen Universe', 782–8, 783, 788.
59. Maxwell, *Letters and Papers*, III, 135n.
60. Horace, *Satires*, II.7.86. Maxwell, *Letters and Papers*, III, 137.
61. Stewart and Tait, *Paradoxical Philosophy*, 96.
62. Campbell and Garnett, *Life of Maxwell*, 650.
63. Maxwell, *Scientific Papers*, II, 763.
64. Campbell and Garnett, *Life of Maxwell*, 650.
65. *Ibid.*, 756.
66. Maxwell, *Letters and Papers*, III, 194.
67. *Ibid.*, III, 710. See John Gay, *The Beggar's Opera*, II, xiii, air 35.
68. See Tait, *Scientific Papers* I, 273–4.
69. Campbell and Garnett, *Life of Maxwell*, 650–1. Cf. Knott, *Life of Tait*, 243. See also Maxwell, *Letters and Papers*, II, 335.
70. Maxwell, *Scientific Papers*, II, 756, 757.
71. *Ibid.*, II, 757.
72. Shelley, *Poetical Works*, 240.
73. Planck, 'Maxwell's Influence', 57.
74. Knott, *Life of Tait*, 255. Cf. Campbell and Garnett, *Life of Maxwell*, 648.
75. Cited in Serres (ed.), *History of Scientific Thought*, 496.

76. Maxwell, *Letters and Papers*, II, 331–2, 336. See also II, 360–1.
77. Maxwell, *Scientific Papers*, II, 462.
78. *Report of the BAAS at Belfast 1874*, lxvii, lxvii–lxviii; lxviii–lxix.
79. Maxwell, *Letters and Papers*, III, 118, 119.
80. I Corinthians 15.55, 54.
81. Knott, *Life of Tait*, 254. Campbell and Garnett, *Life of Maxwell*, 648. See Tait, *Lectures on Recent Advances*, 18–20.
82. Tait, *Lectures on Recent Advances*, 7.
83. Maxwell, *Letters and Papers*, II, 312, see plates VII (opp. 319) and X (opp. 331).
84. See also Snyder, *Puns and Poetry in Lucretius*.
85. Wheelwright, *Heraclitus*, 100.
86. Campbell and Garnett, *Life of Maxwell*, 499.
87. Serres, *The Birth of Physics*, 28–37.
88. Maxwell, *Letters and Papers*, III, 598n.
89. *Ibid.*, III, 601, 556.
90. Maxwell, *Scientific Papers*, II, 761.
91. Maxwell, *Letters and Papers*, III, 603.
92. *Ibid.*, III, 602. Maxwell, *Scientific Papers*, II, 761.
93. Lucretius, *De Rerum Natura*, Bk I, 485–6. Munro (trans.), II, 12.
94. *Ibid.*, Bk I, 6–7; II, 1.
95. Campbell and Garnett, *Life of Maxwell*, 630–1, 631.
96. Thomson, 'Sorting Demon of Maxwell', 126. See also, e.g., Maxwell, *Letters and Papers*, I, 555.
97. Maxwell, *Scientific Papers*, II, 462.
98. Munro (trans.), 35 [Bk II, 282–324].
99. Joyce, *Finnegans Wake*, 1.
100. Lucretius, *De Rerum Natura*, Bk II, ll. 221–4. Munro (trans.), II, 33.
101. Maxwell, *Letters and Papers*, III, 602–3. Cf. Maxwell, *Scientific Papers*, II, 761.

Works cited

UNPUBLISHED MATERIALS

BAAS Papers, Bodleian Library, Oxford, Dep. BAAS 94, It. 159, It. 186.
Herschel Papers, Cambridge University Library, Add. 7617 B, Portfolio F. No. 24.
Huxley Papers, MSS 79, Imperial College, London.
Maxwell Papers, Add. 7655, Cambridge University Library, Cambridge.
Sylvester Papers, w.1, St John's College, Cambridge.
Tait, P. G., Scrapbook, James Clerk Maxwell Foundation, Edinburgh (Edinburgh University Library, Mic. M.134).
Tyndall Papers, Royal Institution of Great Britain, London, RI MS JT.

PUBLISHED MATERIALS

Abbott, Edwin. *Flatland: A Romance of Many Dimensions*. London: Seeley, 1884.
[Anon.] 'Man of the Day, No. 43: Professor John Tyndall, F.R.S.', *Vanity Fair* 7 (6 April 1872): 111.
[Anon.] 'Professor Tyndall's Lectures at the Royal Institution, on Electrical Phenomena and Theories', *Nature* 2 (21 July 1870): 243–4.
Archibald, Raymond Clare, 'Unpublished Letters of James Joseph Sylvester and Other New Information concerning his Life and Work', *Osiris* 1 (January 1936): 85–154.
Bain, Alexander. *Logic. Part Second: Induction*. London: Longmans, Green, Reader and Dyer, 1870.
'Professor Tait on Bain's Logic', *Nature* 3 (15 December 1870): 125–6.
Ball, W. W. Rouse. *A History of the Study of Mathematics at Cambridge*. Cambridge University Press, 1889.
Barton, Ruth. '"Huxley, Lubbock, and Half a Dozen Others": Professionals and Gentlemen in the Formation of the X Club, 1851–1864', *Isis* 89.3 (September 1998): 410–44.
'John Tyndall, Pantheist: A Rereading of the Belfast Address', *Osiris*, 2nd ser., 3 (1987): 111–34.
'Scientific Authority and Scientific Controversy in *Nature*: North Britain against the X Club', in *Culture and Science*, ed. Henson, Cantor *et al.*, 223–35.

Beer, Gillian. 'Alice in Space'. Vice-Chancellor's Distinguished Lecture series. University of Sydney, 3 April 2003.
Open Fields: Science in Cultural Encounter. Oxford University Press, 1996.
Bell, E. T. *Men of Mathematics.* New York: Simon and Schuster, 1986.
Benjamin, Walter. *Charles Baudelaire: A Lyric Poet in the Era of High Capitalism*, trans. Harry Zohn. London: Verso, 1976.
Bennett, E. T. 'Some Account of *Macropus Parryi*, a Hitherto Undescribed Species of Kangaroo from New South Wales'. *Transaction of the Zoological Society of London* 1 (1833–5), 295–300.
Bonney, T. G. 'Tyndall's Hours of Exercise in the Alps', *Nature* 4 (13 July 1871): 198–9.
Bouchelle, Joan Hoiness, ed. *With Tennyson at the Keyboard: A Victorian Songbook.* New York: Garland, 1985.
Braude, Stephen E. *The Limits of Influence: Psychokinesis and the Philosophy of Science.* London: Routledge, 1986.
Bristed, Charles Astor. *Five Years in an English University.* 2nd edn. New York: Putnam, 1852.
Brough, Robert B., ed. *Comic Almanack.* London: David Bogue, 1853.
Brown, Daniel. *Gerard Manley Hopkins.* Tavistock: Northcote House, in association with the British Council, 2004.
Hopkins' Idealism: Philosophy, Physics, Poetry. Oxford: Clarendon Press, 1997.
Browne, Janet. 'Darwin in Caricature: A Study in the Popularisation and Dissemination of Evolution', *Proceedings of the American Philosophical Society* 145.4 (December 2001): 496–509.
Burns, Robert. *The Poems and Songs of Robert Burns*, ed. James Kinsley. 3 vols. Oxford: Clarendon Press, 1968.
Campbell, Lewis, and William Garnett. *The Life of James Clerk Maxwell.* London: Macmillan, 1882.
Cantor, Geoffrey. *Quakers, Jews, and Science: Religious Responses to Modernity and the Sciences in Britain: 1650–1900.* Oxford University Press, 2005.
Cantor, Geoffrey, and Sally Shuttleworth, eds. *Science Serialized: Representations of the Sciences in Nineteenth-Century Periodicals.* Cambridge, MA: MIT Press, 2004.
Carlyle, Thomas. *Sartor Resartus: The Life and Times of Herr Teufelsdröckh.* Edinburgh: Canongate, 2002.
Carpenter, William Benjamin. 'Spiritualism and its Recent Converts', *Quarterly Review* 131 (October 1871): 301–53.
Carroll, Lewis. *Alice's Adventures in Wonderland and Through the Looking-Glass.* Harmondsworth: Penguin, 1998.
The Lewis Carroll Picture Book, ed. Stuart Dodson Collingwood. London: T. Fisher Unwin, 1899.
Cayley, Arthur. *Collected Mathematical Papers of Arthur Cayley*, ed. A. Cayley. Vols. I and II. Cambridge University Press, 1889.
Ceram, C. W. *Archaeology of the Cinema.* London: Thames and Hudson, 1965.

Cercignani, Carlo. *Ludwig Boltzmann: The Man Who Trusted Atoms.* Oxford University Press, 1998.
Chasles, Michel. ['On Descriptive Geometry'], in *The History of Mathematics: A Reader*, ed. Fauvel and Gray, 544–5.
Chitty, Susan. *That Singular Person Called Lear: A Biography of Edward Lear, Artist, Traveller and Prince of Nonsense.* New York: Atheneum, 1988.
Christiansen, Rupert. *The Victorian Visitors: Culture Shock in Nineteenth-Century Britain.* London: Chatto and Windus, 2000.
Clifford, W. K. 'On the Space-Theory of Matter', *Proceedings of the Cambridge Philosophical Society* 2 (1864–76): 157–8.
  'The Unreasonable', *Nature* 7 (17 February 1873): 282.
  'The Unseen Universe', *Fortnightly Review* 23 (1875): 776–93.
Cohen, Daniel J. *Equations from God: Pure Mathematics and Victorian Faith.* Baltimore: Johns Hopkins University Press, 2007.
Coleridge, Samuel Taylor. *Poetical Works*, ed. Ernest Hartley Coleridge. London: Oxford University Press, 1912.
Colley, Ann C. 'Edward Lear's Anti-Colonial Bestiary', *Victorian Poetry* 30.2 (Summer 1992): 109–20.
Colston, James. *The Edinburgh and District Water Supply: A Historical Sketch.* Edinburgh: Printed for Private Circulation, 1890.
Cowen, Ruth. *Relish: The Extraordinary Life of Alexis Soyer, Victorian Celebrity Chef.* London: Weidenfeld and Nicolson, 2006.
Craik, Alex D. D. *Mr Hopkins' Men: Cambridge Reform and British Mathematic in the 19th Century.* London: Springer Verlag, 2008.
Crary, Jonathan. *Techniques of the Observer.* Cambridge, MA: MIT Press, 1990.
Crilly, Tony. *Arthur Cayley: Mathematician Laureate of the Victorian Age.* Baltimore: Johns Hopkins University Press, 2006.
Crookes, William. 'Experimental Investigation of a New Force', *Quarterly Journal of Science* 8 (1871): 339–49.
  'Some Further Experiments on Psychic Force', *Quarterly Journal of Science* 8 (1871): 484–92.
  'Spiritualism Viewed by the Light of Modern Science', *Quarterly Journal of Science* 7 (July 1870). Rpt in *Crookes and the Spirit World*, ed. Medhurst and Barrington, 15–21.
Dallas, Eneas Sweetland. 'Student Life in Scotland', *Cornhill* 1 (1860): 366–79.
Darwin, Charles. *The Descent of Man.* Harmondsworth: Penguin, 2004.
Darwin, Erasmus. *The Botanic Garden: A Poem in Two Parts.* 4th edn. 2 vols. London: J. Johnson, 1799.
Davie, George. *The Democratic Intellect: Scotland and her Universities in the Nineteenth Century.* 2nd edn. Edinburgh University Press, 1964.
Dawson, Gowan. *Darwin, Literature and Victorian Respectability.* Cambridge University Press, 2007.
Dawson, Principal. 'Thoughts on the Higher Education of Women', *Nature* 4 (26 October 1871): 515–16.

Deleuze, Gilles. *The Logic of Sense*, trans. Mark Lester, ed. Constantin Boundas. London: Athlone Press, 1990.
Pure Immanence: Essays on a Life, trans. Anne Boyman. New York: Zone, 2001.
Deleuze, Gilles, and Felix Guattari. *Kafka: Towards a Minor Literature*, trans. Dana Polan. University of Minnesota Press, 1986.
1000 Plateaus: Capitalism and Schizophrenia, trans. Brian Massumi. Minneapolis: University of Minnesota Press, 1985.
Descartes, René. *The Philosophical Writings of Descartes*, trans. John Cottingham, Robert Stoothoff, Dugald Murdoch. 2 vols. Cambridge University Press, 1984.
Desmond, Adrian. *Huxley: The Devil's Disciple*. London: Michael Joseph, 1994.
Douglas, Mrs Stair. *The Life and Selections from the Correspondence of William Whewell*. London: C. Kegan Paul, 1881.
Duncan, David. *Life and Letters of Herbert Spencer*. London: Methuen, 1908.
Earwaker, J. P. 'The New Psychic Force', *Nature* 4 (3 August 1871): 278–9.
Ede, Lisa. 'Edward Lear's Limericks and their Illustrations', in *Explorations in the Field of Nonsense*, ed. Wim Tigges, 103–16. Amsterdam: Rodopi, 1987.
Edgeworth, Maria. *Harry and Lucy Concluded; being the Last Part of Early Lessons*. 4 vols. London: Hunter, 1825.
Ewald, William B. *From Kant to Hilbert: A Source Book in the Foundations of Mathematics*. 2 vols. Oxford: Clarendon Press, 1996.
Fauvel, John, and Jeremy Gray, eds. *The History of Mathematics: A Reader*. London and Milton Keynes: Macmillan with The Open University, 1987.
Fisch, Menachem. *William Whewell Philosopher of Science*. Oxford: Clarendon Press, 1991.
Flood, Raymond, Adrian Rice and Robin Wilson, eds. *Mathematics in Victorian Britain*. Oxford University Press, 2011.
Forbes, Edward. 'The Fate of the Do-do', *The Literary Gazette, and Journal of Belles Lettres* (1847): 493–4.
Fowler, Don. *Lucretius on Atomic Motion: A Commentary on De Rerum Natura 2.1–332*. Oxford University Press, 2002.
Fraser, George. 'The New Psychic Force', *Nature* 4 (3 August 1871): 279–80.
Freud, Sigmund. *The Interpretation of Dreams*, trans. A. A. Brill. New York: Macmillan, 1913.
Jokes and their Relation to the Unconscious, trans. James Strachey. London: Routledge and Kegan Paul, 1960.
Friedman, Michael. *Kant and the Exact Sciences*. Cambridge, MA: Harvard University Press, 1992.
Galton, Francis. *Hereditary Genius: An Inquiry into its Laws and Consequences*. London: Macmillan, 1892.
Gardiner, Brian G. 'Edward Forbes, Richard Owen and the Red Lions', *Archives of Natural History* 20.3 (1993): 349–72.
Garnett, William. 'James Clerk Maxwell, F.R.S', *Nature* 21 (13 November 1879): 45.
Gay, Hannah, and John W. Gay. 'Brothers in Science: Science and Fraternal Culture in Nineteenth-Century Britain', *History of Science* 35 (1997): 425–53.

Geikie, Archibald. *A Long Life's Work: An Autobiography.* London: Macmillan, 1924.
  Memoir of Ramsay. London: Macmillan, 1895.
Gill, Stephen. *Wordsworth and the Victorians.* Oxford University Press, 1998.
Gilman, Daniel Coit. *The Launching of a University.* New York: Dodd, Mead, & Co., 1906.
Gold, Barri J. *ThermoPoetics: Energy in Victorian Literature and Science.* Cambridge, MA: MIT Press, 2010.
Gooday, Graeme. *The Morals of Measurement: Accuracy, Irony, and Trust in Late Victorian Electrical Practice.* Cambridge University Press, 2004.
  'Precision Measurement and the Genesis of Physics Teaching Laboratories in Victorian Britain', *The British Journal for the History of Science* 23.1 (March 1990): 25–51.
  'Spot-watching, Bodily Postures and the "Practised Eye": the Material Practice of Instrument Reading in Late Victorian Electrical Life', in *Bodies/Machines*, ed. Morus, 165–96.
  'Sunspots, Weather, and the Unseen Universe: Balfour Stewart's Anti-materialist Representations of "Energy" in British Periodicals', in *Science Serialized*, ed. Cantor and Shuttleworth, 111–48.
Goodwin, Harvey. *Elementary Course of Mathematics, designed Principally for Students of the University of Cambridge.* 4th edn. Cambridge: Deighton Bell, 1853.
Gopnik, Alison, and Andrew N. Meltzoff. *Words, Thoughts, and Theories.* Cambridge, MA: MIT Press, 1997.
Grandville, J. J. *Un Autre Monde.* Paris: H. Fournier, 1844.
Graves, Robert Perceval. *Life of Sir William Rowan Hamilton.* 3 vols. Dublin: Hodges, Figgis, 1882–9.
Hamilton, William. *Lectures on Metaphysics and Logic*, ed. H. L. Mansel and J. Veitch. 4 vols. Edinburgh: Blackwood, 1859.
  ed. *The Works of Thomas Reid, D.D., Now Fully Collected with Selections from his Unpublished Letters.* 3rd edn. Edinburgh: MacLaughlan and Stewart, 1852.
Hankins, Thomas L. 'Triplets and Triads: Sir William Rowan Hamilton on the Metaphysics of Mathematics', *Isis* 68.2 (June 1977): 175–93.
  Sir William Rowan Hamilton. Baltimore: Johns Hopkins University Press, 1980.
Harman, P. M. *Metaphysics and Natural Philosophy: The Problem of Substance in Classical Physics.* Sussex: Harvester Press, 1982.
  The Natural Philosophy of James Clerk Maxwell. Cambridge University Press, 1998.
  ed. *Wranglers and Physicists: Studies on Cambridge Mathematical Physics in the Nineteenth Century.* Manchester University Press, 1985.
Harrison, W. H. *The Humourist: A Companion for the Christmas Fire.* London: Ackermann, 1832.
Haugrud, Raychel A. 'Tyndall's Interest in Emerson', *American Literature* 41.4 (January 1970): 507–17.

Hay, D. R. 'Description of a Machine for Drawing the Perfect Egg-oval', *Transactions of the Royal Scottish Society of Arts* 3 (1851): 123–7.
Hayley, William. *The Life and Posthumous Writings of William Cowper*. 3 vols. Chichester: J. Johnson, 1803–4.
Heath, T. L. *Appollonius of Perga: Treatise on Conic Sections*. Cambridge University Press, 1896.
Helmholtz, Hermann von. 'Axioms of Geometry', *The Academy* 1 (12 February 1870): 128–31.
Hendry, John. *James Clerk Maxwell and the Theory of the Electromagnetic Field*. Bristol: Adam Hilger, 1986.
Henson, Louise, Geoffrey Cantor, *et al.*, eds. *Culture and Science in the Nineteenth-Century Media*. London: Ashgate, 2004.
Herschel, John. *Essays from the Edinburgh and Quarterly Reviews; with Addresses and Other Pieces*. London: Longman, Brown, Green, Longman and Roberts, 1857.
Hevly, Bruce. 'The Heroic Science of Glacier Motion', *Osiris*, 2nd ser., 11 (1996): 66–86.
Hilts, Victor L. 'A Guide to Francis Galton's English Men of Science', *Transactions of the American Philosophical Society*, 65, pt 5 (1975): 1–85.
Hinton, Charles. *Speculations on the Fourth Dimension: Selected Writings of Charles H. Hinton*, ed. Rudolf v. B. Rucker. New York: Dover, 1980.
Hopkins, Gerard Manley. *The Journals and Papers of Gerard Manley Hopkins*, ed. Humphry House, and Graham Storey. London: Oxford University Press, 1959.
  The Letters of Gerard Manley Hopkins to Robert Bridges, ed. Claude Colleer Abbot. London: Oxford University Press, 1955.
  Poems of Gerard Manley Hopkins, ed. W. H. Gardner. 3rd edn. London: Oxford University Press, 1948.
Horace, *Odes I: Carpe Diem*, ed. and trans. David West. Oxford: Clarendon Press, 1995.
Horn, D. B. *A Short History of the University of Edinburgh: 1556–1889*. Edinburgh University Press, 1967.
Howard, Jill. '"Physics and Fashion": John Tyndall and his Audiences in Mid-Victorian Britain', *Studies in the History and Philosophy of Science* 35 (2004): 729–58.
Howarth, O. J. R. *The British Association for the Advancement of Science: A Retrospect 1831–1931*. 2nd edn. London: BAAS, 1931.
Hume, David. *An Enquiry Concerning the Principles of Morals*. London: A. Millar, 1751.
  A Treatise of Human Nature, ed. L. A. Selby-Bigge. Oxford: Clarendon Press, 1888.
Huxley, Henrietta. *Poems of Henrietta Huxley with Three of Thomas Henry Huxley*. [Eastbourne]: Privately printed, 1899.
Huxley, Leonard. *Life and Letters of Thomas Henry Huxley*. 2 vols. New York: Appleton, 1900.
Huxley, T. H. 'Professor Tyndall', *The Nineteenth Century* 35 (January 1894): 1–11.

Hyman, Susan. *Edward Lear's Birds*. London: Weidenfeld and Nicolson, 1980.
Ingleby, Clement. 'The Antimonies of Kant', *Nature* 7 (6 February 1873): 262.
'The "Hexameter", Πᾶσα δόσις ἀγαθή . . . κ.τ.λ.', *Nature* 21 (27 November 1879): 81.
'Neologisms', *Nature* 4 (14 September 1871): 385.
'Prof. Clifford on Curved Space', *Nature* 7 (13 February 1873): 282–3.
'Transcendent Space', *Nature* 1 (13 January 1870): 289.
'Transcendent Space', *Nature* 1 (17 February 1870): 407.
'The Unreasonable', *Nature* 7 (20 February 1873): 302–3.
Ireland, Samuel. *Graphic Illustrations of Hogarth*. London: Faulder, 1794.
James, Frank A. J. L. 'Reporting Royal Institution Lectures: 1826–1867', in Cantor and Shuttleworth, eds., 67–79.
James, I. M., 'James Joseph Sylvester, F.R.S. (1814–1897)', *Notes and Records of the Royal Society of London* 51.2 (July 1997): 247–61.
Jenkin, Fleeming. 'The Atomic Theory of Lucretius', *North British Review* 9 n.s. (March 1868): 211–42.
Papers Literary, Scientific, &c. by the late Fleeming Jenkin, F.R.S, LL.D, ed. Sidney Colvin and J. A. Ewing. London: Longmans, Green, 1887.
Jenkins, Alice. *Space and the 'March of Mind': Literature and the Physical Sciences in Britain 1815–1850*. Oxford University Press, 2007.
Johnson, James, ed. *The Scots Musical Museum*. Edinburgh: James Johnson, 1787–1883.
Joyce, James. *Finnegans Wake*. London: Faber and Faber, 1939.
Ulysses: The Corrected Text, ed. Hans Walter Gabler. London: Bodley Head, 1986.
Kant, Immanuel. *Critique of Judgement*, trans. James Creed Meredith. Oxford University Press, 1952.
Critique of Pure Reason, trans. Norman Kemp Smith. London: Macmillan, 1929.
Kant's Critique of Practical Reason, and other Works on the Theory of Ethics, trans. Thomas Kingsmill Abbot. 6th edn. London: Longmans, Green, 1909.
Kim, Seahwa, and Cei Maslen. 'Counterfactuals as Short Stories', *Philosophical Studies: An International Journal for Philosophy in the Analytic Tradition* 129.1 (May 2006): 81–117.
Kitchener, F. E. 'Instruction in Science for Women', *Nature* 4 (30 November 1871): 81.
Knott, Cargill Gilston. *Life and Scientific Work of Peter Guthrie Tait*. Cambridge University Press, 1911.
Kolmogorov, A. N., and A. P. Yushkevich, eds. *Mathematics of the 19th Century: Geometry, Analytic Function Theory*, trans. Roger Cooke. Basel: Birkhäuser, 1996.
Kuhn, Thomas. *The Structure of Scientific Revolutions*. University of Chicago Press, 1962.
Lamb, Horace. 'Maxwell as Lecturer', in J. J. Thomson *et al.*, *Maxwell: A Commemoration Volume*, 142–6.
Lamont, Peter. 'Spiritualism and a Mid-Victorian Crisis of Evidence', *The Historical Journal* 47.4 (2004): 897–920.

Lanier, Sydney. *The Science of English Verse*. New York: Charles Scribner's Sons, 1880.
Lear, Edward. *The Complete Nonsense of Edward Lear*, ed. Holbrook Jackson. London: Faber, 1947.
  Selected Letters, ed. Vivien Noakes. Oxford: Clarendon Press, 1988.
Lecercle, Jean-Jacques. *Philosophy of Nonsense: The Intuitions of Victorian Nonsense Literature*. London: Routledge, 1994.
  Philosophy through the Looking-Glass: Language, Nonsense, Desire. La Salle, IL: Open Court, 1985.
Levere, Trevor H. 'Coleridge and the Sciences', in *Romanticism and the Sciences*, ed. Andrew Cunningham and Nicholas Jardine. Cambridge University Press, 1990.
Lewis, David. 'Causation', *The Journal of Philosophy* 70.17 (1973): 556–67.
  Counterfactuals. Cambridge, MA: Harvard University Press, 1973.
  On the Plurality of Worlds. Oxford: Blackwell, 1986.
Lightman, Bernard. 'Scientists as Materialists in the Periodical Press: Tyndall's Belfast Address', in *Science Serialized*, ed. Cantor and Shuttleworth, 199–238.
  Victorian Popularizers of Science: Designing Nature for New Audiences. University of Chicago Press, 2007.
  ed. *Victorian Science in Context*. University of Chicago Press, 1997.
Locke, John. *An Essay Concerning Human Understanding*, ed. Peter H. Nidditch. The Clarendon Edition of the Works of John Locke. Oxford: Clarendon Press, 1975.
Locker, Frederick. *London Lyrics*. London: John Wilson, 1868.
Lockyer, Norman. 'Lectures to Ladies', *Nature* 1 (11 November 1869): 45–6.
  'Science Lectures for the People', *Nature* 4 (1 June 1871): 81.
  'Science for Women', *Nature* 4 (23 November 1871): 57–8.
  'The Scientific Education of Women', *Nature* 2 (16 June 1870): 117–18.
  The Spectroscope and its Applications. 2nd edn. London: Macmillan, 1873.
  'Tait and Tyndall', *Nature* 8 (25 September 1873): 431.
Lockyer, Norman, and Winifred Lockyer. *Tennyson as a Student, and Poet of Nature*. London: Macmillan, 1910.
Lodge, Oliver. *Advancing Science, Being Personal Reminiscences of The British Association in the Nineteenth Century*. London: Ernest Benn, 1931.
Lucretius Carus. *De Rerum Natura*, trans. H. A. J. Munro. 2 vols. 2nd edn, rev. Cambridge: Deighton Bell, 1866.
  De Rerum Natura. Trans. William Emery Leonard. New York: Heritage Club, 1957.
  Of the Nature of Things, in Six Books. 2 vols. London: Daniel Browne, 1743.
  De Rerum Natura. London: Tonson, 1712.
MacLeod, Roy M. 'The X-Club a Social Network of Science in Late-Victorian England', *Notes and Records of the Royal Society of London* 24.2 (April 1970): 305–22.
Mahon, Basil. *The Man Who Changed Everything: The Life of James Clerk Maxwell*. Chichester: Wiley, 2004.

Main, Bert. 'History of Discovery of Anura', ed. C. G. Glasby, G. J. B. Ross, P. L. Beesley, 1–6. *Fauna of Australia*. IIA, Amphibia and Reptilia. Canberra: AGPS, 1993.
Marshall, Gail. *Shakespeare and Victorian Women*. Cambridge University Press, 2009.
Masson, David Mather. *Edinburgh Ladies' Educational Association: Introductory Lectures of the Second Session [by] Prof. Masson, Prof. Tait, Prof. Fraser*. Edinburgh: 1868.
Mathews, William. 'Mechanical Properties of Ice, and their Relation to Glacier Motion', *Nature* 1 (24 March 1870): 534–5.
Max Müller, Friedrich. *Lectures on the Science of Language; delivered at the Royal Institution of Great Britain in April, May, and June, 1861*. 2nd edn. London: Longman, Green, Longman, and Roberts, 1862.
Maxwell, James Clerk. 'A Lecture on Thomson's Galvanometer', *Nature* 6 (16 May 1872): 46.
The Scientific Letters and Papers of James Clerk Maxwell, ed. P. M. Harman. 3 vols. Cambridge University Press, 1990–2002.
The Scientific Papers of James Clerk Maxwell, ed. W. D. Niven. 2 vols. Cambridge University Press, 1890.
'Tait's *Recent Advances in Physical Science*', *Nature* 13 (13 April 1876): 461–3.
A Theory of Heat. 2nd edn. London: Longmans, Green, 1872.
'To the Chief Musician upon Nabla', *Nature* 4 (10 August 1871): 291.
A Treatise on Electricity and Magnetism. 2 vols. Oxford: Clarendon Press, 1873.
McMillan, N. D., and J. Meehan. *John Tyndall: Xemplar of Scientific and Technological Education*. Dublin: ETA Publications, 1980.
Medhurst, R. G., and M. R. Barrington, eds. *Crookes and the Spirit World*. London: Souvenir Press, 1972.
Menninghaus, Winfried. *In Praise of Nonsense: Kant and Bluebeard*, trans. Henry Pickford. Stanford University Press, 1999.
Moktefi, Amirouche. 'Geometry: The Euclid Debate', in Flood, Rice, and Wilson, eds, *Mathematics in Victorian Britain*, 321–38.
Monastyrsky, Michael. *Riemann, Topology, and Physics*. 2nd edn, trans. Roger Cooke, James King, Victoria King. Basel: Birkhäuser, 1999.
Morrell, Jack, and Arnold Thackray. *Gentlemen of Science: Early Years of the British Association for the Advancement of Science*. Oxford: Clarendon Press, 1981.
Morus, Iwan Rhys, ed. *Bodies/Machines*. Oxford: Berg, 2002.
ed. *When Physics became King*. University of Chicago Press, 2005.
Newton, Isaac. *The Principia: Mathematical Principles of Natural Philosophy*, trans. I. Bernard Cohen and Anne Whitman, assisted by Julia Budenz. Berkeley: University of California Press, 1999.
Nietzsche, Friedrich. *Beyond Good and Evil*, ed. Rolf-Peter Horstmann and Judith Norman, trans. Judith Norman. Cambridge University Press, 2002.
Noakes, Vivien. *Edward Lear: The Life of a Wanderer*. Rev. edn. Phoenix Mill: Sutton Publishing, 2004.

Nye, Mary Jo, ed. *The Cambridge History of Science*, Vol. v: *The Modern and Mathematical Sciences*. Cambridge University Press, 2003.
O'Gorman, Francis. 'John Tyndall as Poet: Agnosticism and "A Morning on Alp Lusgen"', *The Review of English Studies*, n.s., 48.191 (August 1997): 353–8.
Olson, Richard. *Scottish Philosophy and British Physics: 1750–1880*. Princeton University Press, 1975.
Owen, Richard. 'Oken, Lorenz', *Encyclopaedia Britannica*. 8th edn. XVI (1858). 498–9.
Paley, William. *Natural Theology or, Evidences of the Existence and Attributes of the Deity, Collected from the Appearances of Nature*. 13th edn. London: Fauldner, 1810.
Paradis, James G. 'Satire and Science in Victorian Culture', in *Victorian Science in Context*, ed. Lightman, 143–75.
Paris, John Ayrton. *Philosophy in Sport made Science in Earnest; Being an Attempt to Illustrate the First Principles of Natural Philosophy by the Aid of Popular Toys and Sports*. 3 vols. London: Longman, Rees, Orme, Brown and Green, 1827.
Philosophy in Sport made Science in Earnest: Being an Attempt to Implant in the Young Mind the First Principles of Natural Philosophy by the Aid of the Popular Toys and Sports of Youth. 7th edn. London: John Murray, 1853.
Parshall, Karen Hunger. *James Joseph Sylvester: Jewish Mathematician in a Victorian World*. Baltimore: Johns Hopkins University Press, 2006.
James Joseph Sylvester: Life and Work in Letters. Oxford: Clarendon Press, 1998.
Pater, Walter. *The Renaissance: Studies in Art and Poetry*. 4th edn. London: Macmillan, 1899.
Peirce, C. S. *The Essential Peirce: Selected Philosophical Writings*. II (*1893–1913*), ed. the Peirce Edition Project. Bloomington: Indiana University Press, 1998.
Pesic, Peter. *Sky in a Bottle*. Cambridge, MA: MIT Press, 2005.
Planck, Max. 'Maxwell's Influence on Theoretical Physics in Germany', J. J. Thomson et al., *Maxwell: A Commemoration Volume*, 45–65.
Plateau, Joseph. *Statique Expérimentale et Théorique des Liquides soumis aux Seules Forces Moléculaires*. 2 vols. Paris: Gauthier-Villars, 1873.
Podmore, Frank. *Modern Spiritualism: A History and Criticism*. 2 vols. London: Methuen, 1902.
Poe, E. A. *The Works of the Late Edgar Allan Poe*. New York: J. S. Redfield, 1850. vol. II, *Poems and Miscellanies*.
Pollock, Frederick. 'The Ice Flower', *The Reader* 5.III (11 February 1865): 168.
Personal Remembrances of Sir Frederick Pollock, Second Baronet, Sometimes Queen's Remembrancer. 2 vols. London: Macmillan, 1887.
Prickett, Stephen. *Victorian Fantasy*. Sussex: Harvester, 1979.
Reichenbach, Hans. *The Philosophy of Space and Time*, trans. Maria Reichenbach and John Freund. New York: Dover, 1957.
Reid, Ian. *Wordsworth and the Formation of English Studies*. Aldershot: Ashgate, 2004.
Rendu. *Theory of the Glaciers of Savoy by M. Le Chanoine Rendu*, ed. G. Forbes. London: Macmillan, 1874.

Report of the Third Meeting of the British Association for the Advancement of Science; held at Cambridge in August 1833. London: John Murray, 1834.

Report of the Thirty-eighth Meeting of the British Association for the Advancement of Science; held at Norwich in August 1868. London: John Murray, 1869.

Report of the Fortieth Meeting of the British Association for the Advancement of Science; held at Liverpool in September 1870. London: John Murray, 1871.

Report of the Forty-first Meeting of the British Association for the Advancement of Science; held at Edinburgh in August 1871. London: John Murray, 1872.

Report of the Forty-third Meeting of the British Association for the Advancement of Science held at Bradford in September 1873. London: John Murray, 1874.

Report of the Forty-fourth Meeting of the British Association for the Advancement of Science held at Belfast in August 1874. London: John Murray, 1875.

Richards, Joan L. 'The Geometrical Tradition: Mathematics, Space, and Reason in the Nineteenth Century', in *Cambridge History of Science*, ed. Nye, 449–67.

Mathematical Visions: The Pursuit of Geometry in Victorian England. Boston: Academic Press, 1988.

Riemann, Bernhard. 'On the Hypotheses which Lie at the Bases of Geometry', trans. W. K. Clifford. *Nature* 8 (1 May 1873): 14–17; (8 May 1873): 36–7.

Rodwell, G. F. 'On Space of Four Dimensions', *Nature* 8 (1 May, 1873): 8–9.

Rowlinson, J. S. 'The Theory of Glaciers', *Notes and Records of the Royal Society of London* 26.2 (December 1971): 189–204.

Rudy, Jason. *Electric Meters: Victorian Physiological Poetics*. Athens, OH: Ohio University Press, 2009.

Ruskin, John. *Fors Clavigera*, letter no. 34, October 1873; rpt Rendu, *Theory of the Glaciers of Savoy*, 163.

Russell, D. A., and M. Winterbottom, eds. *Ancient Literary Criticism: The Principal Texts in New Translations*. Oxford: Clarendon Press, 1972.

Schaffer, Simon. 'A Science whose Business is Bursting: Soap Bubbles as Commodities in Classical Physics', in *Things That Talk: Object Lessons from Art and Science*, ed. Lorraine Daston, 147–94. New York: Zone Books, 2004.

Scott, Walter. *Marmion; a Tale of the Flodden Field*. 10th edn. Edinburgh: Archibald Constable, 1821.

Serres, Michel. *The Birth of Physics*, trans. Jack Hawkes, ed. David Webb. Manchester: Clinamen Press, 2000.

Hermes: Literature, Science, Philosophy. Baltimore: Johns Hopkins University Press, 1982.

ed. *A History of Scientific Thought: Elements of a History of Science*. Oxford: Blackwell, 1995.

Shairp, J. C., P. G. Tait and A. Adams-Reilly. *Life and Letters of J. D. Forbes*. London: Macmillan, 1873.

Shakespeare, William. *As You Like It*, ed. Agnes Latham. London: Methuen, 1975.

A Midsummer Night's Dream, ed. Harold F. Brooks. London: Methuen, 1979.

The Tempest, ed. Frank Kermode. London: Methuen, 1964.

Shaw, George, and Frederick P. Nodder. *The Naturalist's Miscellany*, vol. VI. London: Nodder, 1795.

Shelley, P. B. *The Complete Poetical Works of Percy Bysshe Shelley*, ed. Thomas Hutchinson. London: Oxford University Press, 1945.
Siegel, Daniel M. *Innovation in Maxwell's Electromagnetic Theory: Molecular Vortices, Displacement Current, and Light.* Cambridge University Press, 1991.
Silver, Daniel S. 'The Last Poem of James Clerk Maxwell', *Notices of the American Mathematical Society* 55.10 (November 2008): 1266–70.
Smith, Crosbie. *The Science of Energy: A Cultural History of Physics in Victorian Britain*. London: Athlone Press, 1998.
  'North British Network (act. c. 1845–c. 1890)'. *Oxford Dictionary of National Biography*, Oxford University Press, Sept. 2010 [www.oxforddnb.com/view/theme/96081, accessed 1 Oct. 2011].
Smith, Crosbie, and M. Norton Wise. *Energy and Empire: A Biographical Study of Lord Kelvin*. Cambridge University Press, 1989.
Smith, James and Horace. *Rejected Addresses: or, The New Theatrum Poetarum*. 18th edn. London: John Murray, 1833.
Smith, Jonathan, *Charles Darwin and Victorian Visual Culture*. Cambridge University Press, 2006.
  'Trinity College Annual Examinations', *Teaching and Learning in Nineteenth-century Cambridge*, ed. Jonathan Smith and Christopher Stray, 122–38. Boydell Press and Cambridge University Library, 2001.
Snow, C. P. *The Two Cultures and the Scientific Revolution: The Rede Lecture, 1959*. Cambridge University Press, 1959.
Snyder, Jane McIntosh. *Puns and Poetry in Lucretius' De Rerum Natura*. Amsterdam: B. R. Grüner, 1980.
Spencer, Herbert. *An Autobiography*. 2 vols. London: Williams and Norgate, 1904.
  First Principles. London: Williams and Norgate, 1862.
  'The Nebular Hypothesis', *The Westminster Review* 70 (1858): 185–225.
  'Prof. Tait and Mr. Spencer', *Nature* 9 (2 April 1874): 420–1.
Stewart, Balfour. 'Mr Crookes on the "Psychic" Force', *Nature* 4 (27 July 1871): 237.
Stewart, Balfour, and P. G. Tait. *Paradoxical Philosophy: A Sequel to The Unseen Universe*. London: Macmillan, 1878.
  The Unseen Universe, or Physical Speculations on a Future State. London: Macmillan, 1875.
  The Unseen Universe, or Physical Speculations on a Future State. 4th edn. London: Macmillan, 1876.
Stimson, Dorothy. *Scientists and Amateurs: A History of the Royal Society*. New York: Henry Schuman, 1948.
Struik, Dirk J. *A Concise History of Mathematics*. 4th edn, rev. New York: Dover, 1987.
Stuart, James. 'The Scientific Education of Women', *Nature* 2 (30 June 1870): 165.
Sylvester, James Joseph. 'A Plea for the Mathematician', *Nature* 1 (30 December 1869): 237–9; (6 January 1870): 261–3.
  Address to the Mathematical and Physical Section of the British Association. Exeter, August 19th, 1869, 'Proof' copy, 3. Papers of the BAAS collection, Bodleian Library, Oxford.

'The Ballad of Sir John de Courcy'. Trans. 'Syzygeticus', *Gentleman's Magazine* (February 1871): 313–16.
The Collected Mathematical Papers of James Joseph Sylvester, ed. Henry F. Baker. 4 vols. Cambridge University Press, 1904–12.
Fliegende Blätter: Supplement to The Laws of Verse. London: Grant, 1876.
The Laws of Verse or Principles of Versification Exemplified in Metrical Translations. London: Longmans, 1870.
'The Lily Fair of Jasmin Dean', *The Eagle* 15 (December 1888): 251–3.
'Note on Sonnet to Pritchard', *Nature* 33 (15 April 1886): 558.
'Sonnet to the Savilian Professor of Astronomy in the University of Oxford', *Nature* 33 (1 April 1886): 516.
Spring's Début: A Town Idyll in Two Centuries of Continuous Rhyme. u.d., Baltimore: John Murphy.
Tait, Peter Guthrie. 'Clerk-Maxwell's Scientific Work', *Nature* 21 (5 February 1880): 317–20.
'The Dynamical Theory of Heat', *North British Review* 40 (February 1864): 40–69.
'Energy', *North British Review* 40 (May 1864): 337–68.
'Energy, and Prof. Bain's Logic', *Nature* 3 (1 December 1870): 89–90.
'Herbert Spencer versus Thomson and Tait', *Nature* 9 (26 March 1874): 402–3.
'Imagination in Science', *Nature* 3 (16 March 1871): 395.
Lectures on Some Recent Advances in Physical Science. 2nd edn. London: Macmillan, 1876.
'Note on the History of Energy', *Philosophical Magazine* 29 (1865): 55–7.
On Amphicheiral Forms and their Relations', *Proceedings of the Royal Society of Edinburgh*, 9 (1877): 391–2.
'On the Conservation of Energy', *Philosophical Magazine* 26 (1863): 429–31; 26: 144–5.
'On the History of Thermo-dynamics', *Philosophical Magazine* 28 (1864): 288–92.
'Reply to Prof. Tyndall's Remarks on a Paper on "Energy" in *Good Words*', *Philosophical Magazine* 25 (1863): 263–6.
Scientific Papers. 2 vols. Cambridge University Press, 1898–1900.
'Sensation and Science', *Nature* 4 (6 July 1871): 177–8.
'Sensation and Science [II]', *Nature* 6 (4 July 1872): 177–8.
Sketch of Thermodynamics. Edinburgh: Edmonston and Douglas, 1868.
'Tyndall and Forbes', *Nature* 8 (11 September 1873): 381–2.
Tait, Peter Guthrie, and W. Thomson. 'Energy', *Good Words* 3 (1862): 601–7.
Temple, George. *100 Years of Mathematics*. London: Duckworth, 1981.
Tennyson, Alfred. *Poetical Works*. London: Oxford University Press, 1953.
Thomas, W. K., and Warren U. Ober. *A Mind For Ever Voyaging: Wordsworth at Work Portraying Newton and Science*. Edmonton: University of Alberta Press, 1989.
Thompson, Silvanus P. *Life of William Thomson, Baron Kelvin of Largs*. 2 vols. London: Macmillan, 1910.

Thomson, Allen. 'Address by Dr. Allen Thomson, President of the Section', *Report of the BAAS at Edinburgh 1871*. Notes and Abstracts: 114–21.

'Section D, Biology; Opening Address by the President', *Nature* 4 (10 August 1871): 293–7.

Thomson, J. J. 'James Clerk Maxwell', in J. J. Thomson et al., *Maxwell: A Commemoration Volume*, 1–44.

Thomson, J. J. et al. *James Clerk Maxwell: A Commemoration Volume: 1831–1931*. Cambridge University Press, 1931.

Thomson, William. 'Address by the President, Sir William Thomson', *Report of the BAAS at Edinburgh 1871*. lxxxiv–cv.

'The Sorting Demon of Maxwell', *Nature* (5 June 1879): 126.

Tigges, Wim, ed. *Explorations in the Field of Nonsense*. Amsterdam: Rodopi, 1987.

Todhunter, I. *William Whewell: An Account of his Writings with Selections from his Literary and Scientific Correspondence*. 2 vols. London: Macmillan, 1876.

Tullius, Cicero M. *Of the Nature of the Gods*. London: R. Franklin, 1741.

Turner, Frank M. *Contesting Cultural Authority: Essays in Victorian Intellectual Life*. Cambridge University Press, 1993.

'The Victorian Conflict between Science and Religion: a Professional Dimension', *Isis* 69.3 (September 1978): 356–76.

'Victorian Scientific Naturalism and Thomas Carlyle', *Victorian Studies*, 18.3 (March 1975): 325–43.

Tyndall, John. *Essays on the Use and Limit of the Imagination in Science*. London: Longmans, Green, and Co., 1870.

The Forms of Water in Clouds and Rivers, Ice and Glaciers. London: Kegan Paul, 1872.

Fragments of Science for Unscientific People. 8th edn. 2 vols. London: Longmans, Green, 1879.

The Glaciers of the Alps. London: John Murray, 1860.

Heat considered as a Mode of Motion. London: Longmans, Green, 1863.

Hours of Exercise in the Alps. New York: Appleton: 1871.

Mountaineering in 1861: A Vacation Tour. London: Longman, Green, Longman, and Roberts, 1863.

New Fragments. London: Longmans, 1892.

'On a New Series of Chemical Reactions Produced by Light', *Proceedings of the Royal Society* 17 (1868–9): 92–102.

'On the Atmosphere as a Vehicle of Sound', *Philosophical Transactions of the Royal Society of London* 164 (1874): 183–244.

'On the Bending of Glacier Ice', *Nature* 4 (5 October 1871): 447.

'Principal Forbes and his Biographers', *Contemporary Review* 22 (August 1873): 484–508.

'Remarks on an Article Entitled "Energy" in *Good Words*', *Philosophical Magazine* 25 (1863): 220–4.

'Remarks on the Dynamical Theory of Heat', *Philosophical Magazine* 25 (1863): 268–87.

Sound. A Course of Eight Lectures delivered at The Royal Institution of Great Britain. London: Longman, Green, and Co., 1867.
'Tyndall and Tait', *Nature* 8 (18 September 1873): 399.
Valdés, Mario J., and Etienne Guyon. 'Serendipity in Poetry and Physics', in *The Third Culture: Literature and Science*, ed. Elinor S. Shaffer, 28–39. Berlin: De Gruyter, 1997.
Veitch, John. *Hamilton*. Edinburgh: Blackwood, 1882.
Venn, J. A., ed. *Alumni Cantabrigienses*, vol. II. Cambridge University Press, 1944.
Walker, J. J. 'The "Hexameter", [Πᾶσα δόσις ἀγαθή ... κ.ῑ.λ.]', *Nature* 21 (20 November 1879): 57.
Warwick, Andrew. *Masters of Theory: Cambridge and the Rise of Mathematical Physics*. University of Chicago Press, 2003.
Weber, R. L. (compiler). *A Random Walk in Science*, ed. E. Mendoza. London: The Institute of Physics, 1973.
Wheelwright, Philip. *Heraclitus*. Princeton University Press, 1959.
Whewell, William. *Astronomy and General Physics Considered with Reference to Natural Theology*. London: Pickering, 1833.
The Elements of Morality, including Polity. London: John W. Parker, 1845.
Of a Liberal Education in General and with Particular Reference to the Leading Studies of the University of Cambridge. London: John W. Parker, 1850.
Of a Liberal Education in General: and with Particular Reference to the Leading Studies of the University of Cambridge. London: John W. Parker, 1845.
'On the Connexion of the Physical Sciences. By Mrs Somerville'. *The Quarterly Review* 2 (1834): 54–68.
The Philosophy of the Inductive Sciences. 2 vols. London: Parker, 1840.
The Philosophy of the Inductive Sciences. 2nd edn. 1847. 2 vols. Rpt. London: Frank Cass, 1967.
Wilde, Oscar. *Lord Arthur Savile's Crime, and Other Stories*. London: Methuen, 1925.
Oscar Wilde, ed. Isobel Murray. The Oxford Authors. Oxford University Press, 1989.
Wilson, David B. 'The Educational Matrix: Physics Education at Early-Victorian Cambridge, Edinburgh and Glasgow Universities', in *Wranglers and Physicists*, ed. P. M. Harman. 12–48.
Wilson, George, and Archibald Geikie. *Memoir of Edward Forbes, F.R.S.* London: Macmillan, 1861.
Wordsworth, William. *The Poems*, ed. John O. Hayden. 2 vols. Harmondsworth: Penguin, 1977.
The Prelude: A Parallel Text, ed. J. C. Maxwell. Harmondsworth: Penguin, 1971.
Zerffi, George Gustavus. *Spiritualism and Animal Magnetism*. London: Robert Hardwicke, 1871.

Index

'architectural atoms' (Smith and Smith), 155
Aristotle, 80
Astraea, 8, 203–4, 205, 206
astronomy, 8–9, 129, 213, 242
atoms, *see* Lucretius; clinamen; Maxwell; Thomson, William; Tyndall

Babbage, Charles, 3, 4, 5
Babel, Tower of, 155
Bacon, Francis, 36, 64, 114, 119, 147
Bain, Alexander, *Logic*, 152–3
bard, the, 9, 114
baroque, the, 50, 260, 261
Beer, Gillian, 112, 143
Beggar's Opera, The (Gay), 250
Benjamin, Walter, 29
Bennett, E. T., 14, 16
Berkeley, George, 251
Boltzmann, Ludwig, 237, 239
Bonaparte, Charles Lucien, 12, 17, 91, 94
Boole, George, 191
Brewster, David, 3, 5, 92, 135
'British Ass, The' (Nicolson), 99
British Asses, 93, 117, 176, 177–8, 182
British Association for the Advancement of Science (BAAS), 3–5, 92–4, 175–6, 180, 182
 committees, 107
 meetings, 88
  1831 York, 3, 4, 92
  1833 Cambridge, 3–4
  1835 Dublin, 89
  1839 Birmingham, 90
  1846 Southampton, 100
  1850 Edinburgh, 96–8
  1851 Ipswich, 90, 91
  1867 Dundee, 114
  1868 Norwich, *see* Tyndall, *other science writings*: Norwich 1868
  1869 Exeter, *see* Sylvester, *papers and addresses*: Exeter 1869
  1870 Liverpool, *see* Maxwell, *lectures and addresses*: Liverpool 1870; Tyndall, 'On the Scientific Use of the Imagination'
  1871 Edinburgh, 97, 104, 111, 149, *see also* Maxwell, 'To the Chief Musician upon Nabla: *A Tyndallic Ode*'; Tait, Edinburgh 1871
  1873 Bradford, *see* Maxwell, *poetry*: 'Molecular Evolution'; Maxwell, *scientific works*: 'Discourse on Molecules'
  1874 Belfast, *see* Tyndall, Belfast 1874; Maxwell
  1876 Glasgow, 118–19
 sections, 4
  A (Mathematics and Physics), 4, 99
  D (Natural History), 11, 90, 101,
 see also British Asses; Red Lion Club
bubbles, soap, 79
 Paris on, 33, 36
 Plateau on, 36
 see also Maxwell, toys
Burns, Robert, 42
 'Comin' thro the Rye', 53–4, 174
Busk, George, 103

calculus, 45, 50, 53, 187, 221, 227, 229
Campbell, Lewis, 30, 35, 39, 70, 82
Carlyle, Thomas, 160, *Sartor Resartus*, 122, 147–8, 159–60, *see also* Tyndall
Carroll, Lewis, 9, 143
 Alice in Wonderland, 94
 Through the Looking-Glass, 41, 43, 76, 80
 Maxwell on, 41
 'Syzygies' game, 213–14, 215
Cauchy, Augustin, 207–8
causality, 9, 24, 25, 27–30, 65–6, 70, 154, 207, 238, 249, 262, *see also* counterfactuals; force; Laplace; Maxwell
Cayley, Arthur, 143, 191–2, 202, 220, 224, 225, 234
Chasles, Michel, 189, 221

Index

children, didactic scientific literature for, 32, *see also Philosophy in Sport*
Cicero, 170, 260
Clifford, William Kingdon, 196–7, 249
   'On the Space-theory of Matter', 193, 229, 248
   The Unseen Universe, review of, 241, 242, 248
Coleridge, Samuel Taylor, 1, 3, 6, 78
   'Constancy to an Ideal Object', 62
   'Dejection; an Ode', 153
   'The Eolian Harp', 159
continuity
   principle of the unity of nature, 154, 197–8, 238, 243, 244, 250
   in mathematics (local), 184, 247, *see also* Sylvester
counterfactuals, 27–31, 56, 243, 251
Cowper, William, 38
Crary, Jonathan, 52
Crookes, William
   radiometer, 138–9
   spiritualism, 135, 136–9
   Thallium, 136
Cruikshank, George, 36, 50
curvature of space, 229, 248–9
'Cygne, Le' (Baudelaire), 204

Darwin, Charles, 106, 151, 162–3, 164, 239
   'Darwinism', 20
   natural selection, 148, 151, 165,
   see also Gould
Darwin, Erasmus, 12
Davie, George, 45, 47
Dawson, Gowan, 150
de Biran, Maine, 70
de Blainville, Henri, 95
Deleuze, Gilles, 177, 180, and Félix Guattari, 19
Democritus, 253
de Morgan, Augustus, 242
Descartes, René, 18, 200
   on dreams 76, 77, 82
Design, Argument from, 11, 63–7, 86, 161, *see also* Paley; Maxwell; Tyndall; Whewell
determinism, *see* causality; force; Laplace; Lear; materialism, scientific
dodo, 94–5
dreams, *see* Descartes; Freud; Hamilton, Sir W.; Maxwell

Edinburgh, University of, 6, *see also* Forbes, J. D.; Hamilton, Sir W.; Kelland; Maxwell
Edinburgh Academy, 44
Eisenstein, Gotthold, 191, 234
Eliot, George, 126
Emerson, Ralph Waldo, *see* Tyndall
energy, *see* factional differences
English literature, discipline of, 4, 5–6

ether
   electromagnetic, 71
   luminiferous, 153–4, 156
   plethora of ethers, 242–3, 249
   spiritualist, 132, 250,
   see also Maxwell
Euclid, 32, 46, 195, *see also* Sylvester
evolution, 164, *see also* Darwin; Lear; Oken; Sylvester; Tyndall

factional differences and battles between the North Britons and Metropolitans, 100, 103–4, 108, 238, 252–5
   atoms and molecules, 150–1, 167, 172–3, *see also* Lucretius; Maxwell; Tyndall
   force and energy, 106–7, 118–19, 119–21, 254–5
   glacier motion, 104–6, 107–8, 153
   mathematics, 115
Faraday, Michael, 126, 133, 135
Faucit, Helen, 216
Fechner, G. T., 258
Fichte, J. G., *see* Tyndall
Flatland: A Romance (Abbott), 143
Forbes, Edward, 90–2, 94, 99
   poetry, 91
   'A Yawn from a Red Lion', 91–3
   'Soliloquy Composed by a R[ed] L[ion]', 100
   'The Fate of the Do-do', 94–6
Forbes, James David, 5, 44–5, 46, 55
   on glacier motion, 104, 105, 106, 122
force, *see* Hamilton, Sir W.; Laplace; Maxwell; Newton; Spencer; Tait; Tyndall
Frankland, Edward, 102
Fraunhofer, Joseph von, 129
Freud, Sigmund, 77, 80

Galilei, Galileo, 187
Galton, Francis, 82, 266
Gamow, George, 143
Gassiot, John Peter, 102
Gauss, C. F., 142, 195, 221, *see also* Sylvester
Geikie, Archibald, 99
geometry, *see* curvature of space; Euclid; Riemann; Sylvester; topology
Gilman, Daniel Coit, 212–13, 218
glacier formation, theories of, *see* Forbes, J. D.; Mathews; regelation; Tait; Tyndall
Gold, Barri J., 119
Gopnik, Alison, 29
Gould, John, 10, 17, and Darwin, 23
Gouwen, Gilliam van der, 259, 260
Grandville, J. J., 19, 50
Granville, Augustus, 3
Gray, John Edward, 95

Hamilton, Sir William, 59–60, 73, 84, 108, 246
  Dissertations to Reid, 46–7, 71
  on dreams and the subconscious, 76–7, 78–9, 180
  'the Duality of Consciousness', 60–1, 68
  on force and perception, 70–1, 86
  lectures, 59, 65–6, 71, 76–7, 78
  teaching style, 45–6
  Whewell, debates with, 46–7, 50
Hamilton, William Rowan, 10, 30–2, 89, 91, 144, 197
  on Newton, 7, 31
  quaternions, 113, 196–7
  and Wordsworth, 1–3
  poetry: 'Ode to the Moon under Total Eclipse', 7–9
    'The Enthusiast', 30–1
    'The Tetractys', 113
    'To Poetry,' 3, 7, 89, 234
  scientific works: 'Introductory Lecture on Astronomy', 7, 31,
    see also Hamiltonian; Nabla; quaternions
Hamiltonian, The, 121, 238
Harcourt, W. Vernon, 92
Hay, David Ramsay, 82
heat death, of scientific thought, 178–9, of the universe, 238–9, 257
Hegel, G. W. F., 69, 163, 255
Helmholtz, Hermann von, 171–2, 193, 195
Heraclitus, 88, 256–7
Herschel, John, 1, 3, 151
  'Man the Interpreter of Nature', 64–5, 73
Hinton, Charles, 143, 144
Hirst, Thomas Archer, 93, 102–3, 123, 127
holotype, the, 17–18, 24
Home, Daniel Dunglas, 135, 137
Hooker, Joseph, 103, 108
Hopkins, Gerard Manley, 20, 232–3
  'The Wreck of the *Deutschland*', 231–2
Hopkins, William, 55–6, 105, 106
Horace
  Art of Poetry, 181
  Odes: Ode 1.22, 179–80, Ode 1.34, 180
  Ut pictura poesis, 52, 217, 220, 221,
    see also Sylvester
Huggins, William, 136, 138, 174
Hume, David, 27–8, 59, 68, 175, 177, 180, 250
Huxley, Henrietta, 163
Huxley, T. H., 19, 91, 93, 100, 102–3, 104, 106, 190, 198, 234
  1870 President's address to the BAAS at Liverpool, 148–9,
  see also Euclid

Imagination
  and Fancy, 62, 156
  and mathematics, 142–3, 144
  and physics, 142
  public, 157–8
  Romantic, 153
  scientific, 7, 153–7
  see also Maxwell; Tait; Tyndall
Ingleby, Clement, 144, 185, 196–7, 210, 215

Jenkin, Fleeming, 44, 101, 103, 104
  'The Atomic Theory of Lucretius', 168, 172
Jouffroy, Théodore-Simon, 77
Joule, James Prescott, 102, 106, 110
Joyce, James, 162, 261

Kant, Immanuel, 60, 65, 77, 236
  Critique of Judgement, 180, 'Free Beauty', 83
  Critique of Pure Reason, 83
  see also nebular hypothesis; space; Sylvester
Kirchhoff, Gustav, 129, 237
Kelland, Philip, 44
Klein, Felix, 192, 249
knots, 244, 247
  amphicheiral, 245–7
Knott, C. G., 39, 106, 108, 118, 241, 245

L'Esprit Nouveau (Le Corbusier and Ozenfant), 212
Lacan, Jacques, 69
Lanier, Sydney, 200
  The Science of English Verse, 208–10, 211, 213, 222
Laplace, Pierre Simon
  'Laplace's Demon', 253, *see also* nebular hypothesis
Lear, Edward, 9, 15
  and determinism, 22, 24
  and Edward Stanley, 10
  and evolution, 23–4, 29–30
  Limericks: 14–15, 19–20, 22–3, 24–7, 43, 54
    and logic, 25–6
    'There was an Old Man of Dundee', 14
    'There was an Old Man of Peru', 25, 26
    'There was an Old Person of Hurst', 26
    'There was an Old Person whose habits', 24
    'There was a Young Person of Crete', 26–7
  natural history illustrations, 10, 11, 15, 16–18
    for Darwin's *Voyage of the Beagle*, 23
    Eagle Owl (*Bubo Maximus*), 14, 18
    Lear's Macau, 12, 17

Index

Macropus Parryi, 16, 17, 18
Psittacidae, 10, 11–12, 19
Study of Rose-ringed Parrakeet, 20
Toco Toucan, 23
nonsense alphabets, 12–14, 16
nonsense botanies, 12–13, 22, 27
nonsense drawings, 15, 16, 19, 23, 24
Lecercle, Jacques, 9, 15, 16, 25, 26, 40, 43
lectures, science
   BAAS committee on, 111–12
   popular, 37, 97, 115, 117, 128
   for women, 112
   for working-men, 116–17
   see also Maxwell; Pepper; Tait; Tyndall
Lewes, G. H., 185, 216
Lewis, David, 27–8, 29
Listing, Benedict, 245–6
Locke, John, 58, 175
Locker, Frederick, 214–15, 230
   'On "A Portrait of a Lady"', 215, 217, 232
Lockyer, Norman, 108, 111–12, 129, 185, 242
Lodge, Oliver, 97, 118
Lubbock, John, 103
Lucretius, *On the Nature of Things*, 88, 170–1, 174–5, 261
   atoms, 155, 244
   clinamen, 166, 167, 261
   H. A. J. Munro's edition, 166, 167–9
   Hymn to Venus, 258–60
   see also Deleuze; Jenkin; Maxwell; Tyndall

Mallock, W. H., 251
Masson, David, 116, 124
materialism, 153
   scientific, 147–50, 153, 154, 156, 164–5, 176, 252
Mathematical Tripos, Cambridge, 4, 45, 47–8, *see also* Maxwell
Mathews, William, 123
Maurice, F. D., 6
Max Müller, Friedrich, 13, 151
Maxwell, James Clerk, 6, 103
   aesthetics, 81–4
   on analogy, 56–8, 61, 87–8, and rhyme, 84–5, 'physical analogy', 56–7
   associationism, 176–8
   atoms and molecules, 149–50, 171–2, 260–1, *see also* Lucretius, *below*
   on bubbles, 79
   on causality, 65, 69
   as Cavendish Professor, 111
   on college rules, 42–3, 44
   death of, 239–40
   delight in curves, 82, 221
   on dreams and the subconscious, 75–9, 180
   and Edinburgh University, 42–3, 44–7, 48
   electromagnetic theory of light, 71, 88, 171
   on force, 69–70
   free will and determinism, 169–70, 172–6
   gas, 28, 57, 130, 139, 168, 173, 237, 260,
   gaseous nonsense, 181
   and Sir William Hamilton, 46, 58, 59, 60–1, 70–2
   on Lucretius, 165–6, clinamen, 165–6, 168–70, 175, 177–8, 179, 180, 257, 260–1
   at Marischal College, 46
   and the Mathematical Tripos, 6, 46–7, 48, 49–51, 52–6, 55–6
   'Maxwell's Demon', 28, 173, 174, 181, 241, 253, 260
   nervous breakdown of, 51
   on nonsense, 80, 177–82, and science, 37–8, 120, 177–8, 179–80, 261, 262
   and Peterhouse College, Cambridge, 42
   on pedantry, 44, 45, 49–51, 56, 81
   on Plateau, 36–7, 80, 255
   and play, 34–7, 255–6, *see also Philosophy in Sport*
   on popular lectures, 117, *see also* poetry: '*Tyndallic Ode*'
   puns, 39, 42, 43, 87–8, 119–20, 157, 175–6, 178, 182, 256
   relation to analogy, 40–1, 57–8, 80, 85, 87–8, 176, 178, 179, 181
   on relation, 83–4
   on religion and science, 250, *see also* poetry: 'A Student's Evening Hymn', and essays on 'Real Analogies' *and* 'Design'
   signature of, 39, 257, 262
   on solipsism, 61, 62, 67, 81, 156
   on space, 70–1, 77, 85
   stereoscope, 35, 255
   toys: devil-on-two-sticks', 34–5, 256;
      dynamical top, 35, 257;
      phenakistoscope, 34, 171;
      soap-bubbles, 79, 255, 262;
      thaumatrope, 34; zoetrope, 35, 171
   and Trinity College, Cambridge, 42
   on Tyndall, 107, 108, 258, 261, and Bain, 152, 153
   on *Villette* (Brontë), 30, 251
   see also curvature of space; lectures; Hamilton, Sir W.; Mathematical Tripos; *Philosophy in Sport*.
essays
   for the Apostles, 39–40, 56
   'Are there Real Analogies in Nature?', 40, 42, 43, 56, 57–8, 63, 66, 67–8, 70–1, 74, 80, 85–7, 181–2
   'Has Everything Beautiful in Art its Original in Nature?', 72, 82–3

Maxwell, James Clerk (cont.)
  'What is the Nature of Evidence of
    Design?', 60, 63, 65, 66, 73, 74
  'Idiotic Imps', 132–3
  'On the Properties of Matter', 70
  Paradoxical Philosophy, review of, 136, 242,
    245, 249, 251, 258
  'Psychophysik', 59, 258, 262
  'Science and Free Will', 173, 242
  lectures and addresses:
  Cavendish Professor, inaugural lecture as,
    50, 111, 150
  'On Colour Vision', 111, 127
  King's College, inaugural lecture to, 69
  Liverpool 1870 BAAS, address to Section A
    at, 74, 87, 112, 113–14, 115, 131, 149, 170,
    172, 178–9, 198–9
  poetry: 37, 56, 96–8, 240
  '*A Paradoxical Ode*', 237–8, 239, 242–52
  'A Problem in Dynamics', 53, 54, 113
  'A Student's Evening Hymn', 50, 62–3,
    66–7, 74, 77, 88
  'A Vision: *Of a Wrangler, of a University,
    of Pedantry, and of Philosophy*', 49–51,
    53, 56
  'British Association, 1874: *Notes of the
    President's Address*' 147–8, 167, 178, 254
  '(Cat's) Cradle Song, By a Babe in Knots',
    37, 237
  'In Memory of Edward Wilson, *Who
    repented of what was in his mind to write
    after section*', 53–5, 150, 170, 173–4
  'Lectures to Women on Physical Science',
    112, 173, 260
  '*Lines written under the conviction that it
    is not wise to read Mathematics in
    November after one's fire is out*', 43–4, 55
  'Molecular Evolution', 164–70, 174–82
  'Profes[sor] Tait's Christmas Dream', 110
  'Recollections of Dreamland', 75–80, 180,
    181
  'Reflex Musings: Reflection from Various
    Surfaces', text, 61, on, 58, 59, 61–2,
    68–75, 79, 81–5, 86–7, 88, 251
  'Report on Tait's Lecture on Force: – B.A.,
    1876', 51, 117–18, 119–21, 169, 238,
    254–5, 258
  'Song of the Cub', 165
  'Song of the Edinburgh Academician', 44
  'The Death of Sir James, Lord of Douglas',
    114
  'The F.R.S.E.', 157–8
  'To F. W. F.', 86
  'To the Chief Musician upon Nabla: *A
    Tyndallic Ode*', text of 98–9, 154

  on, 97, 98–100, 107, 108–9, 110, 112,
    121–41, 149–50, 152, 154–6, 164–5, 170,
    177, 181, 251
  'To the Committee of the Cayley Portrait
    Fund', 220, 221, 225–6
  'Valedictory Address to the D[ea]n',
    42–3
  'Valentine by a Telegraph Clerk ♂ to
    Telegraph Clerk ♀', 260
  see also Carroll; rhyme; Rudy; Tennyson
  scientific works
  'Atom', 170, 172
  'Concerning Demons', 241
  'Discourse on Molecules', 151, 165,
    169–70
  'Ether', 245, 249
  Matter and Motion, 116
  'Observations on Circumscribed Figures
    having a Plurality of Foci', 44–5, 82
  'On a Dynamical Top', 35
  'On Faraday's Lines of Force', 56–7, 67, 87
  'On Physical Lines of Force', 71, 87–8, 171
  'On the Dimensions of Physical
    Quantitities', 33
  'On the Dynamical Theory of Gases',
    168–9
  'On the History of the Kinetic Theory of
    Gases: Notes for William Thomson',
    169, 260
  Theory of Heat, 28, 116, 128
  Treatise on Electricity and Magnetism, 115
Maxwell, John Clerk, 44, 75
Mayer, Robert, 106
metaphor, 40, dead, 41
Metropolitans, 100–1, 102–3, *see also* factional
  differences; materialism; Tyndall
Monboddo, Lord, 78
Monge, Gaspard, 189
Munro, H. A. J., *see* Lucretius

Nabla, 99–100, 114, 121, 246
nablody, 113, 114; *see also* Hamilton, W. R.;
  Maxwell
Nägeli, Karl Wilhelm von, 258
natural theology, *see* Design
Nature, see Lockyer
nebular hypothesis, 149, 161, 164, 166, 181
neologisms, 210, nonsense, 26–7; *see also*
  Sylvester; Whewell
Newton, Isaac, 4, 120, 121, 152, 156, 200
  laws of motion, 83, 252–5, 257; *see also* calculus;
    Hamilton, W. R.
Nietzsche, Friedrich, 36, 68
Night-Thoughts (Young), 145
Nodder, Frederick and Ralph, 16, 17; *see also* Shaw

Index

nonsense, ideas of, 9, 10–11, 15–16, 40, 94; *see also* Lecercle; Maxwell
North British network, 101–2, *see also* factional differences
North British Review, 101

ode, 113, 114–15
Oken, Lorenz, 4, 82, 166–7, 235–6
Olympia (Manet), 18
Owen, Richard, 23, 93, 95, 235
   poetry, 95

Paley, William, 66, 151
Parshall, Karen, 225
Pater, Walter, 236
Peirce, Charles Sanders, 207–8
Pepper, J. H., 37, 126, 133, 155
pessimistic meta-induction from past falsity, 37–8, 96
Philosophy in Sport made Science in Earnest (Paris), 32–3, 36, 79
   Maxwell and, 33, 51
   and puns, 41–2, 51–2
Planck, Max, 237
Plateau, Joseph, 34, *see also* bubbles; Maxwell
Plato, 78, 186, 196
play, and science, *see* Maxwell; *Philosophy in Sport*
Playfair, John, 3
Poe, Edgar Allan
   'The Man of the Crowd', 29
   'The Murders in the Rue Morgue', 234
   'The Rationale of Verse', 209
Pollock, Frederick, 124
   '*Dialogue between Urania . . . and Celestine*', 130
   sensitive flame Valentine, 140
   'The Ice Flower', 124–6
Poncelet, Jean-Victor, 189–90
Pope, Alexander, 200, 240
Presbyterianism, Scots, 63, 100, 101
Prichard, Charles, 200–1
Prigogine, Ilya, and Isabelle Stengers, 121
professionalisation of science, 3, 6, 11, 23, 103–4; *see also* British Association for the Advancement of Science
prosody, *see* Lanier; Poe; Sylvester
puns, 43, 80, 202
   relation to analogy, 40, 41
   visual, 50, 52
   see also Heraclitus; Maxwell; *Philosophy in Sport*; Sylvester; Tait
Pythagoreanism, 50, 63, 77, 114, 117, 151, 179, 236

quaternions, 1, *see also* Hamilton, W. R.; Nabla

Ramsay, Andrew Crombie, 90, 92, 93
Rankine, W. J. Macquorn, 97, 101
Rape of Europa, The (Boucher), 261
Reid, Thomas, 59, *see also* Hamilton, Sir W.
Red Lion Club, 88, 89–94, 96–7, 165, 176, 177–8, 179–80, 181, 182
   Metropolitan, 90–0, 100–1
regelation, 105
   Faraday on, 105, 123
   James and William Thomson on, 105, 123
   and Tyndall, 5, 123
religion and science, *see* design; Maxwell
Reichenbach, Carl von, 132
rhyme, in Lear, 24, 25, 216, in Maxwell, 49, 84–5; *see also* Sylvester
Richards, Joan L., 190
Riemann, Bernhard, 'On the Hypotheses which Lie at the Bases of Geometry', 142, 192, 195, 247–8
   the manifold, 193–4, 195
   Riemannian plane, 194–5, 249
   see also Sylvester
Roscoe, William, 111
Royal Institution, 126, 128, *see also* Tyndall; Maxwell, 'On Colour Vision'
Royal Society, 3, 5, 92, 93
Rudy, Jason, 37
Ruskin, John, 82, 108, 130

Salmon, George, 144
Schaffer, Simon, 79
Scott, Walter, 62, 114
science, poetry and the arts, 2–3, 5–7; *see also* Maxwell; Sylvester
Serres, Michel, 189, 257
Shakespeare, William, 39, 202, 240
   A Midsummer Night's Dream, 8, 219–20
   As You Like It, 215–16, 217, 218, 219
   Richard III, 41
   Romeo and Juliet, 40
   The Tempest, 187
Shaw, George, 17, *see also* Nodder
Shelley, Percy Bysshe
   'Mont Blanc', 153, 161
   Prometheus Unbound, 237, 243–4, 255
Silver, Daniel S., 238, 240, 246
Smith, Crosbie, 101, 104, 107
Smith, William Robertson, 99, 153
Snow, C. P., 39
Soyer, Alexis, 91–2
space
   Hamilton, Sir W., 70
   Kantian, 70
   see also Ingleby; Sylvester

Spencer, Herbert, 103, 118, 164, 167, 239
Spenser, Edmund, *Faerie Queene*, 8, 139, 202
spiritualism, and physics, 132–9, and non-Euclidean geometry, 144
Spottiswoode, William, 103, 114
Stewart, Balfour, 101
  'Mr Crookes on the "Psychic" Force', 136–7
  'The Sun as a Type of the Material Universe', 173, 242
  see also Stewart and Tait
Stewart and Tait, 144, 244, 252–37
  Paradoxical Philosophy, 238, 247
  The Unseen Universe, 198, 238, 239, 240–2
  see also Maxwell
Strickland, Hugh, 94, 95
styles of science, 112–15, 124–6, 128–9, 236–7, 261,
  see also Boltzmann; Maxwell; Nabla
Sylvester, James Joseph, 104
  determinants and matrices, 225–8
  Euclidean geometry, 187, 190, 204, 210
  evolution, 234–6
  invariants, 191, 202, 224, 234
  and T. H. Huxley, 190–1, 194
  Jewish faith, 204–5, 206
  mathematics and music, 225, 236–7
  mathematics and poetry, 201–3, 207–9, 221–3, 236; *see also* prosody, below
    catalecticant and canonizant, 224–5, 227–8
    Continuity, 184, 221, 227; 'continuity of sound', *see prosody*: metric: synectic; 'continuity of mental impression', 199
    matrix, 226–7
    projection, 217, 220, 221
  neologisms, 191, 207, 223, 224–5, 234
  puns, 218–19
  space, and continuity, 183–4, 186
    calculus, 187–8, 207
    imagery of, 183–4, 208, celestial, 184, 197–201, 202–4, 235
    Gauss, 185, 188, 192, 196
    Kant, 184–7, 196–7
    non-Euclidean space, 142–3, Riemann, 143, 185, 188, 192–6, 207, 222
    projective geometry, 188–92
    'transcendental space', 196–7
  working life, 205
  papers and addresses:
    'A Constructive theory of Partitions, arranged in three Acts', 202
    Address to Section A at Exeter in 1869, 112, 142–3, 144, 183–4, 185, 186, 187, 193, 196, 198, 205–6, 221, 235
    'An Essay on Canonical Forms', 228
    'An Inquiry into Newton's Rule', 200, 219–20
    Commemoration Day Address to Johns Hopkins, 204, 236
    'On Newton's Rule', 236–7
    'On the Method of Reciprocants ...' (Inaugural address as Savilian Professor), 202–3
    'On the motion of a free rigid body', 173
    'On the Principles of the Calculus of Forms', 224, 227, 228
    'Probationary Lecture' to Gresham College, 185–6, 188, 197
    'Sketch of a Memoir on Elimination', 225
    'Syzygistic Relations', 228
  works:
    Fliegende Blätter, 194, 212, 214
    The Laws of Verse, 113–14, 183–7
  poetry:
    'Indifference', 199
    'Kepler's Apostrophe', 199, 206
    On Horace, 204, 226, Ode 1.1 (To Maecenas,), 210–11
    'Remonstrance', 199–200
    'Sonnet To the Savilian Professor of Astronomy', 200–1
    Spring's Début: A Town Idyll, 228, 229
    'Tasso to Eleonora', 112
    'The Ballad of Sir John de Courcy', 213
    'The Evening Star', 199
    'The Lily Fair of Jasmin Dean', 199
    'To a Missing Member of a Family Group of Terms in an Algebraical Formula', text of, 203, on 202–6
    'To an Ink-spot upon a Lady's Cheek', 199
    'To Rosalind', 212–13, 214, 215–20, 228, 229–33
    'Urbi et orbi', 200
    see also Maxwell, 'To the Committee of the Cayley Portrait Fund'
  prosody:
    breath, 231–3
    music and verse, 209
    rhyme, 203, 212–13, 217, 222–3
    hypersyzygetic canonico-meio-catalectizant, 223–9
    rhythm, 207
    chromatic, 207, 208
    metric, 207, 211
      anastomosis, 210, 233; symptosis, 210, 211–12; synectic, 207–8, 211; syzygy, 208–10, 211, 212, 213–15, 228–9, 230

Tait, Peter Guthrie, 103, 104
  animus toward Tyndall, 108, *see also* factional differences

on knots, 37, 245, 247
poetry, 158
on puns, 119
on science lectures, public 112, 115–16, popular, 117–20
Scrapbook, 108
smoke-rings, 171
addresses and essays:
Edinburgh 1871, address to Section A at, 100, 112–13, 115, 137
Edinburgh Graduates, Address to, 101
'Energy' (with Thomson), 107
'Energy, and Prof. Bain's Logic', 153
'Force', 116, 117–21, 172
'Imagination in Science', 147, 156–7
'Sensation and Science', 128–9
Tyndall's *Heat as a Mode of Motion*, review of, 107, 128, 140
scientific works:
Lectures on Recent Advances in Physical Science, 115, 116, 255
Life of Forbes, 107
Sketch of Thermodynamics, 107
see also curvature of space; Nabla; Stewart and Tait; Maxwell, '*A Tyndallic Ode*'
Tenniel, John, 19
Tennyson, Alfred, 214, 222
In Memoriam, 240
'Lucretius', 169–70, 198–9, 253–4, 261
'The Brook', 98, 104, 155
Thallium, see Crookes, William
thaumatrope, 33–4, 51–2, see also Maxwell
The Reader: A Review of Literature, Science, and Art, 124
thermodynamics, second law of, 39, 238, 240–2, see also heat death
Thomson, Allen, 137, 138
Thomson, James, 102, 103
Thomson, J. J., 34–5, 237
Thomson, William, 28, 39, 79, 101, 103, 104
Presidential address to the BAAS at Edinburgh 1871, 129, 136
(and Tait) *Treatise on Natural Philosophy*, 115
vortex atom, 171–2, 244–6
see also atom; 'Maxwell's Demon'; regelation
thought experiment, see 'Maxwell's Demon'
topology, 245–6
toys, see Maxwell; *Philosophy in Sport*
Turner, Frank M., 103, 168
Tyndall, John, 97, 102, 104, 108, 238
on atoms and molecules, 147–8, 151, and Lucretius, 167–8
glacier motion, 105–6, 122–3
mountaineering, 121–2
religion, 160–2
rhetoric, 124, 128, 131, 140

and romanticism, 166–7; Carlyle, 108, 159–60, 161; Emerson, 160, 166; Fichte, 147, 160; Goethe, 152, 166
Vanity Fair article and caricature, 122, 145–7
lecture demonstrations: 110–11
ice 'flowers', 124–6
optical, 129–32, 162
sounding flames, 139
theatricality of, 126–7, 129, 134, 140–1
see also Maxwell, 'Profes[sor] Tait's Christmas dream'; Maxwell, '*A Tyndallic Ode*'
poetry: 'A Morning on Alp Lusgen', 158
Royal Institution lectures: 97, 102, 107, 155
'On Chemical Rays of the Light of the Sky', 130
On Electrical Phenomena and Theories, 110, 130
'On Force', 106, 118–19
Heat considered as a Mode of Motion, 107, 122, 124, 128, 145
Sound, 'Sounding Flames, &c.', 139, 140
other science writings:
Belfast 1874, Presidential address to the BAAS at, 107, 119, 162, 172, 253–4
Forms of Water, 107
Fragments of Science for Unscientific People, 145; 'Apology for the Belfast Address', 160; 'Science and the "Spirits"', 134
Glaciers of the Alps, 106, 122
Hours of Exercise in the Alps, 122–3
Norwich 1868, address to Section A of the BAAS at, 147–8, 155, 167, 176
'On a New Series of Chemical Reactions produced by Light', 130–2
'On the Scientific Use of the Imagination', 111, 112–13, 123, 131, 134, 145, 147, 149, 150, 151–7, 159, 166
'Physics and Metaphysics', 162
see also atoms; Carlyle; Lucretius; materialism, scientific; molecules

Vestiges of the Natural History of Creation (Chambers), 23, 164
vortices, see atoms; Maxwell, 'On Physical Lines of Force'

Wagner, Richard, 239
Wallace, A. R., 137, 210
Whewell, William, 3–5, 6–7, 48, 86, 92
consilience, 57, 64
fundamental ideas, 77
hypothetico-deductivism, 85, 180–1

Whewell, William (cont.)
  neologisms, 3–4, 6
  see also Hamilton, Sir W.
Whistler, James McNeill, 222
Wilde, Oscar
  'The Canterville Ghost' 144
  Picture of Dorian Gray, 219
Wilson, George, 91, 94
Wöhler, Friedrich, 148
word-games, *see* Carroll; Sylvester

Wordsworth, William
  The Excursion, 2, 7
  The Prelude (1850), 7
  'Tintern Abbey', 68
  'Yarrow Unvisited', 158
  see also Hamilton, W. R.

X Club, 103, 124

Zerfi, G. G., 135

CAMBRIDGE STUDIES IN NINETEENTH-CENTURY
LITERATURE AND CULTURE

General editor
GILLIAN BEER, *University of Cambridge*

TITLES PUBLISHED

1. The Sickroom in Victorian Fiction: The Art of Being Ill
   MIRIAM BAILIN, *Washington University*
2. Muscular Christianity: Embodying the Victorian Age
   edited by DONALD E. HALL, *California State University, Northridge*
3. Victorian Masculinities: Manhood and Masculine Poetics in Early Victorian Literature and Art
   HERBERT SUSSMAN, *Northeastern University, Boston*
4. Byron and the Victorians
   ANDREW ELFENBEIN, *University of Minnesota*
5. Literature in the Marketplace: Nineteenth-Century British Publishing and the Circulation of Books
   edited by JOHN O. JORDAN, *University of California, Santa Cruz*, and ROBERT L. PATTEN, *Rice University, Houston*
6. Victorian Photography, Painting and Poetry
   LINDSAY SMITH, *University of Sussex*
7. Charlotte Brontë and Victorian Psychology
   SALLY SHUTTLEWORTH, *University of Sheffield*
8. The Gothic Body: Sexuality, Materialism and Degeneration at the *Fin de Siècle*
   KELLY HURLEY, *University of Colorado at Boulder*
9. Rereading Walter Pater
   WILLIAM F. SHUTER, *Eastern Michigan University*
10. Remaking Queen Victoria
    edited by MARGARET HOMANS, *Yale University* and ADRIENNE MUNICH, *State University of New York, Stony Brook*
11. Disease, Desire, and the Body in Victorian Women's Popular Novels
    PAMELA K. GILBERT, *University of Florida*
12. Realism, Representation, and the Arts in Nineteenth-Century Literature
    ALISON BYERLY, *Middlebury College, Vermont*

13. Literary Culture and the Pacific
    VANESSA SMITH, *University of Sydney*

14. Professional Domesticity in the Victorian Novel: Women, Work and Home
    MONICA F. COHEN

15. Victorian Renovations of the Novel: Narrative Annexes and the Boundaries of Representation
    SUZANNE KEEN, *Washington and Lee University, Virginia*

16. Actresses on the Victorian Stage: Feminine Performance and the Galatea Myth
    GAIL MARSHALL, *University of Leeds*

17. Death and the Mother from Dickens to Freud: Victorian Fiction and the Anxiety of Origin
    CAROLYN DEVER, *Vanderbilt University, Tennessee*

18. Ancestry and Narrative in Nineteenth-Century British Literature: Blood Relations from Edgeworth to Hardy
    SOPHIE GILMARTIN, *Royal Holloway, University of London*

19. Dickens, Novel Reading, and the Victorian Popular Theatre
    DEBORAH VLOCK

20. After Dickens: Reading, Adaptation and Performance
    JOHN GLAVIN, *Georgetown University, Washington DC*

21. Victorian Women Writers and the Woman Question
    edited by NICOLA DIANE THOMPSON, *Kingston University, London*

22. Rhythm and Will in Victorian Poetry
    MATTHEW CAMPBELL, *University of Sheffield*

23. Gender, Race, and the Writing of Empire: Public Discourse and the Boer War
    PAULA M. KREBS, *Wheaton College, Massachusetts*

24. Ruskin's God
    MICHAEL WHEELER, *University of Southampton*

25. Dickens and the Daughter of the House
    HILARY M. SCHOR, *University of Southern California*

26. Detective Fiction and the Rise of Forensic Science
    RONALD R. THOMAS, *Trinity College, Hartford, Connecticut*

27. Testimony and Advocacy in Victorian Law, Literature, and Theology
    JAN-MELISSA SCHRAMM, *Trinity Hall, Cambridge*

28. Victorian Writing about Risk: Imagining a Safe England in a Dangerous World
    ELAINE FREEDGOOD, *University of Pennsylvania*

29. Physiognomy and the Meaning of Expression in Nineteenth-Century Culture
   LUCY HARTLEY, *University of Southampton*

30. The Victorian Parlour: A Cultural Study
   THAD LOGAN, *Rice University, Houston*

31. Aestheticism and Sexual Parody 1840–1940
   DENNIS DENISOFF, *Ryerson University, Toronto*

32. Literature, Technology and Magical Thinking, 1880–1920
   PAMELA THURSCHWELL, *University College London*

33. Fairies in Nineteenth-Century Art and Literature
   NICOLA BOWN, *Birkbeck, University of London*

34. George Eliot and the British Empire
   NANCY HENRY, *The State University of New York, Binghamton*

35. Women's Poetry and Religion in Victorian England: Jewish Identity and Christian Culture
   CYNTHIA SCHEINBERG, *Mills College, California*

36. Victorian Literature and the Anorexic Body
   ANNA KRUGOVOY SILVER, *Mercer University, Georgia*

37. Eavesdropping in the Novel from Austen to Proust
   ANN GAYLIN, *Yale University*

38. Missionary Writing and Empire, 1800–1860
   ANNA JOHNSTON, *University of Tasmania*

39. London and the Culture of Homosexuality, 1885–1914
   MATT COOK, *Keele University*

40. Fiction, Famine, and the Rise of Economics in Victorian Britain and Ireland
   GORDON BIGELOW, *Rhodes College, Tennessee*

41. Gender and the Victorian Periodical
   HILARY FRASER, *Birkbeck, University of London*
   JUDITH JOHNSTON and STEPHANIE GREEN, *University of Western Australia*

42. The Victorian Supernatural
   edited by NICOLA BOWN, *Birkbeck College, London*
   CAROLYN BURDETT, *London Metropolitan University*
   and PAMELA THURSCHWELL, *University College London*

43. The Indian Mutiny and the British Imagination
   GAUTAM CHAKRAVARTY, *University of Delhi*

44. The Revolution in Popular Literature: Print, Politics and the People
   IAN HAYWOOD, *Roehampton University of Surrey*

45. Science in the Nineteenth-Century Periodical: Reading the Magazine of Nature
   GEOFFREY CANTOR, *University of Leeds*
   GOWAN DAWSON, *University of Leicester*
   GRAEME GOODAY, *University of Leeds*
   RICHARD NOAKES, *University of Cambridge*
   SALLY SHUTTLEWORTH, *University of Sheffield*
   and JONATHAN R. TOPHAM, *University of Leeds*

46. Literature and Medicine in Nineteenth-Century Britain from Mary Shelley to George Eliot
   JANIS MCLARREN CALDWELL, *Wake Forest University*

47. The Child Writer from Austen to Woolf
   edited by CHRISTINE ALEXANDER, *University of New South Wales*
   and JULIET MCMASTER, *University of Alberta*

48. From Dickens to Dracula: Gothic, Economics, and Victorian Fiction
   GAIL TURLEY HOUSTON, *University of New Mexico*

49. Voice and the Victorian Storyteller
   IVAN KREILKAMP, *University of Indiana*

50. Charles Darwin and Victorian Visual Culture
   JONATHAN SMITH, *University of Michigan-Dearborn*

51. Catholicism, Sexual Deviance, and Victorian Gothic Culture
   PATRICK R. O'MALLEY, *Georgetown University*

52. Epic and Empire in Nineteenth-Century Britain
   SIMON DENTITH, *University of Gloucestershire*

53. Victorian Honeymoons: Journeys to the Conjugal
   HELENA MICHIE, *Rice University*

54. The Jewess in Nineteenth-Century British Literary Culture
   NADIA VALMAN, *University of Southampton*

55. Ireland, India and Nationalism in Nineteenth-Century Literature
   JULIA WRIGHT, *Dalhousie University*

56. Dickens and the Popular Radical Imagination
   SALLY LEDGER, *Birkbeck, University of London*

57. Darwin, Literature and Victorian Respectability
   GOWAN DAWSON, *University of Leicester*

58. 'Michael Field': Poetry, Aestheticism and the *Fin de Siècle*
   MARION THAIN, *University of Birmingham*

59. Colonies, Cults and Evolution: Literature, Science and Culture in Nineteenth-Century Writing
    DAVID AMIGONI, *Keele University*

60. Realism, Photography and Nineteenth-Century Fiction
    DANIEL A. NOVAK, *Lousiana State University*

61. Caribbean Culture and British Fiction in the Atlantic World, 1780–1870
    TIM WATSON, *University of Miami*

62. The Poetry of Chartism: Aesthetics, Politics, History
    MICHAEL SANDERS, *University of Manchester*

63. Literature and Dance in Nineteenth-Century Britain: Jane Austen to the New Woman
    CHERYL WILSON, *Indiana University*

64. Shakespeare and Victorian Women
    GAIL MARSHALL, *Oxford Brookes University*

65. The Tragi-Comedy of Victorian Fatherhood
    VALERIE SANDERS, *University of Hull*

66. Darwin and the Memory of the Human: Evolution, Savages, and South America
    CANNON SCHMITT, *University of Toronto*

67. From Sketch to Novel: The Development of Victorian Fiction
    AMANPAL GARCHA, *Ohio State University*

68. The Crimean War and the British Imagination
    STEFANIE MARKOVITS, *Yale University*

69. Shock, Memory and the Unconscious in Victorian Fiction
    JILL L. MATUS, *University of Toronto*

70. Sensation and Modernity in the 1860s
    NICHOLAS DALY, *University College Dublin*

71. Ghost-Seers, Detectives, and Spiritualists: Theories of Vision in Victorian Literature and Science
    SRDJAN SMAJIĆ, *Furman University*

72. Satire in an Age of Realism
    AARON MATZ, *Scripps College, California*

73. Thinking About Other People in Nineteenth-Century British Writing
    ADELA PINCH, *University of Michigan*

74. Tuberculosis and the Victorian Literary Imagination
    KATHERINE BYRNE, *University of Ulster, Coleraine*

75. Urban Realism and the Cosmopolitan Imagination in the Nineteenth Century: Visible City, Invisible World
    TANYA AGATHOCLEOUS, *Hunter College, City University of New York*

76. Women, Literature, and the English Domesticated Landscape: England's Disciples of Flora, 1780–1870
    JUDITH W. PAGE, *University of Florida*
    ELISE L. SMITH, *Millsaps College, Mississippi*

77. Time and the Moment in Victorian Literature and Society
    SUE ZEMKA, *University of Colorado*

78. Popular Fiction and Brain Science in the Late Nineteenth Century
    ANNE STILES, *Washington State University*

79. Picturing Reform in Victorian Britain
    JANICE CARLISLE, *Yale University*

80. Atonement and Self-Sacrifice in Nineteenth-Century Narrative
    JAN-MELISSA SCHRAMM, *University of Cambridge*

81. The Silver Fork Novel: Fashionable Fiction in the Age of Reform
    EDWARD COPELAND, *Pomona College, California*

82. Oscar Wilde and Ancient Greece
    IAIN ROSS, *Colchester Royal Grammar School*

83. The Poetry of Victorian Scientists
    DANIEL BROWN, *University of Southampton*